TRIBOLOGY

a systems approach to the science
and technology of friction, lubrication and wear

TRIBOLOGY SERIES

Vol. 1 Tribology — a systems approach to the science and technology of friction, lubrication and wear (Czichos)

Vol. 2 Impact wear of materials (Engel)

TRIBOLOGY SERIES, 1

TRIBOLOGY

a systems approach to the science and technology of friction, lubrication and wear

HORST CZICHOS

Bundesanstalt für Materialprüfung (BAM)
(Federal Institute for Testing Materials)
Berlin – Dahlem
and
Technische Fachhochschule Berlin

ELSEVIER SCIENTIFIC PUBLISHING COMPANY
AMSTERDAM - OXFORD - NEW YORK 1978

ELSEVIER SCIENCE PUBLISHERS B.V.
Sara Burgerhartstraat 25
P.O. Box 211, 1000 AE Amsterdam, The Netherlands

Distributors for the United States and Canada:

ELSEVIER SCIENCE PUBLISHING COMPANY INC.
52, Vanderbilt Avenue
New York, NY 10017

First edition 1978
Second impression 1979
Third impression 1986

Library of Congress Cataloging in Publication Data

Czichos, Horst.
Tribology.

(Tribology series ; v. 1)
Bibliographv: p.
Includes indexes.
1. Tribology. 2. System analysis. I. Title.
II. Series.
TJ1075.C94 621.8'9 77-28510
ISBN 0-444-41676-5

ISBN 0-444-41676-5 (Vol. 1)
ISBN 0-444-41677-3 (Series)

Printed in The Netherlands

Preface

This volume gives a systematic unified approach to tribology: the fundamentals of friction, lubrication and wear and their influences on the structure and function of dynamic mechanical systems. Since friction is responsible for a major loss of useful mechanical energy, and wear is a major reason for replacing equipment, a better understanding and utilization of the principles of tribology is particularly important for the conservation of energy and materials in engineering design.

The plan to write this book grew from the participation in co-operative work of the "International Research Group (IRG) on Wear of Engineering Materials" operating under the auspices of the "Organization for Economic Co-operation and Development" (OECD) in which I have been taking part since the beginning of the 1970's. Theoretical considerations and practical experience have shown that there is an urgent need for a general, yet practically useful, framework in order to link the many interdisciplinary aspects of tribology, ranging from contact mechanics, surface physics and hydrodynamics to bearing technology and questions of maintenance and reliability of mechanical equipment. Moreover, it became obvious in recent years, that specific analytical methods of isolating and studying single events and treating friction and wear in analogy to other (bulk) mechanical material properties are only partly useful to the study of complex tribological problems. Indeed, the original definition of tribology: "the science and technology of interacting surfaces in relative motion and the practices related thereto" is obviously that of a dynamic mechanical system. Thus, friction, wear and the many related phenomena are essentially due to dynamic interactions between the moving components of systems. (A compilation of mechanical systems in which friction and wear processes occur is given in the Appendix A.) It was concluded that the application of systems thinking and general systems theory are especially suited for developing a convenient framework of tribology, bringing together its many aspects currently scattered throughout the scientific and technical literature. It follows that a tribo-mechanical system should be analyzed and described in terms of its "structure" (elements of the system, properties of elements, interrelations between elements) as well as in terms of its "function" (inputs, outputs, transfer functions). Whereas in the analysis of electrical systems the interest is concentrated mainly on the functional input-output relations, in tribology the analysis and description of the (dynamic) system's structure is the crucial point.

The organization of the book is straightforward and is outlined in Section 1.4. It is suggested that the reader who is interested mainly in the practical application of systems techniques to tribological problems may turn to Chapter 8 on "Practical Systems Methodology", after reading the introductory Chapters 1 and 2, and may follow then the sequence of Chapters 3,4 and so on. Because of the immense scope of tribology, this volume concentrates mainly on the basic general principles of the subject matter in order to fit the huge amount of material into the length available. Thus, the book is aiming to provide unified interdisciplinary background knowledge for engineers, physicists, chemists and material scientists working in the various specialized fields of tribology. When used as a textbook, the text is suited to the level of senior undergraduate or first-year graduate courses in engineering and the natural sciences.

Some part of the material of this book has been presented as lectures at the IRG-OECD Meetings held at (a) Imperial College, London, July 1973, (b) BAM, Berlin-Dahlem, October 1974, (c) CNRS, Paris, April 1976. At these Meetings helpful comments have been made by Professor R. Courtel, CNRS, Paris, and Professor H. Christensen, SINTEF, Trondheim. Further acknowledgements are due to the former Chairman of IRG-OECD, Professor R.L. Johnson, Rensselaer Polytechnic Institute, Troy, New York, for his continuing interest in the application of systems techniques to tribological problems and to the present Chairman of IRG-OECD, Dr. G. Salomon, The Hague, for fruitful joint work in this field. In particular, I like to express my sincere thanks to Professor G.W. Rowe, University of Birmingham, who reviewed the complete draft of this book making various valuable suggestions.

The central theoretical part of this book, namely Chapter 3, "General Theory of Tribology" has been worked out together with Professor J. Mølgaard,

Memorial University of Newfoundland, during his stay at BAM from September 1975 to June 1976 as Fellow of the Alexander-von-Humboldt-Foundation of the Federal Republic of Germany. I am very grateful to Professor Mølgaard for the excellent co-operation and his contribution to this book. Thanks are due also to Ir. A.W.J. de Gee and A. Begelinger, Metaalinstituut TNO, Apeldoorn, for their contribution of practical application examples presented in Sections 8.5.3, 8.5.5, and 8.5.7. In order to make the results of this international joint work easily accessible to the international community, the book has been written in the English language.

The entire manuscript has been graciously read by my colleague Dr. K.-H. Habig who also contributed some useful comments. In addition, my colleague Dr. H.-U. Mittmann reviewed Chapter 6 and Dr. D.H. Buckley and Dr. L.D. Wedeven, NASA Lewis Research Center, Cleveland, reviewed Sections 4.4 and 4.5, respectively, making helpful remarks. The Appendix was checked by P.M. Ku, Southwest Research Institute, San Antonio, who also contributed some supplements. Further I am indebted to many authors and publishers for kindly supplying illustrations and permission to publish them; acknowledgements are presented in the figure captions and in the list of references. Lastly, I acknowledge the valuable assistance of my colleagues at BAM, including the experimental help of the engineers W. Evers, M. Gienau, W. Schrag and J. Schwenzien.

In spite of all the encouraging support, the book would never have been finished without the continuous help of my wife Barbara who not only typed the preliminary drafts but also produced the excellent camera-ready manuscript. In small token for their patience during the time of the preparation of the manuscript, the book is dedicated to my family, my wife Barbara and my son Carsten.

Berlin-Dahlem Horst Czichos

May 1977

Contents

To

Barbara and Carsten

1 Introduction and background

1.1 Dynamics of Physical Systems and the Occurrence of Friction and Wear

The technical function of numerous engineering systems - machines, instruments, vehicles, etc., - depends on processes of motion. According to its basic physical definition, the term motion denotes the change of the position of an object with time. In a larger sense, the study of how things change with time, and of the forces that cause them to do so is the objective of dynamic investigation.

A great many processes in nature and technology depend on the motion and the dynamic behaviour of solids, liquids and gases. For example, processes of motion like the flow of blood in veins or the transmission of forces by moving machine components are essentially necessary for the functioning of countless biological systems and mechanical engineering systems. Moreover, some obviously non-mechanical disciplines, like electronics or optics, are also based on motion processes, in these cases on the motion of sub-microscopic particles, i.e., electrons, ions or light quanta. It appears that processes of motion link the different disciplines as an underlying important principle. Consequently, the analysis of dynamic behaviour has become a keystone to much contemporary science and technology (Ref. 1.1).

A feature common to all processes of motion is the occurrence of effects of "resistance to motion", i.e., the occurrence of friction of some kind or another. The effects of friction are due to physical interactions between bodies or objects moving relatively to each other. As a consequence of friction, the process of motion and the dynamic behaviour of the whole system are influenced or disturbed and some part of the energy of motion is dissipated. Further, if in a dynamic system one moving component consists of a solid body, the effect of friction is accompanied in general by wear: "the progressive loss of substance from the operating surface of a body occurring as a result of relative motion at the surface" (Ref. 1.2). In Figure 1.1, the basic types of motion of solid bodies together with relevant engineering systems and types of resistance to motion and mechanisms of damage are illustrated schematically.

Surface motion	Technical systems	Resistance to motion	Mechanism of damage
← solid / gas	air bearing plane turbine engine	air friction	fluid erosion
← solid / liquid	hydraulic drives ship pipeline	viscous friction	cavitation erosion
← solid / solid	dry bearing wheel and rail brakes	solid friction	wear

Figure 1.1 Types of surface motion and related subjects.

Contemporary technology has developed to a level where the problems of moving surfaces require careful consideration and an integration of all the knowledge relevant to the dynamics of multicomponent systems. For example, in an aircraft literally thousands of parts are subject to friction and wear, for example brakes, bearings, seals, gears, actuators, pumps, flight control systems, instruments, splines and even "static" joints. Whereas, in the past, in the design of dynamic mechanical systems the problems of friction-induced energy losses and wear-induced materials losses have been overlooked to some extent, the future will be seriously concerned with the conservation of energy and materials in engineering design. In order to embrace all aspects of moving surfaces and the transmission and dissipation of energy and materials in mechanical systems, the word "Tribology" was coined by a British committee in 1966 from the word "tribos", which means "rubbing" in classic Greek (Ref. 1.3).

The phenomena of tribology - its fundamental concepts, its basic physical processes and its influences on the function and structure of mechanical engineering systems - are the subject of this book.

1.2 Historical Development of Tribology

> " It is quite difficult to do accurate quantitative experiments in friction, and the laws of friction are still not analyzed very well, in spite of the enormous engineering value of an accurate analysis".
>
> Richard P. Feynman
> - Nobel laureate -
> Feynman lectures, 1963

It appears incomprehensible that such obviously old problems like friction and wear still remain unsolved problems in many aspects. Moreover, in many fields of advanced technology, these problems seem to be more important at present than in former times. Some of the reasons for this fact will be visible from a consideration of the historical development of studies in these fields. Without going into all details, it may therefore be valuable to give a survey of the development of our knowledge on this subject matter. (For further details see Refs. 1.4 - 1.8.) In chronological order, the review deals first with friction and then turns to the subjects of wear and lubrication.

1.2.1 Friction studies

It is likely that even in prehistory man was interested in two aspects of friction. The first practical application seems to have been the use of frictional heat in the lighting of fires. The second practical aspect was to overcome friction in the transportation of materials. It seems probable that man has used sledges for transportation for almost 9,000 years. For the enormous engineering transportation problems connected with the erection of the pyramids, the Egyptians were already using liquid lubricants in order to minimize the work required to transport heavy objects.

It has been said that one of the greatest invention man ever made was that of the wheel. In fact, by proper use of this invention it is possible to reduce the work needed to overcome friction considerably. From the results of archaelogical studies, it might be concluded that the merits of rolling motion compared with sliding had been recognized some 5,000 years in Mesopotamia.

The scientific study of the dynamics of motion and of friction phenomena is, however, much more recent than these early applications might suggest. It starts, as is known, with the work done 360 years ago by Galileo. Galileo made a great advance in the understanding of motion when he discovered the "principle of inertia": if a moving object is unperturbed - if it is not disturbed by friction forces - it continues to move with a constant velocity in a straight line. The next step in the scientific development of dynamics was done by Newton in formulating the celebrated basic laws of classical mechanics.

Supplementing the laws of Galileo and Newton on the mechanics of ideal motion, Amontons in 1699 and Coulomb in 1785 formulated some rules of solid friction based on extensive experiments (Ref. 1.9, 1.10). For the purpose of his friction experiments, Coulomb developed a special measuring equipment named a "tribometer" (see Figure 1.2).

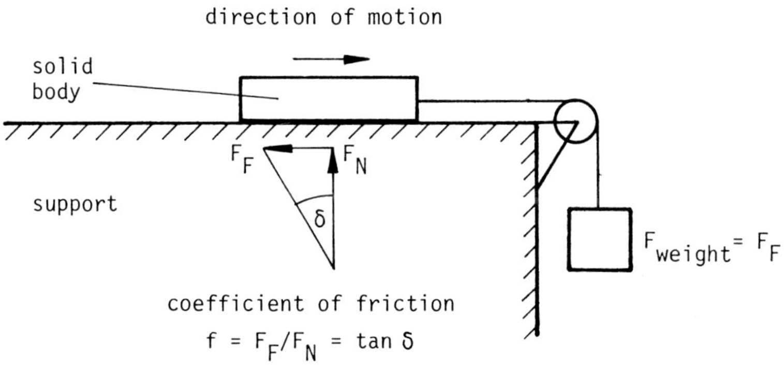

Figure 1.2 Tribometer used by Coulomb for friction measurements.

The experimental results of Amontons and Coulomb showed that the friction force F_F is proportional to the normal load F_N. The ratio of the friction force to the normal load has been termed the "coefficient of friction" $f = F_F/F_N$. The experiments indicated that the value of the coefficient of friction is independent of the geometric contact area. In explaining these results, it has been hypothesized that the surfaces were full of irregularities and that the friction originated in the interlocking of asperities of the contacting surfaces. This theory became known as the "roughness hypothesis of friction". Coulomb also considered the possibility of adhesion processes in friction as suggested by Desagulier in 1724. Desagulier observed strong adhesion between lead spheres when they are pressed together and he considered that similar forces might be involved in friction (Ref. 1.11). Coulomb in his interpretation of the friction process, however, rejected this adhesion hypothesis. Preceding the work of Amontons and Coulomb, Leonardo da Vinci had already found, by the middle of the fifteenth century, that the friction force was proportional to the load (Ref. 1.12). From the results of his experiments, he concluded that the friction force was one quarter of the load and constant for all materials (i.e., f = 0.25 = constant).

Although these findings seem to answer some basic questions of mechanical engineering, from a physical point of view the origin of friction cannot be explained in terms of the classical mechanics of rigid bodies. If the surface asperities are assumed to be rigid, an energy balance shows that the process of merely lifting one sliding surface over the bumps of the counterface consumes no mechanical energy whereas in reality energy is dissipated. It follows that the findings of Amontons and Coulomb can be regarded only as rough empirical rules of limited validity. However, since these rules were formulated at a very early stage of the science of mechanics, these findings have been taken as "laws" of friction. Moreover, the friction behaviour of mechanical systems had been treated in terms of constant coefficients of friction which are assumed to be "intrinsic material properties". As a consequence of:

(i) the development of idealized theories of mechanics in which the effects of friction are neglected entirely,
(ii) the misinterpretation of the coefficient of friction as "intrinsic material constant",

no actual progress beyond the roughness hypothesis was made in the field of friction throughout the nineteenth century and into the twentieth.

As an alternative explanation of the origin of solid friction, the "adhesion theory" of friction was put forward by Hardy (Ref. 1.13) and Tomlinson (Ref. 1.14) in the 1920's. In this theory it is assumed that the friction force is needed to overcome molecular adhesion forces acting between the contacting surfaces. Although this theory appears rather obvious, it requires a friction force proportional to the area of contact. This, however, is in disagreement with the results of numerous friction experiments indicating that the friction force is independent of the contact area.

At this stage of the historical development of the analysis of friction, it was recognized that the physical origins of friction are a very complicated matter which can not be explained by a simple theory but need both accurate experiments and detailed physical analyses. Since, during the 1920's and 1930's, physicists were mainly interested in topics like atomic physics and quantum mechanics it was not until about 1940 that the discrepancies described above were clarified. At this time, Holm (Ref. 1.15), Ernst and Merchant (Ref. 1.16) and Bowden and Tabor (Ref. 1.17) found a great difference between the apparent geometrical area of contact and the "real" area of contact formed by the touching asperities of the two surfaces. In using this observation, together with the assumption that adhesive molecular forces act in the asperity junctions, Bowden and Tabor (Ref. 1.18) could explain the empirical rules of Amontons and Coulomb. It follows that friction has a "dual" nature in that both deformation processes and adhesion processes are involved in friction. This dual molecular-mechanical concept has been generally accepted as a "generalized" theory of friction.

1.2.2 Wear studies

Some of the first systematic studies on the wear behaviour of materials were performed by the great genius Leonardo da Vinci. In his manuscript, found in 1967 in Madrid (Codex Madrid I) (Ref. 1.12), Leonardo described his studies on the wear of simple sliding bearings. He found that wear increases with load and that the direction of wear is not necessarily in a vertical direction but follows the main vector of the load. In order to minimize wear he developed a sliding bearing, suggesting an alloy of 30 percent copper and 70 percent tin to be used as bearing bush.

Apart from the work of Leonardo, the beginning of the study of wear on a scientific basis is far more recent. An early period of systematic wear studies - which has passed almost unnoticed by English-speaking research workers - started in Germany around 1930. Füchsel explained the dry wear of metals as a process of deformation followed by a process of separation (German: "Abblättern") of the deformed metal parts (Ref. 1.19). This theory has been extended by Fink, and Mailänder and Dies (Ref. 1.20, 1.21). They found that the deformed metal parts may oxidize during the wear process, and emphasized the important influence of the environmental atmosphere on the wear of metals. The state of the art of the early period of systematic wear research had been reviewed by Siebel in 1938 (Ref. 1.22). Depending on the nature of external conditions of wear attack, he distinguished between the following different types of wear:

1. Wear during sliding friction of dry surfaces
2. Wear during sliding friction of lubricated surfaces
3. Wear during rolling friction of dry surfaces
4. Wear during rolling friction of lubricated surfaces
5. Wear during oscillating contact
6. Wear by moving solid particles (erosion)
7. Wear by moving fluids (cavitation)

Siebel pointed out that as a consequence of the great variety of different types of wear it seems to be impossible to determine the wear behaviour of a material by one single test and to obtain unique wear values. In a rather modern way of thinking, he suggested close joint research work to be done by physicists, chemists, metallurgists, theorists of elasticity and plasticity, designers, and plant engineers in order to obtain real progress in the most complex field of wear.

The first period of wear research ended in the years of World War II. Thereafter, in the 1950's a certain repetition and extension of this work could be noted, performed mainly in USA and England. For the case of metal/metal wear, Feng, Burwell and Strang were probably the first to recognize the difference between metal transfer from one surface to another and the process of attrition of metal (Ref. 1.23). Kerridge showed that the transfer of metal may be followed by oxidation of the transferred metal and that the last stage of such a process of "mild" wear is given by the attrition of metal oxides (Ref. 1.24). In contrast to mild wear, in processes of "severe" wear the attrition takes place directly at the transferred metal. For certain conditions of sliding dry metal surfaces in a regime of mild wear, Archard found the empirical rule that the wear volume is directly proportional to the load and the sliding distance and inversely proportional to the hardness of the softer of the two interacting metals (Ref. 1.25). At about the same time at which Archard published the empirical rule of dry wear, the main physical mechanisms of wear were classified by Burwell (Ref. 1.26). According to this classification, at least four different main physical mechanisms must be distinguished, namely:

1. adhesive wear
2. abrasive wear
3. corrosive wear
4. surface fatigue wear.

Having briefly reviewed the historical development of the studies of friction and wear up to the end of the 1950's, in retrospect it appears that at that time these subjects were paid increasingly more attention by most of the highly industrialized nations, mainly for economic reasons. Before continuing the chronological review, as a further aspect the science and technology of lubrication - the most important means of reducing friction and wear-induced energy and material losses - should be taken into consideration.

1.2.3 Lubrication studies

The purpose of lubrication is to separate two surfaces moving relative to each other by a fluid film which can easily be sheared without causing any damage to the surfaces. As mentioned above, lubricants were already in use about 5,000 years ago by the Assyrians and Egyptians in facilitating the movement of sledges carrying large statues or buildings blocks of stone.

Apart from this ancient use of the lubricants and the lubrication studies of Leonardo da Vinci, it was only with the advent of accurately made metallic machine parts during the so-called Industrial Revolution that the scientific and engineering aspects of lubrication were consequently studied. Before this period, the scientific interest was concentrated mainly on the development of "pure" theories of fluid mechanics. In this connection an interesting remark of the mathematician John v. Neumann should be quoted (Ref. 1.27). He pointed out that during most of the development of hydromechanics until about 1900, the main interest was virtually the solving of beautiful mathematical problems neglecting terms of internal fluid friction. These approximations, however, had almost nothing to do with real fluids. John v. Neumann characterized the theorist making such analyses as a man who studied the flow of "dry" water.

Most early work on the motion of lubricated machine parts was devoted to journal bearings. In 1883 Petroff proposed the following formula for the frictional force F_F between two coaxial cylinders

$$F_F = \frac{\eta \ v}{h_{av}} \times \text{(wetted area)}$$

where η is the viscosity, v the surface velocity of the shaft and h_{av} the mean effective film thickness. This was the first expression derived for the full-fluid-type of lubrication. (The work of Petroff is discussed in detail in Ref. 1.7.)

The mathematical foundations of all hydrodynamic lubrication theory were laid in 1886 by Reynolds in deriving the famous equations named after him which are the basis on which all subsequent lubrication theory has been based (Ref. 1.28). Fundamental experimental work in this field was performed around the turn of the 20th century by Stribeck at a predecessor institute of the Bundesanstalt für Materialprüfung (BAM), Berlin-Dahlem. Stribeck studied in detail the influence of the different operating variables, like load and velocity, on the lubrication and friction of sliding and rolling bearings (Ref. 1.29). The experimental results of Stribeck were compared by Sommerfeld with his own theoretical friction results which he had derived in integrating Reynolds equation using an ingenious substitution (Ref. 1.30). Biel in 1920 was the first to point out that the type of friction curve measured by Stribeck - the generalized Stribeck curve - is likely to describe the general behaviour of lubricated surfaces as a function of lubricant viscosity, sliding velocity and load (Ref. 1.31). Based on these fundamentals of lubrication theory, criteria for the engineering design of hydrodynamical lubricated bearings have then been developed (Ref. 1.32).

One important question, however, which is left open by the hydrodynamic lubrication theory concerns the limits of full-fluid lubrication. If, for instance, the specific load is very high and the relative velocity is low, it is difficult to build up a sufficiently thick film even with very viscous lubricants. In this lubrication regime, some parts of the sliding surfaces may be covered by lubricant films only one or two molecular dimensions thick. Hardy in 1922 was the first to study in detail this regime of "boundary" lubrication, where the lubricant action is determined by the chemical constitution of the lubricant rather than by its viscosity (Ref. 1.33). Since the work of Hardy, the non-hydrodynamic regimes of lubrication have been studied intensely under different aspects. Further, different chemically active lubricant additives - like extreme pressure additives or antiwear additives - were developed. The development of lubricants with chemically active additives allowed the engineering application of machine elements under conditions which had led to catastrophic failures if lubricated with pure mineral oils. (For instance, the technical application of hypoid gears for rear-axle drive, introduced in 1927 by the Packard Motor Car Company was possible only by using lubricants with chemical additives.)

After World War II, a general trend in mechanical engineering towards higher loads, higher velocities and higher operating temperatures was observed in connection with attempts to reduce the weights of the moving parts in machinery. Since that time there has been a steady increase of interest in the problems of friction, lubrication and wear. Tokens of this increasing interest were, among other things, the appearance of the first international journal entitled WEAR (since 1957) entirely devoted to the science and technology of friction, wear and lubrication. A further new journal, the TRANSACTIONS of ASLE, has been published (since 1958) by the American Society of Lubrication Engineers. In the Federal Republic of Germany a research programme ("Schwerpunkt-Programm") had been launched in 1961 sponsored by the Deutsche Forschungsgemeinschaft (Ref. 1.34). About the same time in England, a Working Group was set up to investigate the present state of lubrication and research and to give an opinion on the needs of industry thereof. In 1966 this Working Group published a report in which for the first time the term "tribology" appeared (Ref. 1.3). The origin of tribology was described in the report "The introduction of a new technology" (1973) as follows (Ref. 1.35).

In the early 1960's there was a steep increase in the reported failure of plant and machinery due to wear and associated causes. At the same time, increased technology and increased capital intensity of plant, and the use of more continuous processes, made breakdowns of such plant and machinery more costly, competitively more serious, and therefore even less desirable than before. This trend was recognized by specialists involved in the subjects of wear, friction and lubrication. The situation seemed to call for more and better education in the subjects and for more and better co-ordinated research.

While trying to establish the reasons for the wide neglect of the subject in the past, despite its technological and economic importance, three principal reasons for this neglect may be mentioned:

(i) The inter-disciplinary nature of the subject which included the disciplines of mechanical engineering, physics, metallurgy and chemistry.

(ii) The fact that only with the advances in technological development in production methods of recent years had attention been focused on the importance and inter-dependence of the constituents of this inter-disciplinary subject.

(iii) The term "lubrication", used in its narrower sense, had not only prevented many people from fully appreciating the economic significance of the subject matter, but it was also a misnomer for the description of the sphere of "transference of force from one moving surface to another" (whether the purpose of the transfer of such forces was associated with high friction, e.g., on brakes, clutches, conveyors or alternatively with low friction, e.g., bearings, slides, etc.)

After consultation with the Editors of the Supplement of the Oxford English Dictionary, the term "Tribology" (Triboscience or Tribotechnology) was recommended for describing the subject matter. Tribology is defined as:

"The science and technology of interacting surfaces in relative motion and of related subjects and practices".

1.3 The Meaning of Tribology

Since its definition in 1966, the term tribology has been widely recognized as a new general concept embracing all aspects of the transmission and dissipation of energy and materials in mechanical equipment including the various topics of friction, wear, lubrication and related fields of science and technology. In order to characterize the significance of tribology, among other things, three points have been emphasized (Ref. 1.36):

(i) The economic aspect of tribology:
It has been estimated that about 30% of the energy generated in the industrial parts of the world is consumed ultimately in friction processes and that in highly industrialized countries, like England, Japan or the Federal Republic of Germany, some \$ 2,000 million per annum are lost as a result of wear processes (Ref. 1.37). Even if these figures are taken as rough estimates only, (Ref. 1.38), they clearly indicate the importance of tribology for the conservation of energy and materials.

(ii) The scientific aspect of tribology:
It is well known that all macroscopic processes in nature are irreversible. Science in its "pure" theories has largely omitted this irreversibility since the laws of "ideal" processes were much easier to develop. Tribology should attempt to investigate the irreversible processes of mechanics in detail and to explain the complex effects of energy and materials dissipation.

(iii) The multidisciplinary aspect of tribology:
Since tribology is defined as "Science and technology of interacting surfaces in relative motion", it includes not only the work of physicists, chemists and materials scientists interested in the surface

properties of materials but also the work of engineers who use "interacting surfaces" for the transmission of motion, forces, work, etc., in various types of machinery. Therefore, tribology is connected with several branches of science and technology, like physics, chemistry, materials science, mechanical engineering, lubrication engineering, etc.

The variety of economic, scientific and technological aspects of tribology is also evident when considering the literature. In the eleven years that have passed since the creation of the term tribology in 1966, the huge number of some 55,000 papers have been published in the field of tribology. According to the BAM Tribology Documentation Service, which annually publishes a compilation (title collection) of tribology papers appearing throughout the world, at present about 8,000 papers are published every year in this field (Ref. 1.39). Due to the multidisciplinary nature of tribology, the papers scatter broadly in purpose and content. Therefore, although some progress in the different specialized subtopics of tribology were obtained, suitable tools are needed in order to link these results and to prepare the way for cooperation between people of widely differing disciplines.

1.4 The Scope of this Volume

The introductory remarks have focussed attention on the problem that tribology, like any multidisciplinary branch of science and technology, poses serious difficulties of communication. The very definition of tribology - "interacting surfaces in relative motion" - indicates already that the behaviour of tribo-technical constructions cannot be predicted from a mere knowledge of single components and their features. Whenever the interplay and interdependence between components is stronger and more complicated than the action of the individual components, a "systems approach" is needed. An understanding of this new outlook is rapidly growing in many areas of contemporary science and technology. In retrospect, it appears that the overlooking of the systems aspects of friction, lubrication and wear problems has been obviously one of the major reasons for the many misinterpretations and the slowness of progress in these fields in the past.

In this volume, we attempt to present a systematic unified approach to tribology based on systems thinking. The book has been written bearing in mind both the specialist who seeks to incorporate his knowledge and practical experience in a general framework, and the general scientist who seeks a systematic overall view of the whole subject. As a result of the unifying attempt, the specialist may find that his special topic is treated in a somewhat unusual manner. It must be borne in mind, however, that in a unifying interdisciplinary treatise a uniform nomenclature and description is necessary which is equally useful for the various specialists in the field, i.e., physicists, chemists, material scientists, lubrication engineers and mechanical engineers. In order to ease the classification of the various aspects of tribology into a general framework, some of the general systems definitions are repeated in almost every chapter.

As a starting point, in Chapter 2, the basic aspects of the system concept are compiled. In applying the system concept, a general theory of tribology in the form of a conceptual framework is outlined in Chapter 3. The theory shows that a complete description of a tribo-mechanical system requires the characterization of the "structure" of the system (system elements, properties of elements, interactions of elements) as well as the characterization of the "function" of the system (inputs, outputs, transfer functions). The theoretical framework of tribology is supplemented in Chapter 4 by a detailed treatise of tribological processes, i.e., contact, friction and wear processes and lubrication modes. The system approach is then used in studying in Chapters 5 and 6 the influence of tribological processes on the structure and function of mechanical systems in connection with tribo-induced changes of the properties of the system elements and material losses as well as stick-slip effects, mechanical efficiency, functional failures and reliability. From these discussions, conclusions for solutions of no-wear conditions and the requirements for a proper functional behaviour of mechanical systems are drawn. In Chapter 7, the various methods of tribometry, i.e. the test, simulation and control methods are described, ranging from laboratory friction and wear tests and simulative tribo-testing to machinery condition monitoring. Finally, in Chapter 8 a tribological systems data sheet suited for a comprehensive data representation is developed and the application of "systems methodology" to practical friction and wear problems is outlined. The Appendix gives a compilation of basic tribo-engineering systems and its components and refers to bibliographic work in the field of tribology provided by the DOCUMENTATION TRIBOLOGY.

2 The system concept

2.1 Introduction

The problem of dealing with complex multidisciplinary subjects like tribology appears to be essentially one of the limitations of "analytical procedures" in science and technology. The theoretical biophysicist Ludwig von Bertalanffy, the founder of "General Systems Theory", has given the following definition of meaning of analytical procedure (Ref. 2.1): An entity; i.e., the object of an investigation, should be resolved into the Parts from which it is combined; hence it could be constituted or reconstituted from the same Parts. These procedures should be understood both in their material and conceptual sense. This basic principle of "classical" science can be applied analytically in a variety of directions, e.g., resolution of causal relations into separate Parts, searching for "atomic units" in science or for "material constants" in engineering.

Application of the analytical procedure to any of these areas depends on two conditions (Ref. 2.1):

(i) The interactions between Parts must be non-existent or, at least, weak enough to be neglected for certain research purposes. Only under these conditions can the Parts be first singled out actually, logically, and mathematically, and then reassembled.

(ii) The relations describing the behaviour of Parts must be linear; only then is the condition of summativity given, i.e., an equation describing the behaviour of the Whole has the same form as the equations describing the behaviour of the Parts; partial processes can then be superimposed to obtain the total process, and so on.

These conditions are not met in the entities called systems, they consist of "Parts in interaction". A system of "organized complexity" may be circumscribed as one in which strong interactions, which are "non-trivial", i.e., non-linear, prevail.

To deal with complex, non-linear systems, additional restraining principles have to be introduced. The economist Kenneth Boulding has suggested a hierarchy of systems based on the following three rules (Ref. 2.2):

(i) Systems belong to classes of different levels of complexity.
(ii) All logical and empirical laws, valid at a low level (system) are also applicable to any of the higher level systems.
(iii) The higher the level, the larger the number of unknown elements and undiscovered laws that make a particular system work.

To illustrate the system concept and the principle of the hierarchical order of systems consider, for example, a traffic system shown schematically in Figure 2.1.

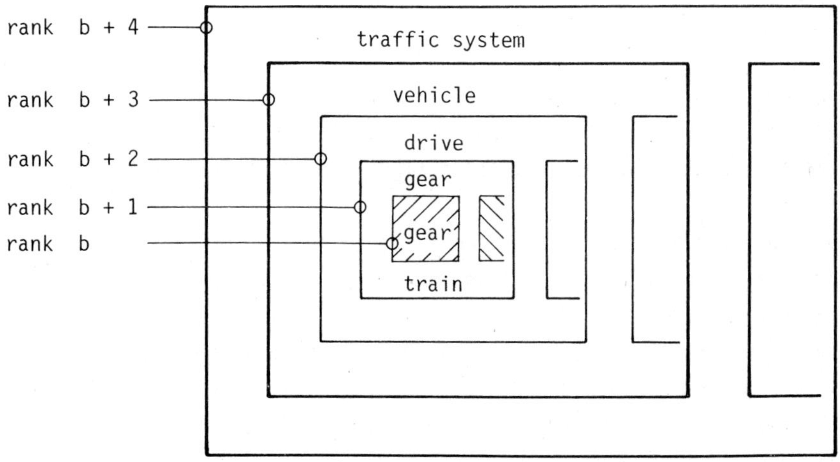

Figure 2.1 Example of a hierarchy of systems.

Depending on the location of a so-called "systems envelope", separating hypothetically the subjects under consideration from their "environment", different levels of complexity or different "ranks" of the system under consideration can be distinguished (Ref. 2.3). The systems of the lowest rank are in general given by the elementary single technical components. (If these elementary parts of technical systems are resolved further, the micro-physical and chemical constituents of the technical systems components are obtained.) Technical systems of the next rank up are obtained in putting together systems of a lower rank by certain technical means or through interactions of lower ranking systems.

In the example shown in Figure 2.1, the elements of the lowest rank, denoted by the symbol "b", are, for instance, technical components like gears, shafts, etc. Through the dynamic interactions of these components, a gear train system of rank b + 1 is formed. This system together with other systems of the rank b + 1 (i.e., the engine system, the clutch system and the power transmission system) form a system of rank b + 2, the drive system. At the next rank up, b + 3, characterizing the whole vehicle, the topmost level of this technical system is reached. At the following level b + 4 man-machine interactions must be taken into consideration. In fact, the whole traffic system of rank b + 4 results through dynamic interactions of vehicle + driver + road + atmospheric conditions.

2.2 Description of a System

After the general survey of the systems concept, a brief simplified compilation of relevant systems definitions and characteristics suited for the purpose of this book is given in this section. (For further details see Refs. 2.4 to 2.7.)

The general definition of a system is contained in the sentence: "A system is a set of elements interconnected by structure and function".

A system is termed open when interchanges of mass and energy with the "outer world" occur, while such interchanges are assumed to be negligible in a "closed" system. In the first case, for example, the transformation of kinetic energy into heat or other forms of energy through irreversible processes leads to "dissipative systems".

The main characteristics of a system are summarized in Figure 2.2, the symbols used have the following meaning.

(I) Structure

The structure of a system is defined by

(a) the set of its elements (A),

(b) the relevant properties of the elements (P),

(c) the coupling of elements, specified as relations between the elements (R).

With these definitions the structure of a system is represented by the set:

$$S = \{ A, P, R \}$$

(II) Inputs, Outputs

Each system can be separated schematically by a hypothetical system envelope (or control surface) from its "environment". The connections between the system and its environment, which are cut by the envelope may be classified as:

(a) Inputs $\{X\}$
and
(b) Outputs $\{Y\}$

(III) Function

The function of a system - utilized for a certain (technical) purpose - is to transform the inputs $\{X\}$ into the outputs $\{Y\}$. The transformation (T) of the inputs into the outputs may be described either in terms of mathematical equations or as a physical analog, or as a verbal description, etc.

Definition: A system is a set of elements interconnected by structure and function

(I) Structure $S = \{A, P, R\}$

(a) Elements

$A = \{a_1, a_2, \ldots , a_n\}$ (n: number of elements)

(b) Properties

$P = \{P(a_i)\}$

(c) Relations

$R = \{R(a_i, a_j)\}$

(II) Inputs $\{X\}$, Outputs $\{Y\}$

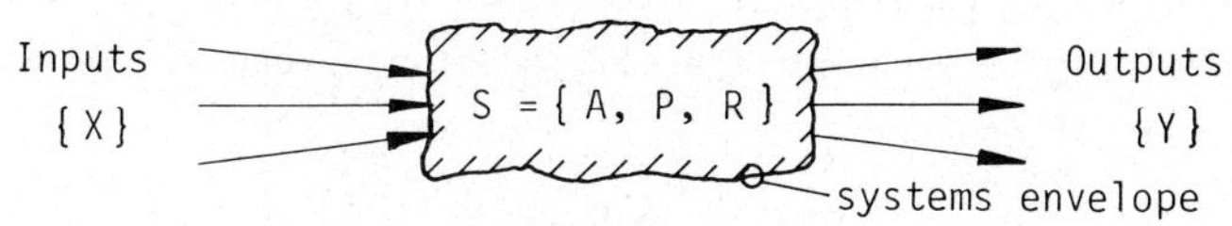

(III) Function

$$\{X\} \xrightarrow{T} \{Y\}$$

Figure 2.2 General description of a system.

For the characterization of the function of a system, mainly three different methods may be distinguished (for further details, see Ref. 2.7):

(a) Dynamic State, Differential Equations
If the inputs and outputs vary with time, the system is said to be in a "dynamic state". This dynamic state may be represented by a set of differential equations, named "equations of motion".

(b) Steady State, Linearity
In certain cases a system may be in dynamic equilibrium, i.e., "steady state". Further, the outputs $\{Y\}$ can often be described as linear superpositions of the inputs $\{X\}$ through an algebraic representation.

(c) Stochastic Processes, "Noise"
In real systems the functional input-output-relations may be influenced by stochastic processes, i.e., dynamic effects of uncertainty and random disturbances, "noise". In such cases, an estimate of the limits of proper function behaviour by means of the theory of probabilities can be attempted.

In describing the behaviour of a system by the terms "structure" and "function" a fundamental difference should be noted:

- A "structural" description of a system is essentially an "internal" one, viz., an attempt to understand the system's behaviour in terms of its elements (the Parts) and their interdependence.
- A "functional" description of a system is "external", characterizing the system's behaviour by its interactions with the environment through its input-output relations.

2.3 Energy Balance, Networks, Analogies

In characterizing the dynamical behaviour of systems, the concept of a generalized energy balance has to be introduced. This means that the system's net power equals zero if we summarize all important processes of storage and transformation of energy:

$$\Delta E = 0$$

In this connection it is important to consider analogies between physical systems (Ref. 2.8), Table 2.1.

The physical variables of a system (inputs and outputs) may be classified conveniently as either "through" or "across" variables. Through variables measure the transmission of something through an element, either an electric current through a resistor or a force through a dashpot. Across variables measure a difference in state between the ends of an element, as the voltage drop across a resistor or the difference in velocity between the ends of a dashpot. Of special importance is the electric analogy of mechanical systems described by the two generalized Kirchhoff's laws:

I Kirchhoff's node law.
The sum of all "through variables" into any node of a system must be zero. (In the case of mechanical systems this is represented by force equilibrium relations according to D'Alembert).

II Kirchhoff's mesh law.
The sum of all "across variables" around a complete loop of the system is equal to zero.

These equilibrium and compatibility relations are valid for through and across variables of electrical systems, mechanical systems, thermodynamic systems, and fluid systems. In many applications one starts directly from Kirchhoff's laws in order to obtain the rules describing the functional behaviour of technical systems.

Discipline	Across variable (effort)	Through variable (flow)
Electrotechnics	electric voltage u	electric current i
Mechanics (displacement)	velocity v	force F
Mechanics (rotation)	angular velocity ω	torque M
Thermodynamics	absolute temperature T	entropy flow $\dot{S}$
Fluidics	pressure p	volume flow $\dot{V}$

Table 2.1 Analogies between variables of physical systems.

It should be mentioned that in the bondgraph technique described below, instead of the force-flow, velocity-effort analogy shown in Table 2.1, a reverse analogy, i.e., a force-effort, velocity-flow analogy is often used. Although mathematically it is of no consequence which analogy is taken, the difference in analogy is important for the comparison of specific results. A comparison of both analogies can be found in Ref. 2.9.

2.4 Bond Graph Methods

The bond graph method, initiated by H. Paynter is a further generalization of the modelling of engineering systems (Ref. 2.10). The original concept central to the bond graph method is that of a multi-energy port representation associated with a "reticulation", i.e., the making of a network of the system (reti, Latin: fishnet). Paynter considered that physical or engineering systems could be represented by a multi-port reticulation in a given region of space, not unlike that used to represent the bonding between atoms in complex chemical molecules. The graphical symbolism of the bond graph method is economical and efficient. The bond graph elements can be represented initially by words and the bonds for power exchange between them as simple lines. For example, Figure 2.3 contains a variable drive, e.g., a Diesel engine with a pump that supplies fluid to a consumer (Ref. 2.9). The energy flux or power in a bond is always the product of two variables - the potential variable called effort and the flow or current variable, simply referred to as flow.

$$\text{Controller} \frac{F}{v} \text{Drive} \frac{M}{\omega} \text{Pump} \frac{p}{\dot{V}} \text{Consumer}$$

Figure 2.3 Word bond graph for a pump drive.

In recent years, the bond graph methods have been developed further, taking into account the effects of energy conversion and dissipation and providing methods for the simulation and electronic data processing directly from a bond graph with the ENPORT program (Ref. 2.11). The ENPORT computer program can provide dynamic responses directly from a suitable bond graph model and specified excitations, without requiring the explicit prior formulation of state equations. Thus bond graph methods offer a unified way to proceed from the basic physical modelling of components, devices, and their connections to analytical and computational results for complex systems involving a variety of types of energy flow.

2.5 Classification of Systems

The abstract formalism of the description of a system reviewed in brief in the foregoing sections should be supplemented by a classification scheme indicating the "type" of system under study. Clearly, physical or engineering systems are composed of materials components whose properties, interrelations, etc., may change with time. Therefore, parameter time is an independent attribute of any of these systems.

Systems can be classified from many different points of view. No classification is complete and perfect for all purposes. The different methods of classification are discussed in detail in Ref. 2.5. According to Norbert Wiener, the founder of "cybernetics", the inputs and outputs of systems may be broadly classified into the three categories "matter", "energy", and "information" (Ref. 2.12). (See also Ref. 2.13, p. 99 f.) Without specifying the internal structure of a system, the most general characterization of a system is then given by a "black-box" diagram as shown in Figure 2.4.

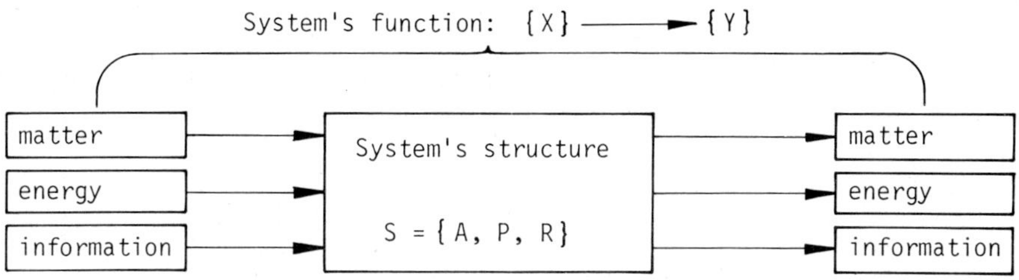

Figure 2.4 General black-box diagram representation of a system.

In using the broad classification of the inputs and outputs of a system, from a functional point of view the entity of physical and engineering systems can be formally described by a matrix representation as shown in Table 2.2. In this scheme the three basic categories of inputs and outputs are sub-divided into some of their sub-quantities. Every point of intersection of inputs and outputs represents a certain class of systems and, correspondingly, a certain field of science or technology. The systems and topics characterized by Table 2.2 range from the class at the upper left:

matter (solid) ⟶ matter (solid)

(corresponding fiélds of science and technology are transportation technology and manufacturing)

to the class at the lower right:

information (digital) ⟶ information (digital)

(corresponding fields of science and technology are the many branches of computer science)

Inputs			Outputs										
			a	b	c	d	e	f	g	h	i	j	k
matter	a	solids											
	b	liquids											
	c	gases											
	d	particles											
energy	e	mechanical											
	f	thermal											
	g	electric											
	h	magnetic											
	i	nuclear											
information	j	analog											
	k	digital											

Table 2.2 Matrix classification scheme of systems.

In refining the matrix representation as illustrated in Table 2.2 still further, it is also possible to obtain sub-classification schemes for the various groups of systems.

The system concept as reviewed in this Chapter is used in the next Chapter as basis for the development of a unified general theory of tribology in the form of a conceptual framework. The reader who is mainly interested in the practical application of systems techniques may first read Chapter 8 on "Practical Systems Methodology" and may then turn back to Chapter 3.

3 General theory of tribology

3.1 Application of Systems Concepts to Tribology

The review of the system concept given in Chapter 2 shows that a system should be analyzed in terms of its "structure" (elements, properties of elements, interrelations between elements) as well as in terms of its "function" (inputs, outputs, transfer functions). In developing further the methods of the application of the system concept to tribology (see Refs. 3.1 to 3.4) we will attempt in the following to outline a simplified general theory of tribology (Ref. 3.5).

The systems approach to tribology is not intended to replace more traditional analyses of friction and wear processes. Much of the information required for the portrayal of tribological phenomena from a systems standpoint can be obtained through conventional procedures of analysis. For instance, important variables such as force, velocity, temperature, chemical potential and the related variables stress, strain, work, power, heat flow and reaction rates can only be determined by standard methods. In the application of systemanalytical methods to a certain complex subject, systems analysis often takes the form of "model building", i.e., the representation of a system graphically or analytically in a manner which may permit the model to be used for a mathematical study of the systems performance. This is not intented here, nor generally possible at the present level of our understanding of the tribology of mechanical systems. The purpose of this chapter is the development of a logical systemanalytical framework, whereby current tribological knowledge can be better organized, taking into account the various aspects and influencing factors. In this general theory, a detailed mathematical analysis is not necessary for a discussion of the pertinent basic concepts. Thus, in the following, systems thinking is applied rather than a formal systems analysis.

In attempting to apply the system concept to the subject of tribology, a fundamental difference between the behaviour and the description of electrical systems and mechanical systems in which friction and wear processes occur must first be emphasized.

Consider, for example, an electrical transformer and a mechanical gear box, illustrated in Figure 3.1 as block diagrams. The technical design of both systems is to transform certain inputs - voltage u and current i in the electrical system and angular velocity ω and torque M in the mechanical system respectively - into outputs used for the technical purposes. The technical function of both systems can be described formally as a transformation of the inputs into the outputs via a certain transfer function (T).

The dynamic performance of both systems is accompanied by perturbations on their function and structure. In both systems, energy losses due to electro-magnetic or friction resistances, respectively, occur. The fundamental difference between the behaviour of the electrical and mechanical systems is that in the electrical case the structure of the system generally remains constant with time. In this case, the transfer function (T) can be worked out mathematically. This has led to various applications of the powerful systems engineering methods of network theory and related methods characterizing the functional behaviour of electrical systems or related models. In the mechanical case, however, the structure of the system generally changes with time through the action of the tribological processes of friction and wear. Therefore, for mechanical systems a systems description in mere terms of input-output relations is not sufficient: the functional description of the system must be supplemented by a detailed study of the structure of the system and the influences of tribo-induced structural changes on the functional behaviour of the system.

In order to develop a convenient framework for the description of systems with both a dynamic functional behaviour and a dynamic structure state, some simplifications must be introduced: a first simplification should be made by applying the methods of "black-box cutting" or "systems tearing". Black-box cutting has worked exceedingly well in physics and chemistry: a model is cut down to smaller boxes until first principles can be applied. Systems tearing can be achieved by locating the hypothetical systems envelope in a convenient manner. In the following treatment the systems envelope is located, as narrow as possible, around the central parts of the mechanical

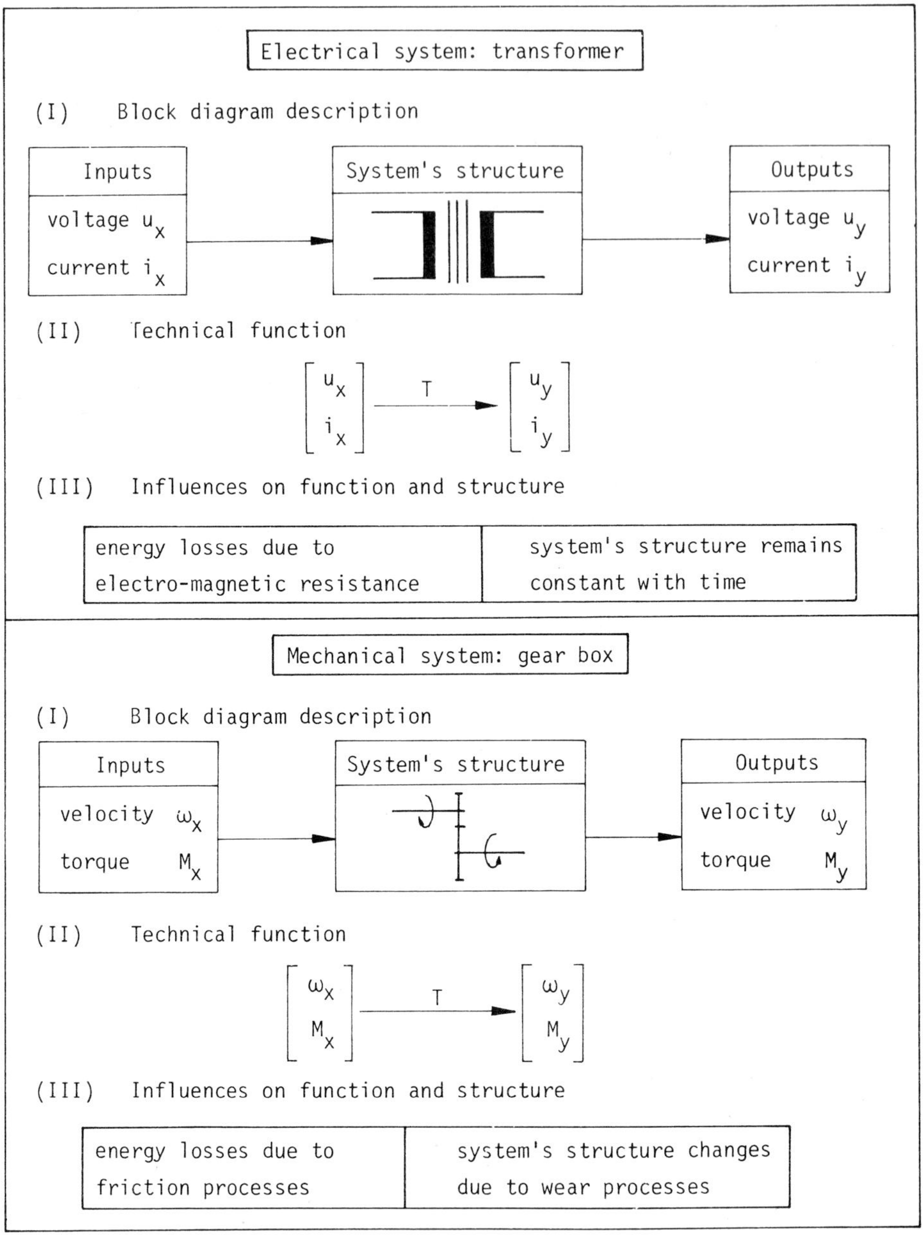

Figure 3.1 Comparison of the characteristics of an electrical and a mechanical system.

system, forming the well-known "interacting surfaces in relative motion". With these reservations in mind, a detailed systems description must involve the following steps:

(I) System's function
- (i) Separate the system from its environment by the choice of a system's envelope,
- (ii) Compile all inputs and outputs
- (iii) Describe the functional input-output relations

(II) System's structure
- (i) Identify the "elements" of the system,
- (ii) Characterize the interrelations and interactions between the elements,
- (iii) Specify the relevant properties of the elements.

Based on this systemanalytical procedure, in the following sections we will attempt to study and compile the basic influencing factors and mechanisms relevant to the function and structure of mechanical systems in which friction and wear processes occur. The theoretical framework obtained will then be used for various analyses and applications in the following chapters of this book.

3.2 Function of Tribo-Mechanical Systems

In the attempt to apply the system concept to the phenomena of friction, lubrication and wear, the question of the functional, technical purpose of the system under consideration has first to be answered. A tribo-mechanical system is defined as an entity whose functional behaviour is connected with interacting surfaces in relative motion. (The words "tribological system" or "tribo-system" may be used synonymously to the word "tribo-mechanical system".) The technical aims realized through moving surfaces may range from aerospace applications to biomechanical joints. However, from a physical point of view, four basically different groups of technical purposes can be distinguished as illustrated by the examples shown in Figure 3.2.

The most general technical purpose of a tribo-mechanical system is the realization of motion through various types of "bearings". The other basic groups are the transmission of mechanical work, the transmission of information - for instance the control of machine functions with cams - and the forming of materials. As a supplement to the basic classification of tribo-mechanical systems into four different groups, illustrated in Figure 3.2, a detailed function-oriented compilation of tribo-mechanical systems is given in the Appendix (A).

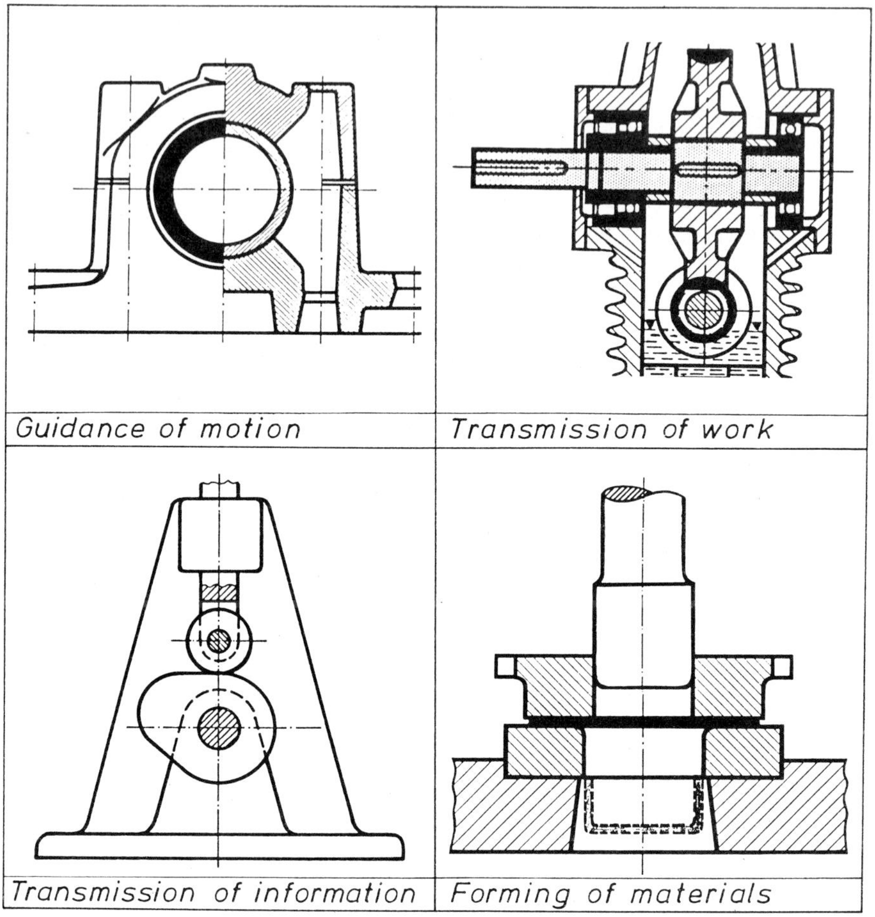

Figure 3.2 Types of tribo-mechanical systems.

In an abstract and highly simplified description, the function of the different tribo-mechanical systems consists, basically, in converting the inputs - for instance, motion, mechanical energy and materials - into outputs, which are used technically. The functional cause-and-effect relations between inputs and outputs are accompanied by loss-outputs of mechanical energy and of materials, denoted summarily by the terms friction and wear losses. From a point of view external to the system, the system may be treated as a black-box with inputs and outputs, as shown schematically in Figure 3.3.

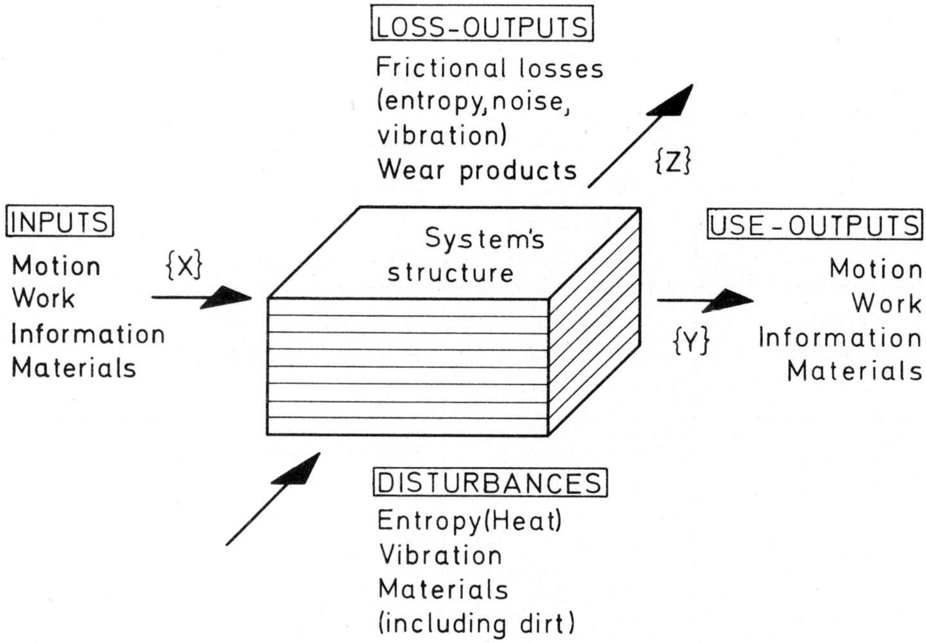

Figure 3.3 Black-box description of tribo-mechanical systems.

The technical function can often be expressed as a relationship between an input and a useful output. Not all the inputs may be considered desirable, some may be regarded as disturbances. Not all the outputs may be desired: such outputs may, from the practical point of view, be considered as losses. In general, part of a functional input may be lost, appearing as a loss output. However, the loss output need not be quite the same type of quantity as the input or useful outputs.

In relating a system to its function, we are concerned with operational variables which can be controlled by a designer or an operator, i.e., variables which we can alter before or during the operation of a system without altering the physical composition of the system in terms of the materials or shapes of the elements. Table 3.1 lists some operational parameters that we may be concerned with, and which we may also refer to as operational variables in the environment. The actual variables used to describe the relationship between a system and its environment will depend to some extent on the system under consideration, but will generally be related to, or derived from, the variables listed.

Input and output quantities	Primary variables	Related and derived variables
Work (Motion)	Force/Torque Position/Dimension	Work rate (Power) Rotational velocity Linear velocity
Thermal (Heat transfer)	Temperature Specific heat Specific entropy	Heat transfer rate Entropy transfer rate
Mass (Mass transfer)	Composition Free energy	Rate of mass transfer Rate of free-energy transfer
Information	Position Force/Torque	Frequency Phase

Table 3.1 Operational inputs and outputs of mechanical systems.

3.3 STRUCTURE OF TRIBO-MECHANICAL SYSTEMS

According to systems theory, the structure of a system is characterized by the elements or components of the system, their relevant properties and their interrelations. The simplest structure of a tribo-mechanical system is given by two solids (1) and (2) exchanging mechanical inputs and outputs through their contact interface. A compilation of examples of the elements or components of tribo-mechanical systems is provided in the Appendix (B).

A few words may be appropriate here on the flow of one or more quantities through a system, implied by the use of the terms "input" and "output", e.g., work or mass. A mass flow is obviously a flow of a material quantity. However, the "flow" of other quantities, such as work, can perhaps be regarded more correctly as an influence of one element upon another. The notion that a quantity "flows" is therefore merely a useful convention, which incidentally helps us to distinguish between the different kinds of process which can occur in a system.

The most sensible manner, in which to handle the different types of quantity is to treat each formally on a plane of its own, linked to the other quantities by transformations between the planes. This leads to a conceptual three-dimensional internal view of a tribo-mechanical system, shown in Figure 3.4. The picture can be understood as a resolution of the systems structure (the central box of Figure 3.3) into greater detail, see also Figure 2.4. It is likely that all the parameters, identified in a study of the function of the system, will be represented on planes for the function, work and thermal processes. Separate planes may be used for each of the chemical species taking part in tribological processes. This includes chemical reaction products.

Figure 3.4 might be taken to represent a very "simple" system of just two solid mechanical elements of the same material in contact, so that the interfacial space at the region of contact is the third element, i.e.,

Element (1) First machine element
Element (2) Second machine element
Element (3) The interfacial volume

The possible transactions and processes between three such elements are, in fact, quite complex even for this elementary system (Ref. 3.6). We do not attempt here to represent all the possible processes in Figure 3.4, but rather to illustrate and classify the various types of processes which may occur.

Consider first the transfer of the technically used inputs $\{X\}$ and outputs $\{Y\}$ of the system. On each conceptual plane, the transfer of such a quantity from one element to another may occur indicated by horizontal arrows. The term "translation" is chosen for these transfers. Consider now the processes of transfers between different planes, referred to as "transformations". The vertical solid arrows in Figure 3.4 represent a transformation in material or chemical species, occurring, for instance, as a result of wear processes within the system. The vertical dotted arrows represent the transformation of work or mechanical energy to heat or entropy, occurring as a result of friction processes within the system. Both translations of mass, e.g., the flow of lubricant through the system and transformations of mass in wear processes are likely to involve work, accompanied by the simultaneous transformation of the work into heat or entropy which appears on the thermal plane.

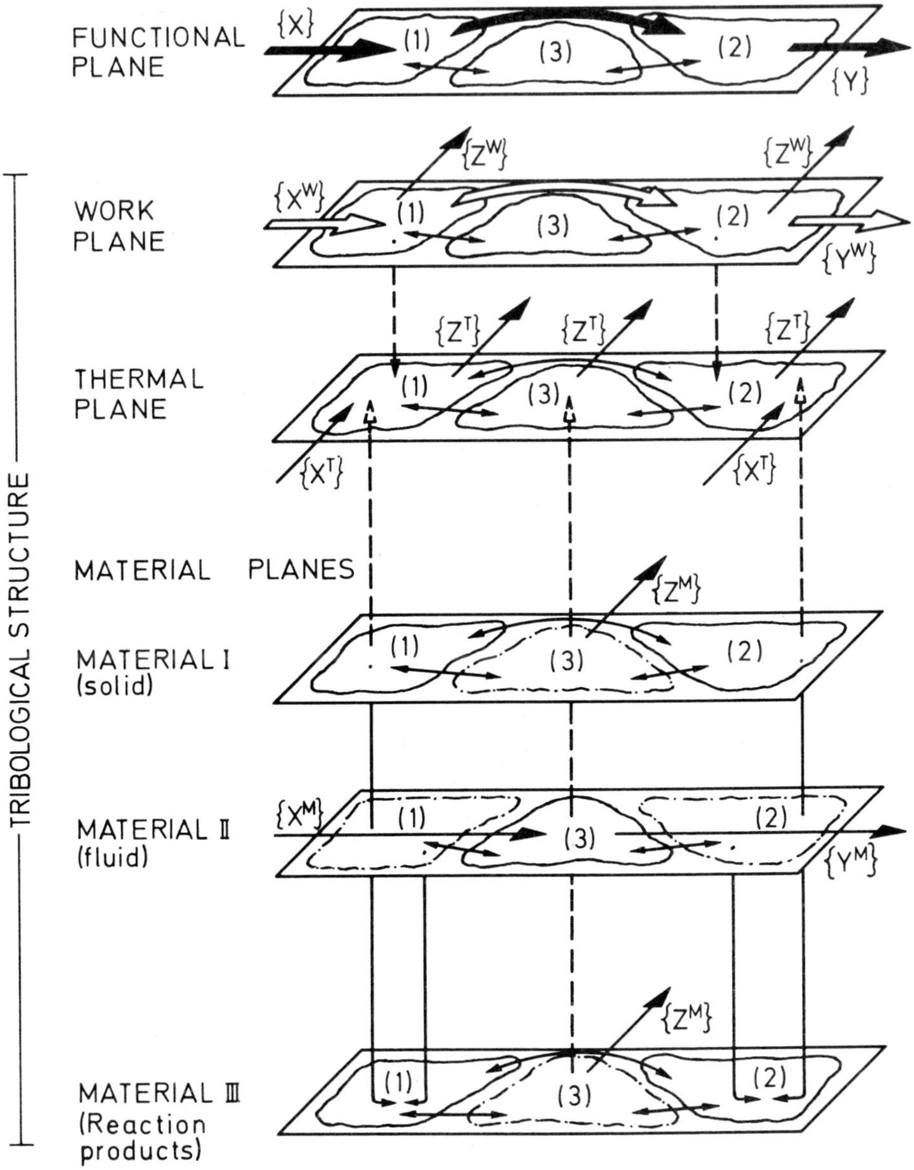

Figure 3.4 The Tribo-process diagram: Conceptual planes of parameters and processes in a tribo-mechanical system.

The abstractions of the tribo-process diagram, Figure 3.4, clearly demonstrate the significance of the systems approach to the analysis and description of mechanical systems in which friction and wear processes occur:

The characteristic feature of tribology is the interaction between the different planes of Figure 3.4, while the conceptual analysis of single separate planes reverts to more conventional analytical simplification.

At this point it is relevant to indicate the aspect of lubrication in relation to the general framework of tribology. Lubrication represents a deliberate attempt to avoid or reduce the effect of friction and wear upon a mechanical system. A lubricant can also act, as it flows away, as a cooling agent removing heat from the location of the friction process. If the sliding or rolling surfaces, i.e., elements (1) and (2) in Figure 3.4, are completely separated by the action of the element (3) at all times, there may be no wear process. In this event, the analysis is simplified, since a materials transfer between the materials planes shown in Figure 3.4 will be of no concern ("no-wear model"). However, if in a lubricated state there is some contact between the surfaces of the elements (1) and (2), or between boundary lubricants on the surfaces, the processes represented on and between the materials planes are of paramount concern. In such cases, the presence of a lubricant may complicate the analysis relative to the unlubricated system, partly because the reaction products present may be complex and difficult to characterize, and partly because transient conditions may be the major concern.

3.4 Tribological Interactions

The separation and representation of the processes within a mechanical system, onto conceptual planes shown in Figure 3.4, allows us to concentrate on one aspect of the system at a time.

3.4.1 Functional plane

The emphasis on the functional plane, "extracted" from Figure 3.4 and shown in Figure 3.5, must be on a representation of the technical purpose of the system and the evalution and representation of the parameters which have most immediate relevance to this technical purpose.

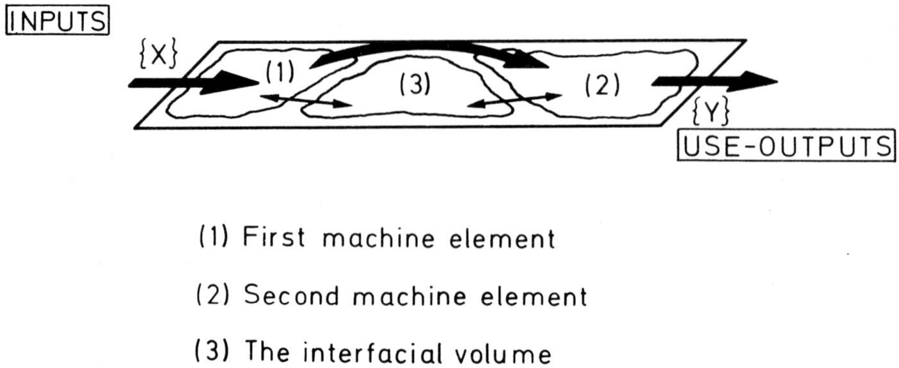

Figure 3.5 The functional plane of the Tribo-process diagram: Functional transactions in a tribo-mechanical system.

We may wish to classify mechanical systems according to their function. Invariably, motion is a characteristic of a tribo-mechanical system. This motion may constitute a transfer of work, materials or information. In some instances, the purpose of a system may be to change a rate of motion or to eliminate it altogether. It is also often desired to restrict motion, i.e., to reduce the number of degrees of freedom a machine element may possess. In other instances, materials are not merely moved, but also changed in state or form. Mechanical devices which produce or transfer information are still common, but are being steadily replaced by devices in which there is little or no mechanical motion (viz. the replacement of the mechanical clock by digital electronic clocks.) These main aspects of the function of mechanical systems are listed in Table 3.2.

In order to represent the function of the system in terms of {X} — {Y} input-output relations, electrical network analogies have been applied to mechanical systems (Ref. 3.7). As explained in Chapter 2, these approaches are based on a classification of the operating variables into "through and across" variables or "flow and effort" variables and the application of the well known Kirchhoff's node and mesh laws. Recently, the bondgraph methods have also been applied to tribological subjects (Ref. 3.8). As indicated above, these applications of theories of electrical analogies are restricted, however, to no-wear, steady state models. Therefore, from a tribological point of view, additional aspects must be taken into account.

Particular characteristics in friction and wear are generally be desirable. These characteristics may be dependent on the operating parameters.

In addition, and perhaps more important, there may be certain limits to be identified for these characteristics, outside which limits the system will not perform its function satisfactorily. The identification of properties which are critical to the function of each element concerned will serve as a guide for the analyses carried out on work, thermal and mass planes, to ensure that such analyses are relevant to the function of the system. The relationship between the values of these properties and the function of the system and the manner in which the operation of the system affects these properties, is, from a practical point of view, the crux of a tribological study.

<table>
<tr><th colspan="2">Inputs and outputs needed for technical function</th><th rowspan="2">Primary technical function of the system</th><th rowspan="2">Examples</th></tr>
<tr><th>Main inputs {X}</th><th>Main outputs {Y}</th></tr>
<tr><td rowspan="4">Motion
+
Work</td><td>Motion</td><td>Guidance of motion
Coupling of motion
Annihilation of motion</td><td>Bearings
Clutches
Brakes</td></tr>
<tr><td>Work</td><td>Power transmission (mech., hydr., pneum.)</td><td>Gears</td></tr>
<tr><td rowspan="2">Information</td><td>Generation of information</td><td>Clocks; Cams and followers</td></tr>
<tr><td>Reproduction of information</td><td>Data transducer (audio, video; tape or record)</td></tr>
<tr><td rowspan="2">Motion
+
Materials</td><td rowspan="2">Materials</td><td>Transportation</td><td>Wheel/rail
Pipeline</td></tr>
<tr><td>Forming of materials</td><td>Wiredrawing</td></tr>
</table>

Table 3.2 Technical functions of mechanical systems.

All this can be seen on the functional plane in a systems analysis. We may then imagine ourselves looking down into the tribological structure depicted below the functional plane in Figure 3.4, in order to ascertain the physical and chemical relationship which determine the functional characteristics.

3.4.2 MECHANICAL WORK PLANE

In Figure 3.6, we concentrate our attention on the processes involving mechanical work in the system. It may be that one function of the system is, in fact, to transmit and transform work, in which case this analysis will duplicate to a large extent the analysis on the functional plane. Although this does occur quite often, it is nevertheless a special case.

A system or element may perform mechanical work upon another system or element. Such mechanical work may be measured as the product of a force times a distance (or the product of related variables). Work may also be transferred between elements by the transfer of material in which work is stored. The transfer of work in either manner is referred to as a translation work. Work may also be transmitted from an element or a system as mechanical vibration, or work may be transformed through the tribological interactions in other forms of energy and either stored, emitted or passed to other conceptual planes of the system.

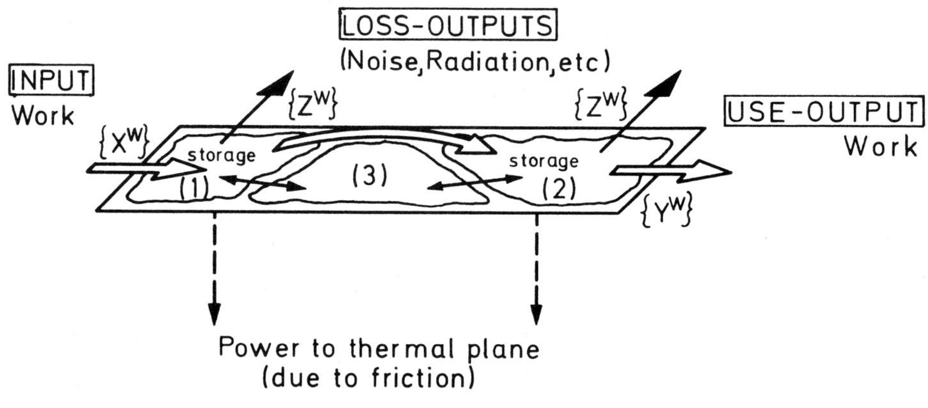

Figure 3.6 The work plane of the Tribo-process diagram: Work transactions in a tribo-mechanical system.

From a generalized energy balance consideration of Figure 3.6, it follows that the input work must be equal to the use-output work + the loss-output energy + the energy stored in the system + the energy transformed to other conceptual planes of the system.

The rate of performance of mechanical work is power. In summarizing the various portions of power, for a system as a whole, a power balance equation is

$$\sum \dot{E}_X^W = \sum \dot{E}_Y^W + \sum \dot{E}_Z + \sum \Delta \dot{E}_s + \sum \dot{E}^{WT}$$

where

$\dot{E}_X^W$: input power

$\dot{E}_Y^W$: use-output power

$\dot{E}_Z$: loss-output energy rate

ΔE_s : stored energy

E^{WT} : thermal energy transformed from mechanical work

Viewing only one element, j:

$$\sum_{\text{all } i} \dot{E}_{ij}^W = \sum_{\text{all } i} \dot{E}_{ji}^W - \sum \dot{E}_j^W + \sum \Delta \dot{E}_{sj} + \sum \dot{E}_j^{WT} + \sum \dot{E}_j^{WM}$$

where the subscript i refers to all elements with which the element j has work transactions and $\dot{E}^{WM}$ to the absorption of mechanical power, as discussed below in the section on materials planes.

The transmission of power $\dot{E}_X \longrightarrow \dot{E}_Y$ in a mechanical system is usually a relatively simple application of mechanics. The question which is less well understood and which often constitutes a basic problem in mechanical systems is the manner in which mechanical power is not transmitted, i.e., "absorbed" by friction or vibration. Thus, the energy or power balance considerations should be supplemented by a discussion of the tribological processes involved in the storage, transformation, emission and dissipation of mechanical work. Clearly, all these phenomena are closely connected with the contact, deformation and friction processes occurring at the interface between elements (1) and (2) of a tribo-mechanical system. In Table 3.3 some of the basic processes are listed. A detailed discussion of the relevant tribological processes is given in Chapter 4.

Work transactions	Processes
Translation of work	Elastic deformation
Transformation of work	Friction mechanisms: elastic hysteresis plastic deformation adhesion
Dissipation of energy	Storage of energy Emission of energy Conduction to other conceptual planes

Table 3.3 Mechanisms and processes of work transactions.

3.4.3 THERMAL PLANE

Figure 3.7 shows the principles of an analysis on a thermal plane of a tribo-mechanical system. We have already referred to the transformation of mechanical work to entropy, i.e., mechanical power to thermal power, which proceeds according to the equation

$$\dot{E}_j^{WS} = \dot{S}_j T_j$$

where $\dot{S}_j$ is the rate of generation of entropy in element j and T_j is the absolute temperature.

The thermal power generation will usually result in the temperature of locations of entropy generation being appreciably higher than their surroundings, resulting in the transfer of the thermal power, by conduction or radiation away from the locations of friction. Heat and entropy can be transferred between systems elements. The transfer of thermal power down a temperature gradient is a unique work process in that the work involved appears as thermal power, i.e., a flow of the same type of quantity as the one being transferred. The net result is that in thermal conduction or radiation, the total thermal-power-flow rate remains constant, i.e.,

$$\dot{Q}_j = \dot{S}_r T_r = \dot{S}_j T_j$$

where $\dot{S}_r$ is the total flow rate of entropy across a surface at a distance r from a source of entropy and at an absolute temperature T_r surrounding a source of entropy S_j at an absolute temperature T_j. $\dot{Q}_j$ is the associated heat-flow-generation rate.

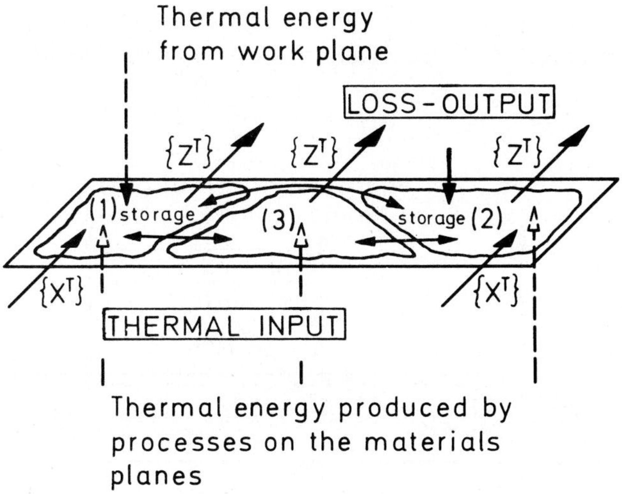

Figure 3.7 The thermal plane of the Tribo-Process diagram: Thermal transactions in a tribo-mechanical system.

Work may also be performed when the potential of a system or an element is changed. This concept can be applied not only to mechanical work, but also to chemical work (change of chemical potential), electrical work (change of electrical potential) and thermal work (change of temperature). All these different forms of work will generally lead to the formation of entropy (Ref. 3.9). Thermal power may be stored, in which case the location of the storage increases in temperature. Vice versa, the temperature drops when stored thermal power is given up.

In general, thermal processes are of interest in tribology in so far as the temperature changes involved affect friction and wear processes.

That is, we are concerned with the relationship between temperature and the mechanical function of the system. This amounts to a concern for the effect of temperature on properties which determine the processes of vibration, friction and wear. These aspects are also discussed in some detail in Chapter 4.

3.4.4 Material Planes

A principle purpose of a representation on the material planes as shown in Figure 3.8 is to facilitate insight into the mass transfer and mass transformation processes which are invariably part of the wear of a mechanical system.

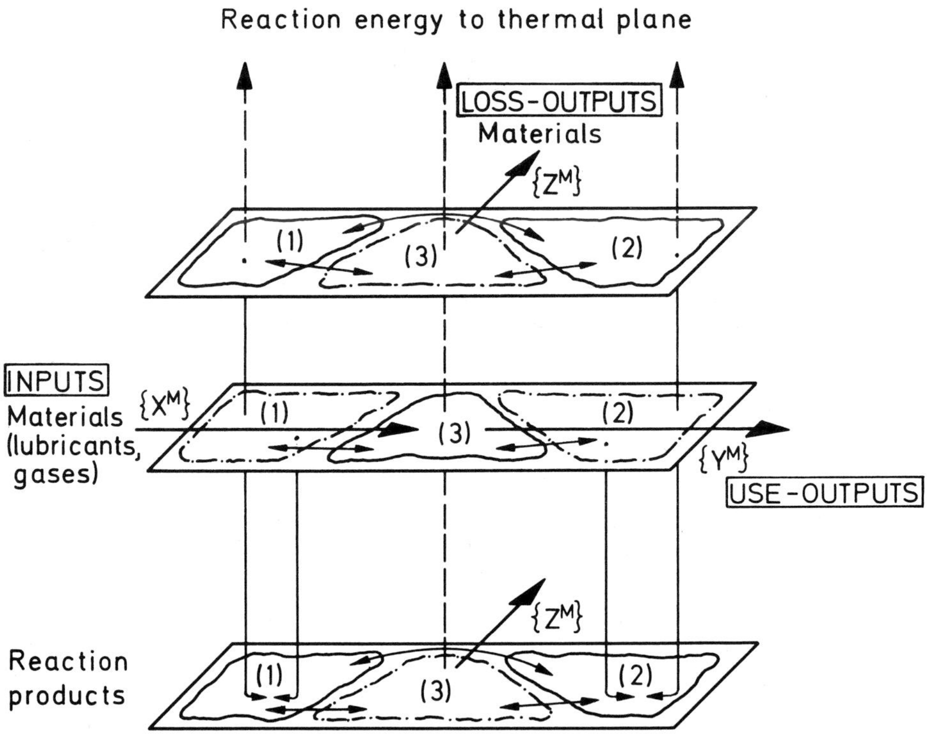

Figure 3.8 Material planes of the Tribo-process diagram: Material transactions in a tribo-mechanical system.

In some cases, the purpose of the mechanical system may be to transport some material, in which case an evaluation of aspects of the transport of the material will be duplicated on the functional plane and on a material plane.

The basic characteristic of wear is the removal of material from one or more machine elements. In some instances, merely a permanent change in shape, through deformation, of a machine element is considered to constistute wear. It may be preferable to call the actual process of the removal of material a process of "attrition", and to apply the term "wear" to cover all the processes in a system which combine to cause attrition.

In wear, material may be deformed within an element. It may also be transferred from one element to another. This process is called translation, which includes not only transfer from one solid element to another, but also attrition, in which case material becomes, as debris, part of the interfacial volume. Furthermore, a material may, through chemical reaction, be transformed to another material.

Using $\dot{m}_{ij}^{a}$ as the symbol for the flow rate of a mass species a from element i to element j, letting $\dot{m}_{j}^{ab}$ refer to a transformation of a mass species a to mass species b within element j, and letting m_{j}^{a} be the mass of species a in element j, then a mass balance equation for species a in element j would be

$$\sum \dot{m}_{j}^{a} = \sum_{\text{all } i} \dot{m}_{ij}^{a} - \sum_{\text{all } k} \dot{m}_{jk}^{a} + \sum_{\text{all } c} \dot{m}_{j}^{ca} - \sum_{\text{all } b} \dot{m}_{j}^{ab} ,$$

where i refers to all elements from which the material is translated to j, k to all elements to which the material is translated from j,c all the chemical species which in chemical reaction become part of material a, and b represents all the chemical species to which the material is transformed. (In order to also distinguish graphically between the different material species, it may be appropriate in certain cases to split up the material planes shown in Figure 3.8 into material sub-planes.)

In attrition, we will expect m_{j}^{a} to be negative for elements which lose mass by attrition. However, the equation illustrates that certain elements may retain a more or less steady mass, i.e., $\dot{m}_{j}^{a} = 0$, while significant translation and transformation processes occur on the element. This is often the case in practice.

For a system as a whole, we may write

$$\sum \dot{m}^{a} = \sum \dot{m}^{a}_{X} - \sum \dot{m}^{a}_{Y,Z} + \sum \dot{m}^{ca} - \sum \dot{m}^{ab} ,$$

where X and Y,Z refer to inputs and outputs respectively.

The processes on and between the material planes are likely to involve entropy changes, generally the production of entropy which will appear as sources of entropy in the thermal plane of the same element. It is possible that work may be performed in certain aspects of mass-translation processes, notably the formation of new free surfaces as in fracture or other forms of the separation of material. The opposite process, the joining of two surfaces, eliminating some free surface, is likely to be accompanied by the formation of entropy. Thus, a repeated sequence of separation-joining-separation and so on will, in effect, convert mechanical power into thermal power and appear as a mechanisms of friction. Limiting our concern to the material planes, we can list some pertinent processes in Table 3.4. The wear mechanisms are discussed in detail in Section 4.4.

Wear mechanism	Processes		
	Within element and plane	Translation	Transformation
Surface fatigue	Cyclic deformation	Fracture	-
Abrasion	Deformation	Fracture	-
Adhesion + fracture	Deformation	Transfer Fracture	- -
Tribochemical reactions	Deformation	Adsorption Fracture	Chemical reactions

Table 3.4 Wear mechanisms and processes.

3.5 Conclusions

The analysis of a tribo-mechanical system from the systems viewpoint as illustrated in the tribo-process diagram, Figure 3.4, clearly shows that the characteristic feature of tribology is the interaction between the different conceptual planes of Figure 3.4. Concluding the discussion on the processes with respect to the different planes of the conceptual framework of Figure 3.4, some remarks on the correlations of properties should be made.

We have already noted above that each of the properties of concern on the work plane is likely to be affected by temperature and also possible by mechanical stress or strain. The same applies to the properties of concern on the material planes. As a general principle, certain properties of the elements may be regarded as dependent properties on certain planes and independent properties on other planes. For example, temperatures are clearly dependent on the processes represented on the thermal plane, but may be regarded as independent properties influencing processes on other planes. A complete analysis must therefore include a correlation of the properties on the different planes for each location of interest in the system. Temperatures consistent with the process on the thermal plane must enter into our consideration of the processes on the other planes. Likewise, stresses, strains and velocities consistent with the processes on the mechanical work plane must enter into our considerations of the processes on the thermal and materials planes.

This analysis not only illustrates the complexity of the tribology of mechanical systems, but also the inevitable interdisciplinary nature of efforts made to deal with such systems.

Thus the work of a mechanical designer or the production engineer takes place, so to speak, on the "functional plane" of Figure 3.4. The mechanical engineer will also be concerned with the analyses on the work and thermal planes. These analyses are also in the province of the physicist. Physicists, along with materials scientists, will be qualified to investigate the processes on the material planes, while chemists and chemical engineers will be closely concerned with the interplay between the material planes, i.e., the chemical reactions causing the transformation of material.

The specialists, however, inevitable seeks to incorporate his knowledge and practical experience in some kind of conceptual framework. Given the complexity of tribological processes, a theory of tribology based on systems thinking is likely to provide the best type of framework available.

The analysis shows that a satisfactory description of a tribo-mechanical system would be one which clearly relates the elements of a system, the processes between the elements and their relevant properties, to the operational variables and the inputs and outputs, in particular in relation to a particular function for the system. The information compiled in the various Figures and Tables of this chapter will be used as a convenient guide-line for the discussions of these aspects in the following chapters.

4 Tribological processes

4.1 Role of Tribo-Processes in Mechanical Systems

The general theory of tribology as outlined in Chapter 3 can be regarded as a general framework linking the various processes and variables connected with friction and wear of mechanical systems. It has been shown that in an abstract manner a tribo-mechanical device can be characterized as a system in which certain interfacial processes of "translation" and "transformation" of functional quantities, e.g., motion, work, materials, occur. The interfacial processes connected with the interaction of material surfaces in relative motion are termed "tribological processes" or, briefly, "tribo-processes". In this connection the following questions arise:

(a) What are the mechanisms by which the use-quantities of a mechanical system are transmitted through the system, i.e., what are the actual physical processes by which, for instance, work is transmitted through the system as described on the conceptual functional plane (see Figure 3.5)?

(b) What are the tribological processes which lead to the generation of loss-outputs, i.e., what are the actual physical mechanisms of power dissipation due to friction (see Figure 3.6) and the generation of wear products (see Figure 3.8)?

(c) How can the dissipative loss-mechanisms (b) be influenced or minimized?

Since the general theory of tribology as developed in Chapter 3 is a conceptual framework only, the answers or solutions to these questions must be worked out by conventional methods of analysis. These details must then be integrated into the framework of the general theory.

Obviously, the answers to the above questions (a), (b), (c) depend on the type of system under consideration. It has been shown above that the different tribo-mechanical systems can be classified according to their function into different groups (see Table 3.2). For any of these different groups, different mechanisms of translation and transformation of the systems quantities are relevant, for instance:

(I) If the technical function of a mechanical system consists in the transmission of motion or mechanical power, as in the transmission of a torque through a bearing shaft, the actual transmission process (a) is given by a bulk elastic deformation of the shaft. No practical devices exist, however, which can transmit power without corresponding parasitic "guidance" - or "support-losses" (b). These losses can be minimized (c) through proper lubrication, as in a hydrostatic or hydrodynamic bearing, for example.

(II) If the technical function of a mechanical system consists in the interfacial transmission of mechanical power, as in a gear train for instance, the transmission of mechanical work (a) is obtained through contact deformation processes. The contact deformation inevitably involves some friction processes (b) which can be influenced (c) by lubrication of an elastohydrodynamic or boundary lubrication mode.

(III) If the technical function of a mechanical system consists of forming materials, as in wiredrawing for instance, an essential part of the materials transmission mechanism (a) consists of plastic deformation processes. In this case, some energy-loss-processes (b) may result through dislocation damping. Also these processes may be influenced considerably by proper lubrication (c).

The inspection of the possible tribological processes of "translation" and "transformation" of quantities within a mechanical system indicates that the following main effects should be studied:

(i) bulk elastic deformation and interfacial contact deformation processes,

(ii) the mechanisms of energy and material dissipation through interfacial friction and wear processes,

(iii) the processes and modes of lubrication as the most effective means of influencing the dissipation mechanisms.

In order to complete the general framework derived in Chapter 3, a survey of the main tribological processes generally recognized as being relevant to mechanical systems is given in this chapter. The treatment does not attempt to enter into specific details of the tribological behaviour of materials but tries to present a unified picture of the important basic phenomena. Since the treatment in this chapter is based on conventional analyses, the survey can be regarded - besides the systems aspects - as a review of the present knowledge of the mechanisms of friction, lubrication and wear from a physical point of view. The integration of the results of this chapter to the general systems concepts is given in Chapter 5 and 6.

4.2 Contact Processes

In a great deal of tribological systems, the transactions of the relevant inputs and outputs occur through the contact interface between tribo-element (1) and tribo-element (2). In these interfacial contact processes, the forces and displacements of the interacting bodies, i.e., the "contact mechanics", as well as the materials interactions i.e., the "contact physics and chemistry" must be taken into account. In this section the discussion is restricted to the contact processes between dry surfaces; the contact processes between lubricated surfaces are treated separately (see Section 4.5).

4.2.1 Contact mechanics

In contact mechanics, a great variety of situations can be considered depending on:

(a) the number of bodies taking part in the contact process,
(b) the macro-geometry of the bodies (two- or three-dimensional problems),
(c) the surface topography (smooth or rough surfaces),
(d) the mechanical properties of the bodies (modules, hardness, etc.),
(e) the deformation mode (elastic, plastic, elasto-plastic),
(f) the contact forces (normal forces F_N, tangential forces F_T),
(g) the type of relative motion (static contact, rolling, sliding, spinning, rolling with slip, etc.),
(h) the velocity of relative motion.

It is not the intention here, nor is it possible within the scope of this volume, to deal with all these situations. (Excellent review articles on various aspects of contact mechanics can be found in the proceedings of a symposium on "The mechanics of the contact between deformable bodies", see Ref. 4.1.) The purpose of this section is to gain a general physical picture of the contact-deformation process and the mechanisms of the transfer of mechanical work through a contacting interface. Obviously, a work transfer can be obtained only through elastic contact-deformation processes. It therefore seems appropriate to consider first the elastic contact of smooth solid bodies under the application of a pure normal load F_N (elastostatics) and then to study the influence of the effects of surface roughness, impulsive loading and elasto-plastic conditions.

Consider as a starting point a semi-infinite elastic solid under an arbitrary load distribution $p\ (x_1,\ x_2)$, as illustrated in Figure 4.1. Integrating over the contact region A_o, it will be recognized that load F_N is given by

$$F_N = \int_{A_o} p dx_1 dx_2$$

The classical problem of elastostatics, shown in Figure 4.1, which was first solved by Boussinesq, can be treated by the Fourier-transform method. (For details, see Ref. 4.2.)

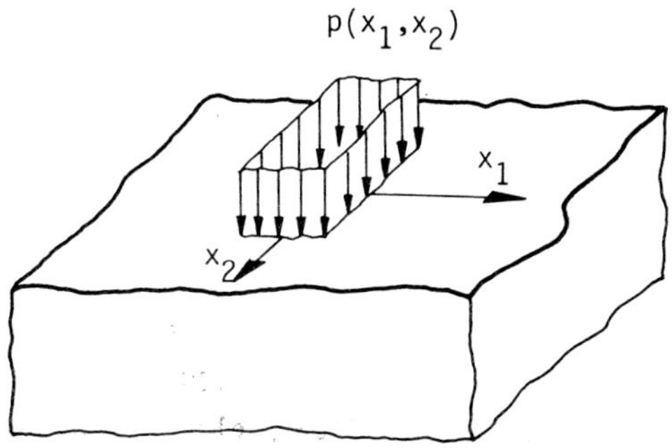

Figure 4.1 Load distribution acting on a semi-infinite body.

In tribology, in a great many contact problems we are concerned with the contact of curved bodies such as cylinders, spheres, as in rolling-element-type bearings, gear-teeth contacts, etc. The classic basis for all of these contact problems of "contraformal surfaces" is given by the famous Hertz theory (Ref. 4.3).

This theory treats the contact of elastic bodies as an elasto-statical problem under the following restrictions:

(i) perfectly elastic materials,
(ii) both bodies have the same elastic constants,
(iii) topographically smooth surfaces,
(iv) the two surfaces are not closely conforming in the contact region,
(v) no relative angular velocity of "spin" about the common normal to the contact surface,
(vi) no tangential (tractive) force is transmitted.

However, the results of the Hertzian theory have also been widely applied successfully to many situations in which the bodies under question are moving. Owing to the great importance of the Hertz equations in tribology, some of the results of the Hertz theory will be reproduced here. For the purpose of this treatment, it is convenient to restrict our attention to the case of the elastic contact of two spherical bodies under a pure normal load F_N.

The contact situation is illustrated in Figure 4.2, where r_1 and r_2 are the radii of curvature of the two bodies and E_1, E_2 and ν_1, ν_2 their elastic moduli and their Poisson's ratios, respectively. Using the abbreviations:

$$\frac{1}{r} = \frac{1}{r_1} + \frac{1}{r_2}$$ r: equivalent radius of curvature

and $$\frac{1}{E} = \frac{1}{2}\left[\frac{1-\nu_1^2}{E_1} + \frac{1-\nu_2^2}{E_2}\right]$$ E: composite elastic modulus

the contact pressure p at a location l within the contact area, and the radius of contact a_H under the action of a normal load F_N is given by

$$p = \frac{3\,F_N}{2\pi a_H^2}\left[1 - \left(\frac{l}{a_H}\right)^2\right]^{1/2} \qquad a_H = \left[\frac{3\,r}{2\,E}\right]^{1/3} F_N^{1/3}$$

For the contact area A_H, it follows that

$$A_H = \pi a_H^2 = \pi \left[\frac{3\,r}{2\,E} \right]^{2/3} F_N^{2/3}$$

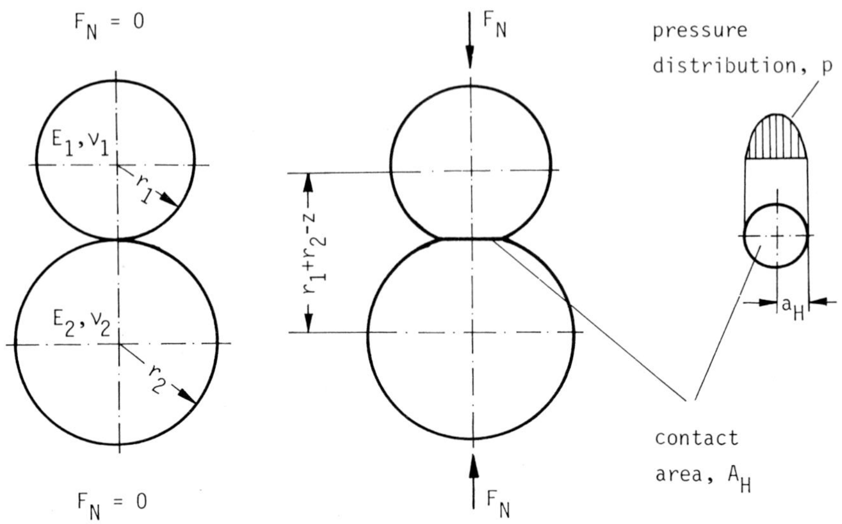

Figure 4.2 Hertzian contact between two spheres.

Considering now the process of a transfer of work perpendicular through the interface, it is necessary to calculate the value of approach z of the two bodies as a consequence of their elastic deformation. From Hertzian theory it follows that

$$z = \left[\frac{9}{4\,E^2 r} \right]^{1/3} F_N^{2/3}$$

Using the expression it is possible to determine the potential elastic energy ΔU of the two contacting bodies. Since $F_N = \frac{\partial U}{\partial z}$ the potential is given by

$$\Delta U = \frac{2}{5} \left[\frac{9}{4\,E^2 r} \right]^{1/3} F_N^{5/3}$$

If the contact situation is entirely elastic, the potential energy can be transformed into kinetic energy in releasing the contact. In this case, the system of the elastic bodies performs work against its environment. In considering the whole process of contact formation and contact release, it turns out that the quasi-static Hertzian contact situation can be regarded as model of the interfacial transfer of mechanical energy through a system of contacting bodies.

Although the main features of an "ideal" transmission of mechanical energy through an elastic contact interface can be understood on the basis of Hertzian theory, in any contact situation of "real surfaces" the effect of surface topography must be taken into account (Refs. 4.4, 4.5).

According to the German standard DIN 4760, the deviations from an ideal smooth surface can be classified into different groups:

(a) deviations in shape
(b) waviness (large-scale surface deviations)
(c) roughness (small-scale deviations)
(d) lattice structure

In order to describe the roughness of surface, a couple of parameters have been defined in relation to a cross-section of the surface, see Figure 4.3.

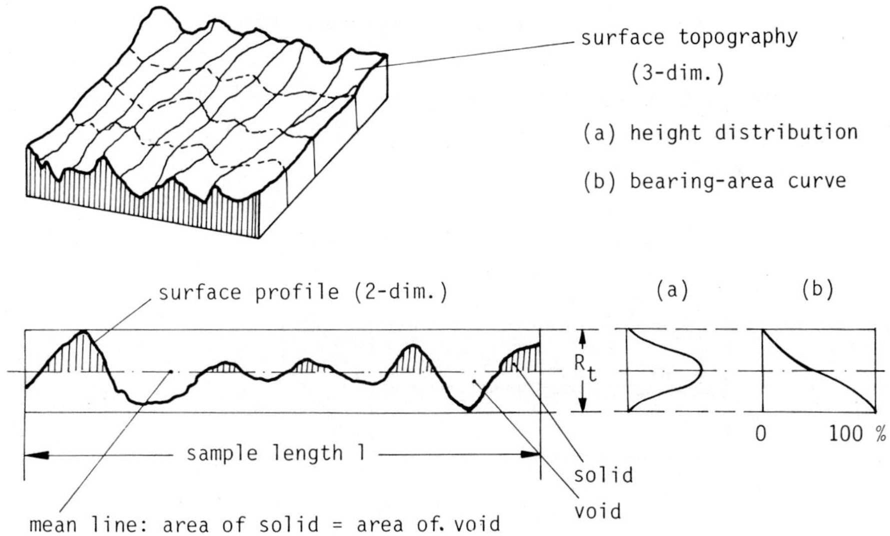

Figure 4.3 Characteristics of surface roughness.

The most commonly used height parameter in surface profilometry is the R_a value or center-line-average value (c.l.a.). It is simply the average deviation of the profile from the reference mean line. Other height parameters include the maximum peak-to-valley height R_t within the sample length l, the average deviation of the five highest peaks and five lowest valleys, R_z, and many more, all of which are well documented in the literature (Refs. 4.6, 4.7).

Another approach to describe the topography of surfaces is to regard the profile, as detected by a profilemeter, as an electrical signal and to analyze it statistically (Ref. 4.8). It is then possible to classify the profile in terms of the following functions:

(I) The probability distribution of ordinate heights. For many engineering surfaces finished by grinding or abrasion processes it was found that the heights of the profiles were distributed in a Gaussian manner.

(II) The auto-correlation function of the profile defined as:

$$Z\,(\Delta) = \lim_{l \to \infty} \frac{1}{l} \int_{-\frac{1}{2}l}^{+\frac{1}{2}l} y(x) \cdot y(x+\Delta)\,dx$$

where y(x) is the height of the profile at a given co-ordinate x and y(x +Δ) is the height at an adjacent co-ordinate (x +Δ). Peklenik has analyzed the surfaces produced by many different techniques and has proposed a system of classifying surfaces based upon the nature of the auto-correlation functions of their profile (Ref. 4.9).

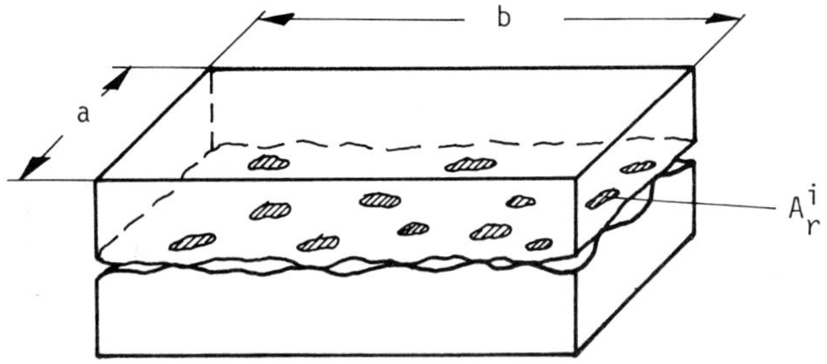

$$A_o = a \cdot b \gg A_r = \sum_{i=1}^{n} A_r^i \quad \text{(n: number of contacts)}$$

Figure 4.4 Nominal and real area of contact.

Now, returning to contact mechanics, in order to avoid ambiguity, for the contact of rough surfaces two terms must be distinguished, as illustrated in Figure 4.4:

(i) the nominal area of contact A_o, i.e., the apparent area of overlap of the contacting solids,

(ii) the real area of contact A_r, i.e., the sum of the separate microscopic areas at which the asperities are in contact.

Since it is obvious that the real area of contact, rather than the nominal area, is crucial for the interfacial transmission of work, a great many studies have been devoted to the estimation and experimental determination of the real area of contact A_r (Refs. 4.10, 4.11).

A comprehensive study of elastic deformation using model arrangements was made by Archard (Refs. 4.12, 4.13). Although these models are oversimplifications, they illustrate that multiple-point contact under elastic deformation can give a real area of contact almost linearly proportional to F_N, i.e.,

$$A_r = \text{constant} \left[\frac{F_N}{E} \right]^C \qquad 4/5 \leq C \leq 44/45$$

(depending on the model)

Another extension of the Hertzian theory of elastic contact between spheres to the case of rough surfaces with a Gaussian height distribution was made by Greenwood and Tripp (Ref. 4.14). Their model is defined by three parameters

σ^* : mean deviation of the asperity,

β : mean radius of the curvature of the asperity-tips assumed to be constant for all asperities,

n : density of asperities per unit area.

They showed mathematically:

(i) the total number of microcontacts is almost proportional to the load F_N,

(ii) the average size of a microcontact is almost load independent,

(iii) the real area of contact A_r is proportional to the number of contacts,

thus the real area of contact A_r, i.e., the sum of the micro-contact areas, is proportional to the load F_N. While at low loads there are considerable differences between the contact areas A_r and A_0(Hertz), at higher loads in the rough surfaces contact situation an effective radius can be defined which approaches the Hertzian radius and like it varies as the one-third power of the load.

In the discussion so far, mainly reversible quasi-static elastic contact deformations have been considered which can be regarded as models for the transmission of work in a tribo-mechanical system. However, in contact situations under pure normal loads some irreversible interfacial processes already exist which lead to a dissipation of mechanical energy. Unfortunately, for the irreversible processes in contact deformation there exists no theory as there is for the reversible elastic case. Therefore, in the following only some short remarks on dissipative processes in contact mechanics will be made (a physical treatise on the origins of energy dissipation is given in Section 4.3).

The following processes may contribute to an energy dissipation in contact mechanics:

(a) INTERFACIAL SLIP, I.E., PARTIAL MOTION OF THE CONTACTING SURFACES OF THE BODIES IN THE INTERFACE

If in a Hertzian contact the two contacting bodies have different elastic properties, an interfacial shear stress results which in turn influences the pressure distribution. Depending on the physical nature of the surfaces, the interfacial tangential shear may cancel out through local micro-slip processes, thus dissipating energy (see also Section 4.3.4). Also if the tangential stresses are "frozen in" during the duration of the contact as a consequence of interfacial adhesion forces (see Section 4.2.2), it is likely that in the release of contact, energy is dissipated.

(b) THE GENERATION OF ELASTIC WAVES IN THE INTERFACE

In any contact situation at least two transient states are involved: the formation of the contact and the separation of the contact. In these transient states due to changes in the load or the pressure distribution, there may be a pronounced effect of micro-slip and a generation of elastic waves. It has been shown experimentally that under impulsive loading of Hertzian contacts, gross slip effects occur and energy applied during gross slip is dissipated (Ref. 4.15).

(c) VISCOELASTIC DAMPING

The above discussions of contact-deformation processes on the basis of the Hertzian theory refer to elastic solids as given approximately by hardened ball-bearing steel. There are, however, other groups of technically important materials, for instance elastomers, which cannot be treated in this manner. Unlike a hard material like steel, the elastomer structure is composed of flexible chains which are in a constant state of thermal motion. The term "viscoelasticity" is commonly applied to materials which are neither ideal elastic solids nor viscous liquids (see Section 4.5.2), but in fact possess characteristics which are typical of both. The elastic behaviour may be represented by an ideal spring and viscosity by an ideal dashpot. Combining two basic mechanical models, known as the Voigt and Maxwell models consisting simply of spring and dashpot in parallel and series respectively, the contact-deformation behaviour of viscoelastic materials may be modelled. As a general feature, in the sequence of contact formation and separation of such materials, mechanical work is dissipated through hysteresis damping effects (Ref. 4.16).

(d) PLASTIC CONTACT PROCESSES

As early as 1901, Stribeck pointed to the limits of the Hertzian theory set by plastic deformation processes at higher loads (Ref. 4.17). An elastic-plastic transition criterion, i.e., the so-called plasticity index ψ is given by Greenwood and Williamson (Ref. 4.18) as

$$\psi = \left[\frac{E}{H}\right]\left[\frac{\sigma^*}{\beta}\right]^{1/2}$$

E : composite elastic modulus

H : hardness

σ^* : mean deviation of surface asperity height

β : mean asperity radius

A similar expression has been derived by Whitehouse and Archard (Ref. 4.19). (For a comparison of both models see Ref. 4.20.) If $\psi < 0.6$, the contact should be elastic under all practical loads. If $\psi > 1$, a part of the contact will involve plastic flow. As most surfaces have values ψ in excess of unity, there will be some plastic contact deformation of asperities in almost any contact situation. This is in line with the view of Bowden and Tabor (Ref. 1.4) who argued that, because surfaces contain asperities of small-radius-of-curvature, plastic flow would occur at the true areas of contact even under very small loads. They expressed the real area of contact by the relation

$$A_r = \frac{F_N}{p_y}$$

where p_y is the yield pressure for the weaker material. It is therefore likely that in almost any contact situation there is some dissipation of mechanical energy due to plastic asperity-deformation processes. According to Nicholas (Ref. 4.21), in general, four main mechanisms of energy dissipation in plastic deformation can be named:

(i) The moving dislocations require a high kinetic energy, which is set free if the dislocations are locked.
(ii) The dislocations dissipate energy continuously through the thermo-elastic damping and scattering of acoustic waves.
(iii) Dislocations are generated and annihilated during the deformation.
(iv) Point defects are generated and annihilated during the deformation.

4.2.2 CONTACT PHYSICS AND CHEMISTRY

In the contact of physical bodies, besides the contact "mechanics", interfacial molecular material interactions and "surface forces" must be taken into account. Since these processes depend on the physical and chemical nature of the surfaces in contact, a description of the surface characteristics is needed.

From the view of solid state physics, surfaces of crystalline bodies represent an abrupt termination of the periodicity of the crystalline lattice. Compared with the bulk properties of a crystalline solid, the implications of such an abrupt termination of lattice are (Ref. 4.22):

(a) structural reorientations (crystallinity and energetics)
(b) surface/environment interactions (physical adsorption, chemisorption, oxidation or film formation)
(c) changes in physical properties

For technical surfaces, due to the manufacturing process, additional aspects must be taken into account like:

(d) a work-hardened layer

(e) texturing of the surface

(f) local internal stress fields

In order to describe the composition of real surfaces, Schmaltz as early as 1936 distinguished between an "inner surface layer", bulk material and worked layer, and an "outer surface layer", contaminants, adsorbed layers, oxide layers (Ref. 4.23). The composition of a surface in a cross-sectional view is shown schematically in Figure 4.5. Clearly, the actual composition of a given surface and the structure and thickness of the different surface layers, as shown schematically in Figure 4.5, depends on the preparation of the surface and on the environment, apart from the bulk material properties.

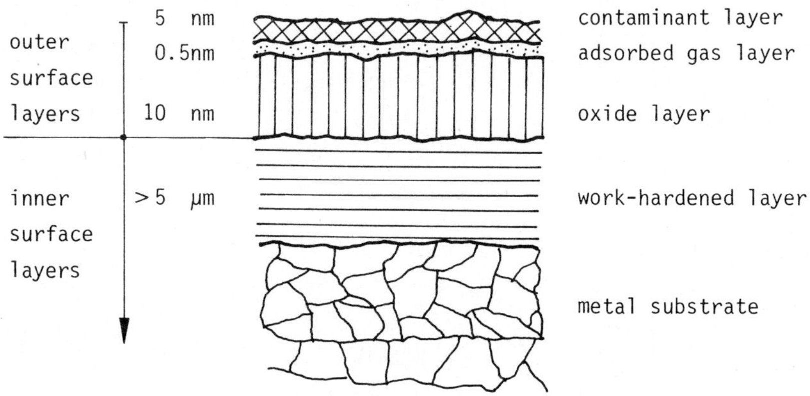

Figure 4.5 Schematic representation of the layered composition of a metallic surface.

Consider now the contact process of two solid bodies under a pure normal load F_N. As described in the foregoing section, the contacting surfaces will touch only within discrete micro-spots whose sum forms the real area of contact A_r. In the process of contact formation, besides an elastic deformation process the asperities undergo some plastic deformation. Due to this deformation processes, intimate contact between the partners occurs so that molecular surface forces may act through the interface, which results in "interfacial bonding" and the generation of adhesive "junctions". Thus the different types of molecular forces that can act within the interface of two contacting bodies to produce "adhesion" and the energy of interfacial adhesion bonding must be considered.

Adhesion mechanism	Bonding through:	Validity
van der Waals 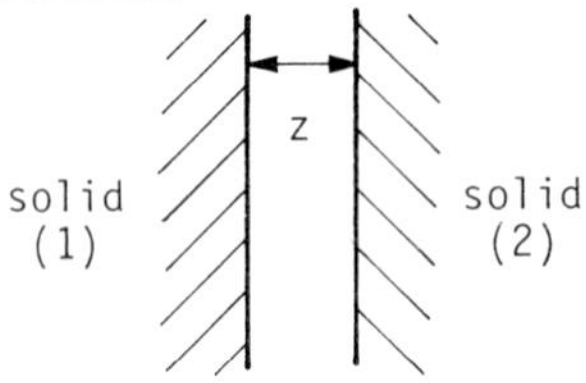	electromagnetic fluctuation fields	z > atomic dimensions
electrostatic 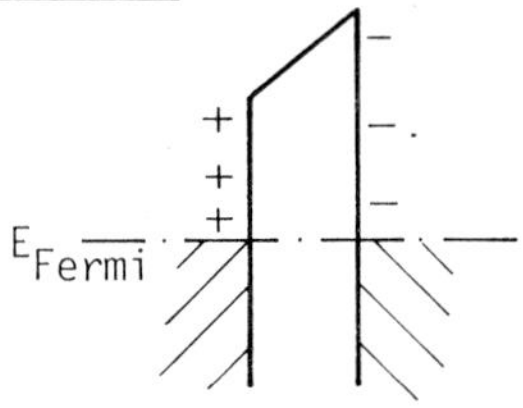	electrostatic fields	z ⩽ tunnel distance (localized charges)
metallic 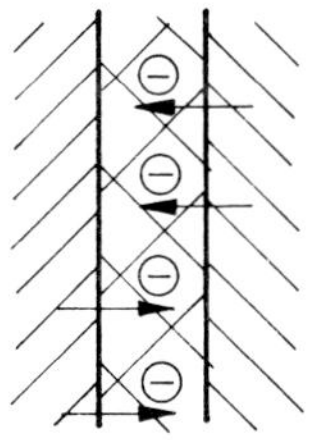	mobile interfacial electrons	z ≈ atomic dimensions (grain boundary analogy)
covalent 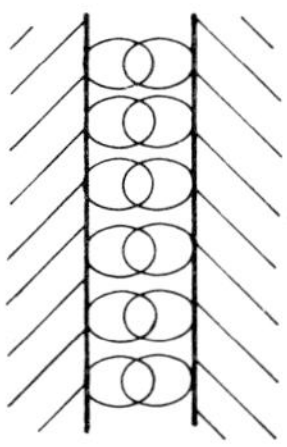	overlap of electron distributions	z ≈ atomic dimensions (dangling bonds)

Figure 4.6 Types of adhesion bonding mechanisms.

Theoretically, the attractive interaction forces between two contacting solids include, at least in principle, all those types of interaction that contribute to the cohesion of solids, such as metallic, covalent and ionic, i.e., primary chemical bonds (short-range forces), as well as secondary van der Waals bonds (long-range forces). For example, two pieces of clean gold placed in contact will form metallic bonds over the regions of atomic contact and the interface will have the strength of bulk gold. With clean diamond, the surface forces will resemble valency forces; with rock salt the surface forces will be partly ionic. All these are essentially short-range forces. Long-range van der Waals forces act in the adhesion between soft rubber-like materials and between polymeric solids. In Figure 4.6 some characteristics of the adhesion bonding mechanisms are shown.

Due to the great importance of adhesion in tribological processes, some features of the action of van der Waals forces and metallic bonds - the main reasons for an adhesion of polymers and metals - will be described in some detail.

If two solids are brought together they will first experience attractive forces due to van der Waals interactions. These forces arise from the fluctuating charge distributions in the atoms or molecules of the two bodies, creating momentary and continuously shifting dipoles and correspondingly induced dipoles in neighbouring atoms or molecules. As a first approximation for a solid body, it may be assumed that the van der Waals force between any pair of atoms or molecules is additive so that the total force may be calculated by summation for all pairs of atoms in the two bodies. On the basis of this assumption, Casimir was the first to derive the following formula for the force of adhesion F_A in the interface A of two mirror planes (Ref. 4.24):

$$F_A = \frac{\hbar c \pi^2}{240} \cdot \frac{A}{z_o^4}$$

$\hbar = \frac{h}{2\pi}$ (h : Planck's constant)

c : speed of light

z_o : distance between the surfaces

In the calculation of the van der Waals forces, two cases must be distinguished:

(a) non retarded forces occurring between interacting dipoles which are in phase,

(b) retarded forces occurring if the initial dipole and the induced dipole are no longer in phase.

For these two cases the following expressions for the adhesion force F_A between a sphere of radius r at a distance of nearest approach z from a flat surface can be derived

(a) non-retarded forces $$F_A = \frac{k_1}{6} \cdot \frac{r}{z^2}$$

(b) retarded forces $$F_A = \frac{2\pi k_2}{3} \cdot \frac{r}{z^3}$$

where k_1 and k_2 are the Hamaker constants for non-retarded and retarded forces respectively. The validity of these formulae has been proved in experiments with clean smooth mica surfaces. It was found that at $z > 100$ nm, the formulae for the retarded forces is valid whereas at $z < 20$ nm the measured van der Waals attractive force F_A agrees well with the theory for non-retarded interactions (Ref. 4.25). If the van der Waals surface forces are described in terms of surface energy γ, typical values of about 0.02 to 0.04 Jm^{-2} are found.

These results lead to an important generalization: since van der Waals interactions occur between any type of atoms or molecules, it can be assumed that the van der Waals adhesion forces act between all materials that are brought into contact down to distances of a few nm.

Below 1 nm, at atomic dimensions, short range surface forces of different type come into action which depend on the nature of the materials in contact. As an important example of the action of short-range surface forces, the adhesion of clean metals will be discussed.

In order to estimate the strength of metallic adhesion bonding, consider first the bulk cohesion of metals. Due to the complicated electronic structure of most metals, at present it is not yet possible to calculate the cohesion strength of metals from first principles. Estimations of the cohesion bond strength of simple metals can be attempted using the so-called "jellium model". According to Ziman (Ref. 4.26), it can be seen that in the simple picture of a jellium model cohesion depends, basically, on having an "electron gas" of fairly high density, not on the details of the crystal structure. The electron gas behaves like a ubiquitous liquid glue that will bind together more or less any arrangement of ions that is sufficiently densely packed.

The application of the jellium model to the case of adhesion at a bimetallic interface is illustrated in Figure 4.7. In this model, just the overlap of the metal-vacuum electron-number densities n_1 and n_2 of the two contacting partners within the true area of contact is considered.

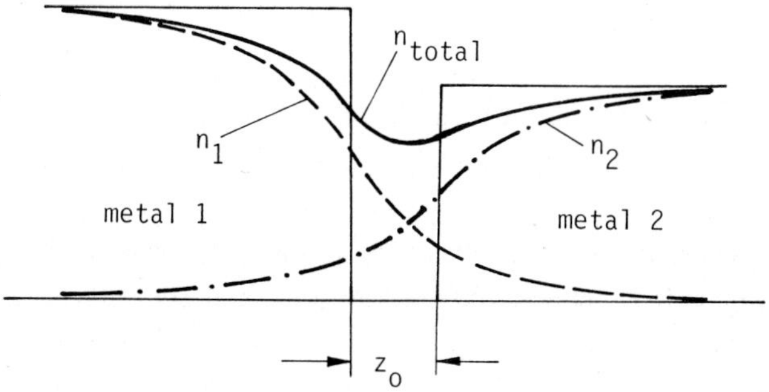

Figure 4.7 Jellium model of a bimetallic adhesion interface.

A first preliminary calculation of the adhesive interaction of a bimetallic interface energy was performed by Ferrante and Smith (Ref. 4.27), which has recently been extended (Ref. 4.28). The theory proceeds from an expression for the total energy of the system in terms of electron number density, including exchange and correlation energies, and is applied to the close-packed planes of simple metals. In this theory the strength of adhesion bonding depends, basically, on the free-electron density. As interfacial free-electron density in turn depends on how strongly the electrons are localized inside the contacting partners, in addition to the free-electron jellium formalism, the electronic structure of metals and the character (s, p, d) of electrons free for interfacial adhesion bonding must be considered. This has led to the hypothesis that for like metals the strength of adhesion bonding should increase within the periodic table from left to right in the order of transition metals, noble metals, B-subgroup metals (Ref. 4.29). For the adhesion bonding of unlike metals, further aspects such as the differences in Fermi levels, electron work functions and structure and density of electron states of both partners must be considered. Here, the density of states in the metal on one side of the interface may determine the total number of electrons impinging on the junction, and the density of states in the other metal may determine the number of final states available for the passed adhesive bonding electrons. It may be hypothesized that strong adhesion bonds will occur if one partner can act as an "electron donor" and the other as an "electron acceptor" (Refs. 4.29, 4.30). These considerations are, however, rather speculative because at present there is no direct way of studying the actual interfacial electron densities or determining the interfacial bond strength.

Although adhesion is essentially the result of interfacial molecular bonding forces, it should be borne in mind that the net adhesion effect between solids is influenced by at least three further important factors:

(i) The elasto-plastic contact behaviour of the partners, determining the size of the real area of contact in which the actual surface forces are able to act.

(ii) The influence of elastic stresses in contact separation, eventually peeling junctions apart, particularly if the contacting partners are lacking in ductility.

(iii) The presence of surface films and surface asperities which may reduce the adhesion forces.

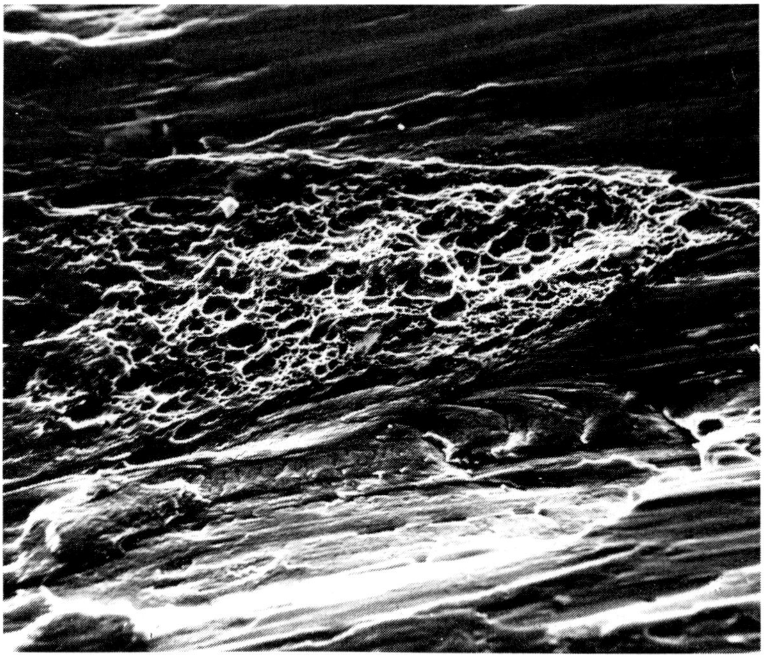

20 μm

Figure 4.8 SEM micrograph of an adhesive junction after contact separation (copper surface).

Experimentally, the only way of studying adhesive interactions between two solid bodies which contact under a pure normal load F_N is to destroy the bonding and to measure the force F_A necessary for the separation of the surfaces. According to Bowden the ratio

$$a = \frac{F_A}{F_N}$$

is termed the coefficient of adhesion.

The typical appearance of a broken metal/metal adhesion junction can be seen in Figure 4.8 in a scanning electron microscope (SEM) micrograph, having the appearance of ductile fracture.

The early experimental work of Bowden and Rowe (Ref. 4.31) showed clearly that very strong adhesion could be obtained between metal surfaces that had first been cleaned by heating to a high temperature in a vacuum of 130 μPa (10^{-6}Torr). In these experiments, relatively heavy joining loads had been used so that the surfaces underwent plastic deformation. Likewise under plastic contact conditions combined with a disruption of surface contaminant layers, Sikorski studied the correlation of the coefficient of adhesion with various bulk properties of metals in applying the twist-compression bonding method under normal atmospheric conditions (Ref. 4.32). In Figure 4.9 the results of Sikorski are represented showing the mean coefficient of various metals as a function of its Vickers hardness.

These results indicate two basic features:

(i) The adhesion coefficient decreases with increasing hardness (and also with increasing elastic moduli and surface energy).

(ii) The values of the adhesion coefficient are split up for the different types of crystal structure; the adhesion is lowest for hexagonal close-packed metals.

This general feature of the adhesion of metals has also been observed for clean pure metals under conditions of ultrahigh vacuum in extensive experiments by D. Buckley (Refs. 4.33, 4.34). It was also found that, contrary to a previously held view, strong adhesion could be obtained between unlike metals which had no mutual solubility (Refs. 4.35, 4.36). Further, on separation of the surfaces the cohesively weaker metal has generally been observed to transfer to the cohesively stronger.

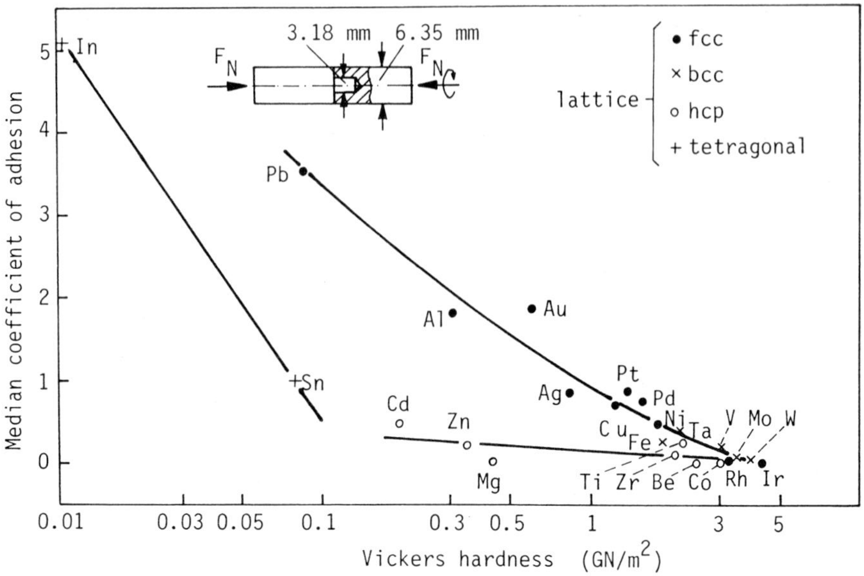

Figure 4.9 Correlation between the median coefficients of adhesion and Vickers hardness for metals of various lattice structures (Ref. 4.32).

In trying to explain these experimental observations in terms of physical properties of metals, two aspects have been emphasized:

(a) The ability of surfaces to deform plastically obviously exhibits a marked influence on adhesion since it influences the true area of contact. As illustrated in Figure 4.10 with face-centred-cubic (fcc) metals there are 12 possible slip systems. Body-centred-cubic (bcc) metals have 12 main operable slip systems and 36 secondary slip systems which require higher shear stresses to produce slip. With hexagonal metals (hcp), however, at room temperature there are only 3 operable basal slip systems. The limited slip behaviour in the hexagonal metals, therefore, seems to indicate that the hexagonal metals might well exhibit lower adhesion behaviour than either bcc - or fcc - metals (Refs. 4.37, 4.38).

(b) The electronic structure of the metals and the interfacial free-electron density is assumed to determine the interfacial bond strength (Refs. 4.27 to 4.29). For example, in calculating the binding energies for a couple of unlike fcc metals using the electron theory of metallic binding, an ordering of binding energies was obtained as shown in Table 4.1. As indicated in Table 4.1 the theoretical findings have been experimentally verified (Ref. 4.39).

Besides the extensive experimental work on the adhesion of metals, relatively few studies have been reported of adhesion between other types of solid such as covalent or ionic materials or van der Waals bonded solids. In addition, little work appears to have been carried out between such solids and clean metals. (Examples of polymer-metal adhesion as studied by optical microscopy and field ion microscopy can be found in Section 7.6.) Tabor and co-workers described a study of the adhesion between solids of different type performed in a vacuum of 13 nPa (10^{-10}Torr) (Ref. 4.40). They found that adhesion of soft metals to a hard solid such as titanium carbide can be high and when the surfaces are separated fragments of metal are found attached to the harder surface. With other hard solids such as saphire or diamond the adhesion of copper is appreciably less, indicating that the bonding between copper and these covalent materials is intrinsically weak.

From the study of the adhesion bonding mechanisms an important general conclusion can be drawn:

Since both partners contribute to the adhesion bonding, the bond strength is essentially an interaction characteristic. Therefore, any measure or characteristic of adhesion cannot be handled in engineering applications like a materials property (as for instance the elastic modulus) but rather as a quantity related to the whole system of contacting partners under consideration. In contrast to material properties like the elastic modulus which can be regarded as a "system-independent" parameter, the adhesion behaviour of a material pair is a strongly "system-dependent" characteristic.

In the discussion of contact processes so far, the contact mechanics as based on Hertzian theory and the interfacial adhesion surface forces have been treated separately. An analysis of the elastic contact between a smooth elastic sphere and a smooth flat surface taking into account the effect of interfacial attractive adhesion forces was made by Johnson, Kendall and Roberts (Ref. 4.41). They showed that the area of contact is always greater than that given by classical Hertzian theory. This is obviously due to a net release of surface energy resulting from the replacement of two surfaces by one solid/solid interface of lower surface energy. The change in surface

Structure	Slip plane	Slip direction	Number of slip systems	
FCC Cu, Al, Ni, Pb, Au, Ag, γ Fe, ...	{111}	⟨110⟩	4 x 3 = 12	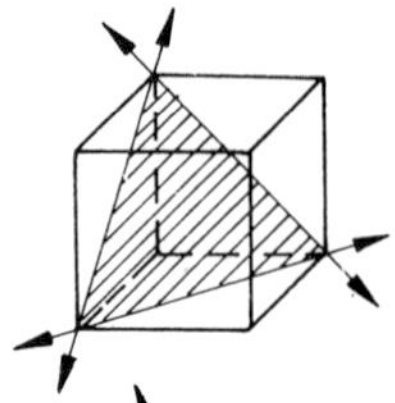
BCC W, Mo, α Fe, β brass,...	{110} {211} {321}	⟨111⟩ secondary slip systems: ⟨111⟩ ⟨111⟩	6 x 2 = 12 12 x 1 = 12 24 x 1 = 24	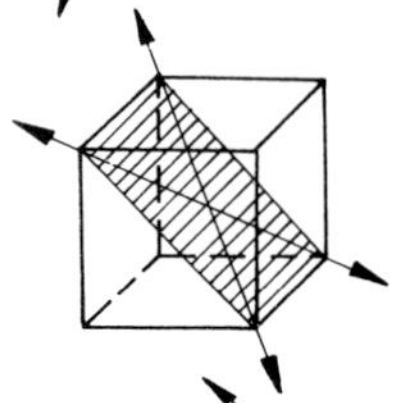
HCP Cd, Zn, Mg, Co, ...	(0001)	⟨1120⟩ (basal slip)	1 x 3 = 3	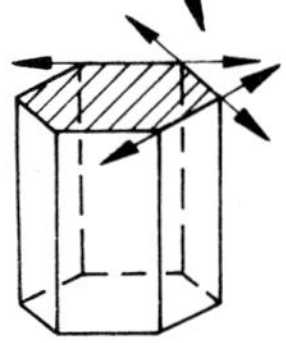

Figure 4.10 Observed slip systems in crystals.

Theoretical	Ordering binding energy, ergs/cm^2	Experimental ordering
Al - Al	525	1
Al - Au	465	2
Al - Cu	455	3
Al - Ag	385	4
Cu - Cu	375	5
Au - Au	345	6
Cu - Ag	335	7
Ag - Ag	290	8

Table 4.1 Ordering of adhesive binding energies (Ref. 4.39).

energy per unit area of contact, $\Delta\gamma$, may be written as

$$\Delta\gamma = \gamma_1 + \gamma_2 - \gamma_{12}$$

where γ_1 is the surface energy of the sphere, γ_2 of the flat and γ_{12} of the interface. The analysis shows that there is a finite tensile force F^*_N required to pull the sphere off the flat. The value of F^*_N is the same whether or not the surfaces are initially pressed together with an external force, and is given by

$$F^*_N = \frac{3}{2}\pi r \Delta\gamma$$

The results of the analysis of Johnson, Kendall and Roberts for the case of a rubber/rubber contact are shown in Figure 4.11. For soft elastic van der Waals solids, (r = 22 mm, γ = 30 mJm^{-2}) the adhesion observed is several mN. For thoroughly clean crystalline solids where γ is at least an order of magnitude greater one would expect adhesive forces, independent of the joining load, of order of 10^{-2}N.

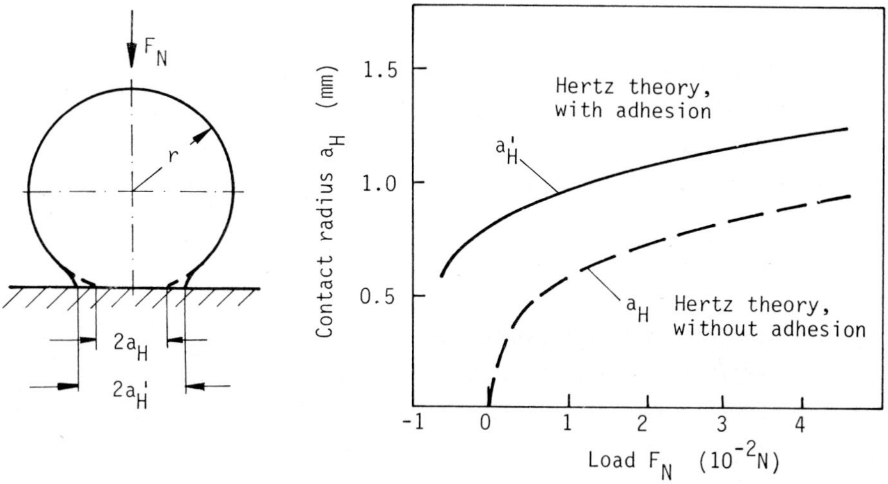

Figure 4.11 Hertzian elastic contact without and with interfacial adhesion (Ref. 4.41).

The analysis illustrated in Figure 4.11 is valid for optically smooth rubber-like materials where overall deformations are large and local micro-asperities are easily accommodated. As a criterion for the influence of surface roughness on the adhesion of elastic solids, Fuller and Tabor (Ref. 4.42) have defined an "adhesion parameter" given by

$$\Theta = \frac{E\, z^{3/2}\, \beta^{1/2}}{\beta \Delta \gamma}$$

where β is the average radius of curvature of individual asperities, z the deviation from the mean height, E the reduced elastic modulus and $\Delta\gamma$ the surface-energy change defined above. Physically, the quantity $\beta\Delta\gamma$ is a measure of the adhesive force, the quantity $E\, z^{3/2}\, \beta^{1/2}$ a measure of the force exerted by the asperities in pushing the surfaces apart. If Θ is small, this implies that the adhesion is dominant, if Θ is large the asperities dominate and the adhesion is small. It is assumed that the transition from strong to weak adhesion accurs if $\Theta > 10$. This corresponds to surface irregularities of dimensions $z \approx 1$ µm for van der Waals solids, like rubber, and $z \approx 5$nm for hard materials such as diamond. This analysis shows that with hard elastic solids, very small surface irregularities are sufficient to reduce adhesion to a very low value. If, on the other hand, the junctions can deform plastically, then adhesion dominating over roughness is more likely to occur.

In summary, in any contact situation the combined influences of contact mechanics and contact physics and chemistry must be taken into account. The analyses, performed for the case of two solid bodies contacting under the action of a pure normal load F_N, show that the following main characteristics and processes must be considered:

(I) Surface characteristics

- (a) Surface topography
  - surface roughness parameters
  - profile distribution curve
  - profile auto-correlation function
  - statistical surface parameters
- (b) Surface composition
  - contaminants
  - physisorbed layers
  - chemisorbed layers

- oxide layers
- worked layers

(II) Contact formation

(a) Contact mechanics
- Hertz theory
- apparent and real area of contact
- deformation mode (plasticity index)
- elastic-asperity-contact deformation
- plastic-asperity-contact deformation

(b) Contact physics and chemistry
- van der Waals forces
- metallic bonding
- effect of other short-range forces
- effect of surface films

(III) Contact separation
- interfacial slip
- generation of elastic waves
- elastic recovery
- transfer of adherent material
- fracture

4.3 Friction Processes

4.3.1 Tangential forces in contact processes

In this section, the origins and mechanisms of dry friction will be considered in order to explain the dissipation processes that occur on the conceptual work plane in a tribological system as illustrated in Figure 3.6. Consider as a starting point the static contact of two bodies under a pure normal force F_N on which a tangential force F_T is superimposed without causing a macroscopic relative motion between the two bodies. In this case, the addition of F_T to F_N modifies the interfacial elastic- and plastic-contact-deformation situation as described in the foregoing section. There are four main aspects that must now be considered:

(a) ELASTIC STRESSES AND DISPLACEMENTS

Mindlin made a detailed study of the idealized case of elastic Hertzian contact under the additional action of a small tangential force (Ref. 4.43). In the case of a tangential force across a circular Hertzian contact area, i.e., the case of two spherical bodies in contact, the tangential traction is everywhere parallel to the direction of the applied force. The contours of constant tangential traction are concentric circles. The magnitude of the traction rises from one-half the average at the centre to infinity at the edge. Obviously, the high traction stresses in some way cause micro-slip between the two contacting materials depending on the shear strength of adhesion bonding between the materials in contact. If the resistance against motion is expressed formally by the coefficient of friction f, presumably the tangential component of traction cannot exceed the product of the coefficient of friction and the normal component of traction. If a_H denotes the radius of the Hertzian contact circle, it follows that no slip occurs within a circle of radius

$$a' = a_H \left[1 - \frac{F_N}{fF_T} \right]^{1/3}$$

whereas in that part of the Hertzian contact zone in which the radius is between a' and a_H slip occurs. The situation is illustrated in Figure 4.12.

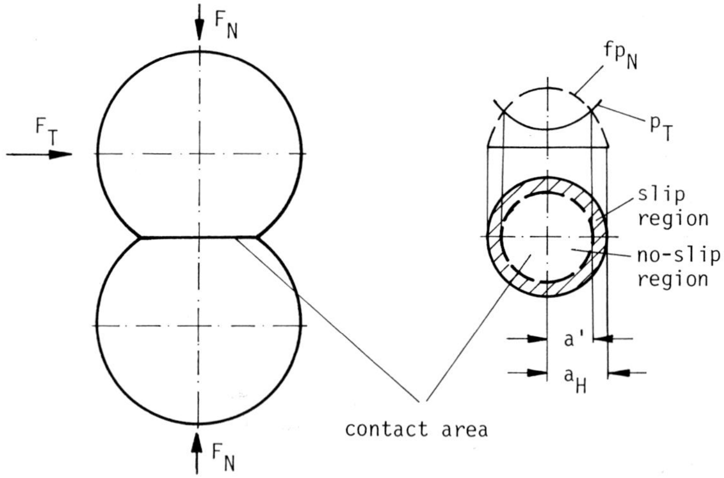

Figure 4.12 Slip and no-slip regions of a Hertzian contact under combined normal and tangential forces.

(b) JUNCTION GROWTH

Consider now the possibility of plastic deformation of a static contact under the action of a normal force F_N and a superimposed small tangential force F_T. According to the plasticity theory of bulk materials, in the two-dimensional case the von Mises yield criterion is given by

$$p^2 + 3\tau^2 = p_y^2$$

p : normal pressure
τ : shear stress
p_y : yield pressure

It has been shown in Section 4.2 that in the contact of clean rough metallic surfaces under a pure normal load, plastic deformation of the touching asperities occurs and adhesive junctions are formed. For an ideal two-dimensional model of an intermetallic junction, by analogy the yield criterion may be expected to have the form (Ref. 4.44):

$$p^2 + C\,\tau^2 = p_o^2$$

where C is a constant with a value of about 10, and p_o is the static contact pressure.

Since $$p = \frac{F_N}{A_r}, \quad \tau = \frac{F_T}{A_r}, \quad p_o = \frac{F_N}{A_{ro}}$$

it follows that

$$\left[\frac{F_N}{A_r}\right]^2 + C\left[\frac{F_T}{A_r}\right]^2 = \left[\frac{F_N}{A_{ro}}\right]^2$$

which leads to

$$A_r = \left[\frac{F_N^2}{p_o^2} + C\,\frac{F_T^2}{p_o^2}\right]^{1/2}$$

or

$$\frac{A_r}{A_{ro}} = \left[1 + C\,\frac{F_T^2}{F_N^2}\right]^{1/2}$$

A plot of this expression is shown in Figure 4.13.

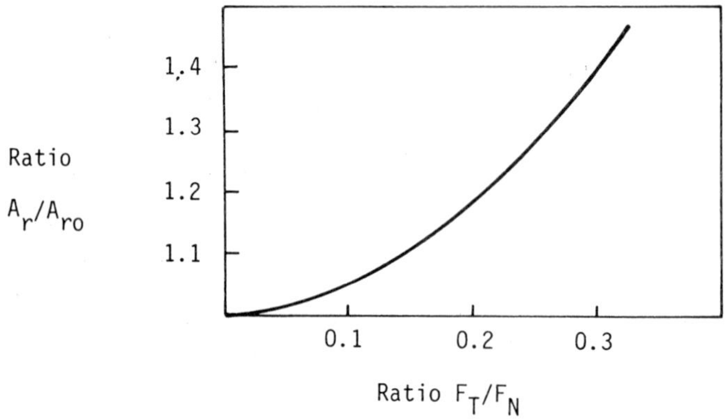

Figure 4.13 Junction growth under combined normal and tangential forces.

It can be seen that there may be a considerable increase in the contact area due to the combined action of F_N and F_T. The validity of this results has been proved experimentally for clean metals, i.e., high interfacial adhesion bonding. It is possible that in this case the real area of contact A_r increases until it is equal to the nominal geometric area of contact A_o. If the interfacial adhesion bonding is lowered, for instance through lubrication of the surfaces, similar junction growth is observed until gross tangential sliding occurs as indicated in Figure 4.13.

(c) DISPERSAL OF CONTAMINANTS

Most surfaces under normal atmospheric conditions possess a surface composition as shown in Figure 4.5. The topmost, i.e., the "outer" surface layers are formed by contaminants, like oxides and moisture. Through these contaminants, the short-range surface forces existing at a solid surface are saturated under atmospheric conditions. If two surfaces are brought into contact in the direction normal to their interface, in general only (week) long-range forces of the van der Waals type can act adhesively, since the (strong) short-range forces are already saturated and the thickness of the contaminant layers of both surfaces is in general larger than the range of the short-range surface forces. If, however, a tangential force is superimposed there may be a dispersal of the contaminant films through the combined action of junction growth and microscopic shearing of contaminants. Consequently, short-range forces come into action and lead to strong local adhesive bonds. For metals it has been proved experimentally that through the superimposition of tangential forces and a relative motion at the interface, adhesion may increase more than tenfold.

(d) MICRODISPLACEMENTS BEFORE SLIDING

It has been observed that when a tangential force F_T is applied to a static contact, a small relative displacement of about 0.1 to 1 μm between the components in contact in the direction of F_T occurs almost instantaneously (Ref. 4.45). These microdisplacements before gross sliding are in some cases reversible and in others irreversible. This indicates that elastic or plastic processes on a small scale take place before gross sliding occurs.

4.3.2 SLIDING FRICTION

If, in a dry static contact situation, the load applied tangential to the contact surface exceeds a certain value, gross sliding between the components in contact occurs, Figure 4.14. In this situation, the following macroscopic rules have in general been observed experimentally:

(i) When tangential motion between two contacting bodies occurs, the friction force F_F always acts in a direction opposite to that of the relative velocity of the surfaces.

(ii) The friction force F_F is proportional to the normal force F_N

$$F_F = f \cdot F_N$$

Through this relationship it is possible to define a coefficient of friction

$$f = \frac{F_F}{F_N}$$

(iii) The friction force is independent of the apparent geometric area of contact.

These rules, known as the "Amontons-Coulomb laws" of dry sliding friction have been used so far as guiding rules in engineering applications.

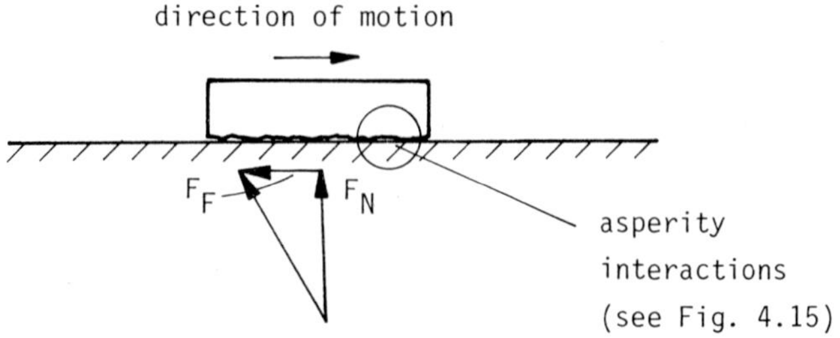

Figure 4.14 A solid body in sliding motion.

In Section 1.2.1 the danger of misinterpretations of these "laws" was emphasized. Therefore, a physical picture of the mechanisms of solid friction is needed, from which the influencing factors and the relevant properties of the sliding components can be compiled. As explained above, a general treatment must include both an explanation of the forces necessary to overcome friction and a physical picture of the processes of energy dissipation, i.e., the mechanisms of the transformation of mechanical work to other forms of energy as described schematically in Figure 3.6.

To begin with, let us summarize the basic results of the contact process which may also be relevant to dry sliding friction.

The results of Section 4.2 show that the following points are relevant in any static contact situation:

(a) Two solid surfaces in contact over a nominal geometric area A_0 are actually touching only in discrete spots, called micro-contacts.

(b) The sum of the micro-contacts forms the real area of contact A_r which carries the normal load F_N.

(c) Depending on the deformation mode, the following relations between the real area of contact A_r and the normal load F_N exist:

- elastic deformation (plasticity index $\psi < 0.6$)

$$A_r \propto \frac{F_N}{E} \qquad E \;:\; \text{composite elastic modulus}$$

- plastic deformation (plasticity index $\psi > 1$)

$$A_r \propto \frac{F_N}{p_y} \qquad p_y \;:\; \text{yield pressure}$$

(d) In the micro-contacts, adhesive junctions are formed through the action of surface forces:
- long-range forces, acting between any combination of materials down to separation of about 1nm,
- short-range forces of metallic, ionic or covalent type, depending on the nature of the materials in contact, acting at separations < 1nm.

For the case of a sliding contact, based on the existing knowledge of the topography and composition of solid surfaces, we may postulate the following microscopic view of sliding friction: friction occurs through asperity interactions, i.e., through dissipative processes involved in the joining and separation of microcontacts.

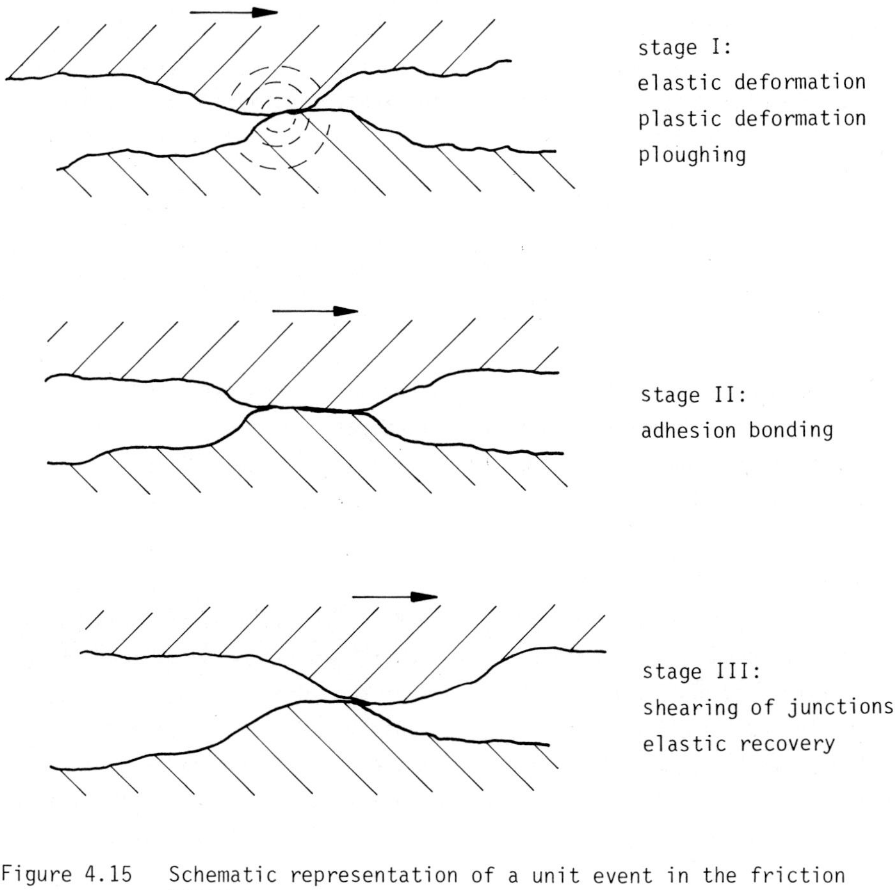

Figure 4.15 Schematic representation of a unit event in the friction process.

In this model, the macroscopic friction force can be expressed as the sum of microscopic friction forces at the individual micro-contacts, and the energy dissipated may be expressed as the sum of corresponding elementary dissipation processes. The elementary processes that occur during the joining and separation of a micro-contact formed between two asperities interpenetrating to a certain depth are illustrated schematically in Figure 4.15.

Obviously, in the different stages of the formation and separation of a micro-contact, the following main processes are involved:

(i) elastic asperity deformation,
(ii) plastic asperity deformation,
(iii) ploughing,
(iv) shearing of adhesive junctions.

Each of the partial processes involves a tangential force necessary to maintain the relative motion as well as a partial process of energy dissipation. Kragelski (Ref. 1.6) in a detailed analysis studied the different causes of friction taking place in the micro-contacts and expressed the total macroscopic friction force F_F as

$$F_F = \sum F_1 + \sum F_2 + \sum F_3 + \sum F_4$$

where

F_1 : resistance caused by elastic displacement of the material
F_2 : resistance caused by plastic displacement
F_3 : resistance due to shearing (ploughing) of the material
F_4 : resistance due to shearing of the (adhesive) film

With this model, from a physical point of view the causes of the friction force can be understood qualitatively (Ref. 4.46). A quantitative calculation of friction forces or friction coefficients is, however, only possible to very rough approximations under special simplified conditions. This is due mainly to our incomplete knowledge of the surface properties of solids. In order to compare the implications of the microscopic model of friction with macroscopic experimental results, some estimations of the magnitude of the friction coefficient for metals in sliding contact will be made in the following based on the theory of Bowden and Tabor (Ref. 1.4). It should be emphasized that these can be regarded only as rough approximations.

According to the microscopic model of friction, the different contributions to friction may be classified broadly into two groups: (a) deformation processes and (b) adhesion processes. In general, the processes belonging to these two groups are not independent of each other (Ref. 4.47). Under certain special conditions, however, one process may dominate so that the other contributions are comparative small and may be neglected to a first approximation. Consider, for example, two special cases of the friction of metals:

(a) PLOUGHING COMPONENT OF FRICTION

This case may be approximately given if a very hard rough surface slides over a soft one. The frictional resistance is then mainly caused by the asperities of the harder surface ploughing through the softer, and the coefficient of friction may be estimated from the forces required for plastic flow of the softer metal. The normal load F_N is balanced by the yield pressure p_{yN} of the metal acting via the real area of contact A_r:

$$F_N = A_r p_{yN}$$

The resistance F_T to tangential motion is balanced by the yield pressure p_{yg} of the metal acting over the cross-sectional area of the groove A_g:

$$F_T = A_g p_{yT}$$

Assuming that the plastic-yielding metal is isotropic, i.e., $p_{yN} = p_{yT}$, the "friction coefficient" is given by

$$f = \frac{F_T}{F_N} = \frac{A_g}{A_r}$$

In this model the value of f is determined by the ratio of the areas A_g/A_r, depending on the asperity shape model. For the case of a conical indenter, as illustrated schematically in Figure 4.16, it follows from simple geometric consideration that

$$f = \frac{2}{\pi}\cot\varphi \qquad \varphi: \text{ semiapex angle of the conical identer}$$

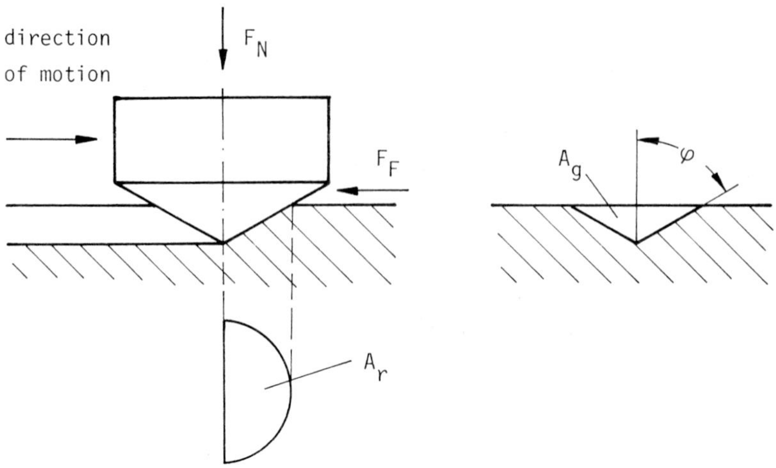

Figure 4.16 Ploughing component for a hard conical identer sliding over a soft metal.

Normal surface asperities seldom have an effective slope exceeding 5^0 or 6^0 (i.e., $\varphi = 85^0$ or 84^0). It follows that

$$f \approx 0.05$$

This value may be regarded, however, only as a lower friction-coefficient limit for the deformation component of friction due to the neglections described above and the experimentally observed fact that a pile-up of material ahead of the grooving path ("bourrelet frontal") occurs in most cases of plastic deformations during sliding (Ref. 4.48).

(b) ADHESION COMPONENT OF FRICTION

This case may be approximately given for a sliding contact of two relatively soft clean metals. Depending on the degree of interpenetration of asperities and on the surface composition, adhesion bonding in the micro-contacts occurs. If sliding is to take place, the friction force is needed to shear the weakest tangential planes at the areas of actual contact. If the mean shear strength of the weakest planes at the contact areas is denoted by τ_s (assumed to be of equal value in all micro-contacts) and their total area is assumed to be approximately equal to the real area of contact, neglecting the effect of junction growth, we may write

$$F_F = \tau_s A_r$$

(For an extension of this approximation see Ref. 4.49.) Since for the plastic contact condition (plasticity index $\psi > 1$) the real area of contact is given by

$$A_r = \frac{F_N}{p_y} \qquad p_y \text{ : yield pressure}$$

it follows that the coefficient of friction is given by

$$f = \frac{F_F}{F_N} = \frac{\tau_s}{p_y}$$

For metals which are not work-hardened, the shear strength τ_s of the interface is approximately equal to the critical shear stress of the metal τ. The yield pressure p_y has been shown to be approximately 5τ. It follows that

$$f \approx 0.2$$

Further, this value can only be regarded as a lower friction-coefficient limit for the adhesion component of friction of (technically) clean metals, since for really clean metals with the help of junction growth, friction coefficients as high as $f \approx 40$ to 100 have been measured. On the other hand, it is possible to lower the friction coefficient down to values of $f \approx 0.02$ with boundary lubricants.

From the microscopic model of friction and the simple estimations based on this model, it is possible to verify the form of the empirical Amontons-Coulomb law of sliding friction. It has been described that the friction processes occur within the real area of contact A_r, and that for both the deformation component and the adhesion component

$$F_F \propto A_r.$$

The results of the contact mechanics of rough surfaces for both elastic-

and plastic-contact conditions show that

$$A_r \propto F_N.$$

If it is assumed that these results also hold for sliding contact conditions, it follows that

$$F_F \propto F_N.$$

Thus, the friction force is proportional to the load and independent of the geometric contact area, in agreement with the empirical Amontons-Coulomb law of sliding friction.

However, the microscopic model of friction as illustrated in Figure 4.15 indicates that friction is not only a composite process but is also sensitive to changes in the deformation and adhesion characteristics of the contacting surfaces. Depending on the operating conditions and the atmospheric environment, i.e., the "third body", interacting with the two sliding partners, in general, changes in the surface topography as well as in the surface composition and the surface strength properties will take place, see Section 5.2. These changes may influence the friction coefficient markedly, see Section 8.5.2.

Thus, as discussed already in Section 4.2 for the static contact situation, there are many properties of materials which influence the nature of the interface in sliding friction. For example, for metals these include surface orientation, lattice spacing, grain boundaries, crystal structure, nature of adhesion bond character and alloying elements. The effect of alloying elements can alter interfacial behaviour by segregation to the surface of metals or by altering bulk properties through crystal transformation kinetics (Ref. 4.50). Since it is not possible to deal with the friction behaviour of the various materials within the scope of this volume, this section may be closed best in quoting the summarizing remarks from the recent excellent review of D. Tabor on the general mechanisms of sliding friction (Ref. 4.51):

"The friction between unlubricated solids is due to two main factors. The first is the adhesion occurring at the regions of real contact. The second arises if the asperities on a hard surface produce grooving or cutting in the other. With metals the ploughing action involves plastic displacement, with brittle solids some cracking or fragmentation, with rubbers and polymers hysteresis losses. Other factors may also play a part in the frictional process, but except in special circumstances these are generally of only tertiary significance.

Solids that deform easily and isotropically will tend to give very high coefficients of friction. On the other hand, brittle solids will give restricted junction growth and lower coefficients of friction. Again, solids with well-defined anisotropy may show anisotropic frictional properties. This is most marked with lamellar solids that are often low friction materials, though this is by no means always the case. With viscoelastic solids the deformation properties are rate- and temperature-dependent, and there is a corresponding variation of friction with sliding speed and temperature.

Although deformation properties are important, the factor that has the largest overall influence on friction is the cleanliness of the surfaces. A small trace of oxygen or some other contaminant can radically reduce the adhesion and friction. It is probable that in practical affairs involving the sliding of unlubricated metals, atmospheric contamination is the main factor responsible for reducing friction and wear and for preventing total seizure. With polymers and brittle materials the situation is not as clearly defined, but, in general, surface contamination has less effect than with metals".

In addition to the discussion of the friction force, i.e., the resistance to sliding motion as discussed in this section, the question on the frictional energy dissipation must be studied. Before discussing the friction-induced energy dissipation processes (see Section 4.3.4), the mechanisms of rolling friction force, i.e. the causes of the resistance to rolling motion are reviewed in the next section.

4.3.3 Rolling friction

Rolling friction, although in general much smaller than sliding friction, is also a very complex phenomenon because of its dependence upon so many factors (Ref. 4.52). Clearly, the contact processes, i.e., the contact mechanics and contact physics and chemistry, as discussed in Section 4.2, are adequately valid in rolling-contact situations. The contact deformation in steady rolling is conveniently approached from the Eulerian point of view.

The nominal point of contact is taken as the origin of coordinates; material of both solids "flows" steadily through a time-invariant strain field in the contact region.

Comparing the interfacial kinematics and stress conditions of sliding friction, it is obvious that the main reasons for sliding friction - like ploughing and shearing of adhesive junctions - are not met in a rolling situation. Therefore, the microscopic model of sliding friction cannot be applied in order to explain the occurrence of resistance to rolling, so that other explanations are needed. As in the case of sliding, any model of rolling friction must explain both the rolling friction force and the rolling-friction energy dissipation.

Rolling friction may be classified into a type, where large tangential forces are transmitted, e.g., the driving wheels of a locomotive, or in another type where small tangential forces are transmitted, often called "free rolling". Consider, as a starting point a Hertzian contact. If a torque is introduced which causes a rolling motion, then in addition to the Hertzian normal stresses, tangential stresses occur, see Mindlin's theory, Section 4.3.1. This results in the division of the contact area into a region of micro-sliding, and regions of adhesion within which the surfaces roll without relative (sliding) motion.

As main contributions to rolling resistance, the following effects can be listed:

(a) micro-slip (i.e., micro-sliding)
(b) elastic hysteresis
(c) plastic deformation
(d) adhesion effects

In the following, these different components of free-rolling friction will be discussed in brief.

(a) MICRO-SLIP EFFECTS

In the theories of micro-slip, different cases are distinguished in the literature, the most important of which are:

(i) Reynolds slip (Ref. 4.53)
Consider a Hertzian contact of two bodies having different elastic constants. If the two bodies roll freely together the pressure which acts on each gives rise, in general, to unequal tangential displacements of the surfaces leading to interfacial slip processes.

(ii) Heathcote slip (Ref. 4.54)

Consider a rolling contact where the rolling element is a ball which rolls inside a grooved track. In this case the profiles may conform closely in the transverse direction to the track on which they roll. Due to surface points lying at appreciably different distances from the axis of rotation, tangential tractions are introduced and micro-slip effects occur.

(iii) Carter-Poritsky-Föppl slip (Refs. 4.55. 4.56, 4.57)

For the two-dimensional case (e.g., two cylinders rolling on each other) with a tangential force in the direction of rolling, the micro-slip area could be calculated. This work led to the important conclusion that, in rolling, the adhesion zone is adjacent to the leading edge of the contact area; in contrast to the static problem where the adhesion zone is centrally placed, see Figure 4.17.

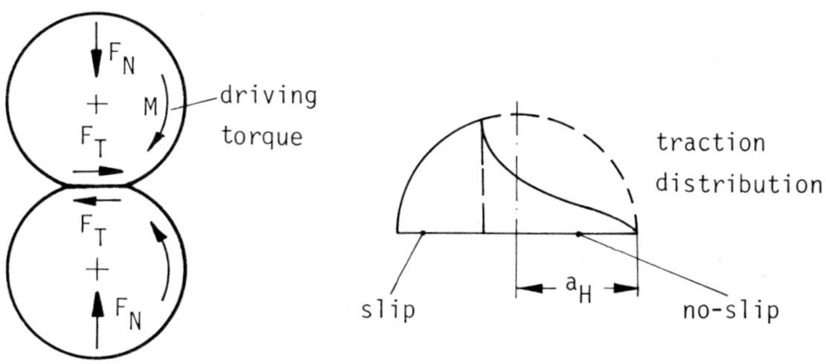

Figure 4.17 Slip and no-slip regions in a rolling contact.

Tabor studied experimentally the slip effects of the Reynolds and Heathcote type using steel balls rolling in grooved rubber tracks (Ref. 4.58). He showed that unless there is a high degree of geometrical conformity, both micro-slip effects contribute only very little to the observed rolling-friction force. Recently it was shown experimentally that above a critical value of the ratio ball radius r_B to groove radius r_G ($r_B/r_G > 0.8$ say), rolling friction is primarily due to a sliding mechanism (Ref. 4.59). In general, however, according to Bowden and Tabor, rolling friction cannot be traced back to sliding friction via the effect of micro-slip since the rolling-friction coefficient is not reduced by lubrication as is the case for the sliding friction coefficient. These experiments indicate that micro-slip may contribute only very little to the rolling-friction coefficient. On the other hand, it is generally recognized that the existence of friction forces at the interface between rolling elastic bodies results in the division of the contact area into regions of micro-slip, and regions within the surfaces roll without relative motion. Obviously, these effects have important engineering implications for the transmission of mechanical work through a rolling contact, as in a friction drive for example.

(b) ELASTIC HYSTERESIS

The elastic hysteresis theory of rolling friction, proposed in 1952 by Tabor (Ref. 4.60), assumes that the rolling resistance is due to the elastic hysteresis losses in the material of the track. In Section 4.2.1 it has been shown that in the case of an ideal Hertzian contact, a certain amount of energy is needed to build up the contact zone. For the formation of an elastically deformed rolling track a certain mechanical power is needed. In the deformation process the surfaces undergo complicated combined compressions and torsions. The elastic deformation energy is recovered mainly in the contact release. As a consequence of relaxation effects, however, the released forces may be smaller than the previous ones; the difference may be interpreted as rolling-friction loss.

The elastic hysteresis losses of materials, being related to the damping and relaxation properties of the materials, are obviously more pronounced for viscoelastic materials than for metals.

It has been found for viscoelastic materials that the rolling-friction coefficient can be related to the relaxation time (or the relaxation spectrum) of the material. Representing a viscoelastic material by a series of parallel linear Maxwell elements which undergo compression and recovery during rolling, it follows that in a simple model at a certain rolling velocity, which corresponds to the retardation time, the rolling-friction coefficient exhibits a maximum (Ref. 4.61).

At low rolling velocities the viscoelastic material will recover at the trailing side of the contact sufficiently quickly to maintain a fairly symmetrical pressure distribution, so that the rolling resistance will be small. At high rolling velocities the material will not recover sufficiently quickly to even make contact on the trailing side. The increase of pressure asymmetries with increasing rolling velocities have been checked experimentally with opto-elastic techniques (Ref. 4.62).

(c) PLASTIC DEFORMATION

It is well known that if in a rolling contact of metallic bodies the contact pressure exceeds a certain value, gross yielding will occur. For freely rolling cylinders yield occurs first at a point beneath the surface when

$$p_H \approx 3p_y$$

where p_H is the maximum Hertz pressure and p_y is the yield stress of the material in simple shear. In this case energy is needed to form plastic deformed rolling track.

An exact analysis of the plastic deformation processes in the rolling of a ball over a plane is very complicated. An approximate solution is due to Eldredge and Tabor (Ref. 4.63). They observed that in the first traversal a plastic deformed rolling track is formed, and found for the rolling-friction force an empirical expression of the form

$$F_F \propto \frac{F_N^{2/3}}{r} \qquad \begin{array}{ll} F_N & : \text{normal load} \\ r & : \text{ball radius} \end{array}$$

They explained this relation by assuming that rolling resistance is mainly due to the plastic deformation in front of the rolling ball. Under well-defined metallographic conditions, i.e., in the case of rolling deformation of metal single crystals, the plastic deformations can be explained on the basis of crystallographic data, such as crystallographic orientation and slip systems (Ref. 4.64).

In repeated rolling-contact cycles, however, the above yield criterion does not hold (Ref. 4.65). During the first contact cycle the surface material is plastically compressed and residual compressive stresses acting parallel to the surface are introduced. During subsequent rolling cycles the material is subject to the combined action of residual and contact stresses. Further yielding is less likely and a steady-state may be reached in which the material is no longer stressed beyond its elastic limit. This process is known as "shakedown" and the maximum load for which it occurs will be called the shakedown limit, given in pure rolling by

$$p = 4p_y$$

If rolling cylinders are subjected to loads in excess of the shakedown limit a new type of plastic deformation occurs. This mode of plastic deformation in rolling contact was first observed by Crook (Ref. 4.66) and studied experimentally by Hamilton (Ref. 4.67) and theoretically by Merwin and Johnson (Ref. 4.68). It consists of forward shearing of the surface of each cylinder relative to its core. The deformation is cumulative in the sense that an equal increment of plastic strain is acquired with each revolution. It follows that, at loads above the shakedown limit, continuous and cumulative plastic deformation is observed, whereas at loads below it, even though some yielding is caused initially, after a few traversals the system shakes down to an elastic cycle of stress. Accordingly, the friction mechanism is changed.

(d) ADHESION EFFECTS

According to Section 4.2.2, in rolling-contact conditions surface forces may act in the interface between the rolling bodies. Comparing the joining and separation of surface elements in rolling contacts and sliding contacts some important differences should be noted. Due to the differences in kinematics, in rolling contact the surface elements approach and separate in a direction "normal" to the interface rather than in a tangential direction. Therefore, as described in Section 4.3.1, junction growth and dispersal of surface contaminants is unlikely in the main part of the contact area. Consequently, at the regions within the rolling-contact interface where no relative motion in a tangential direction occurs, adhesion forces may be mainly of the (weak) van der Waals type. Short-range forces such as strong metallic bonds may act only in micro-contacts within the micro-slip area. If adhesion bonds are formed they are separated at the trailing end of the rolling contact in tension rather than in shear as in a sliding contact. Therefore, in general the adhesion component of rolling friction may be only a small portion of the friction resistance. For the rolling of metals under specific experimental situations, however (e.g., different thin metallic layers on hard balls) an adhesion component can nevertheless be the dominant factor in determining the order of rolling-friction coefficients for different metal pairs (Ref. 4.69).

Also the rolling contact between a glass cylinder and a smooth rubber flat has been successfully treated as an adhesive joint through which two cracks propagate at the same speed, one crack continuously opening to break the interfacial joint and the other closing to reform it (4.70), see Figure

4.18. Rolling friction on this basis is the consequence of adhesive hysteresis - the energy loss associated with forming and breaking an adhesive bond.

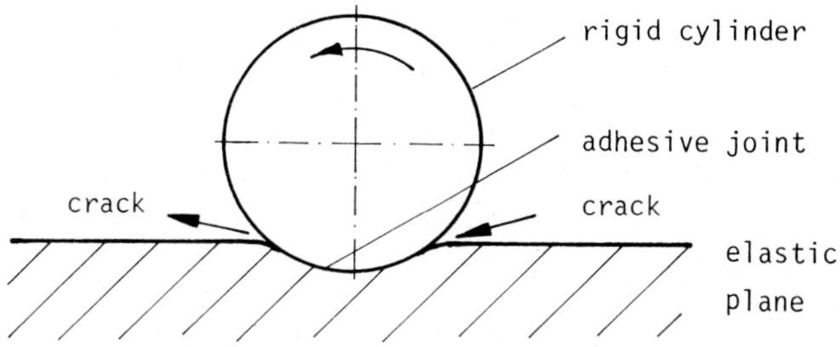

Figure 4.18 The contact between a rigid cylinder and an elastic plane considered as an adhesive joint containing two cracks.

Another type of adhesion theory for rolling friction was proposed by Derjaguin and Smilga (Ref. 4.71). According to this theory, contact partners of different electronic affinity undergo an electro-static charging. During rolling there is a flow a charge leading to energy dissipation.

The discussion of the cause of the rolling friction resistance shows that rolling friction is also a complex composite process determined by a couple of properties of the rolling-contact partners and operating variables. In special cases, i.e., under specific experimental situations, it is possible for one mechanism to predominate. It may then be permitted to relate the rolling-friction data with specific properties of the rolling partners. In the general case, however, the different mechanisms are superimposed. Therefore, all influencing parameters must be taken into account systematically on considering the whole process of energy transmissions and energy dissipations.

4.3.4 Energy transmission and energy dissipation

Having discussed the main friction processes, it is now possible to complete the general framework of the processes on the conceptual "work plane" of a tribo-mechanical system (see Section 3.4.2). According to Figure 3.6, the input work of a tribo-mechanical system can only be partly used as use-output; some part of the work is either stored within the system or emitted as loss into the environment or transferred to the thermal plane.

It was explained above (see Section 4.1) that a transmission of work through a mechanical system can be obtained only through processes of elastic deformation either of a bulk machine element, like the shaft of a sliding bearing, or through the interface of two contacting machine elements, as in the case of contacting gear-teeth. These elastic deformation processes are reversible. In idealized cases, the transmission of work through a mechanical system can be calculated or estimated using conventional techniques of the mechanics of solid bodies.

The processes by which mechanical work is not transmitted through a mechanical system are, of course, essentially irreversible. Since they are the ultimate origin of friction, these processes will be discussed in the following in order to gain a physical picture of the effects that take part in the dissipation of mechanical energy in a tribo-mechanical system. Owing to the great complexity of these irreversible processes it is only possible so far to outline a qualitative picture of the main features of these processes.

The whole course of the "loss" process of mechanical energy in a tribo-mechanical system can be formally divided into three phases:

I. Introduction of work into the contact zone
   → formation of real area of contact

II. Transformation of work within the contact zone
   → elastic deformation
   → plastic deformation
   → adhesion

III. Dissipation of energy
   (a) Storage
      → generation of point defects and dislocations
      → strain energy storage
   (b) Emission
      → phonons (acoustic waves, sound)
      → photons (tribo-luminescence)
      → electrons (exo-electrons, Kramer effect)
   (c) Transformation to thermal plane
      → generation of heat and entropy

Phases (I) and (II) have been described in some detail in the foregoing section in connection with the discussion of friction forces. In the following, the main processes of the dissipation phase (III) will be considered.

(a) STORAGE OF ENERGY

In the middle of the 1950's it became possible to study the molecular array of crystalline bodies and the dislocations and imperfections in this array by using of the Moiré technique in conjunction with an electron microscope. Since on one hand, the plastic deformation and shearing of solids is governed by imperfections and dislocations and, on the other hand, most friction processes involve plastic deformation processes, it appeared obvious to relate the friction process directly to the dislocation effects (4.72). In 1965 Kostetski and Nazarenko tried to explain the Coulomb-Amontons law of sliding friction through a relation between normal forces, friction forces and the dislocation structure of the solids (Ref. 4.73). They obtained experimentally for the sliding of armco iron parallel curves for the increase of friction force and the logarithm of dislocation density with increasing normal load as shown in Figure 4.19. It has now been well established through direct observations with the transmission electron microscope that in sliding contact of metals dislocations are generated (Ref. 4.74).

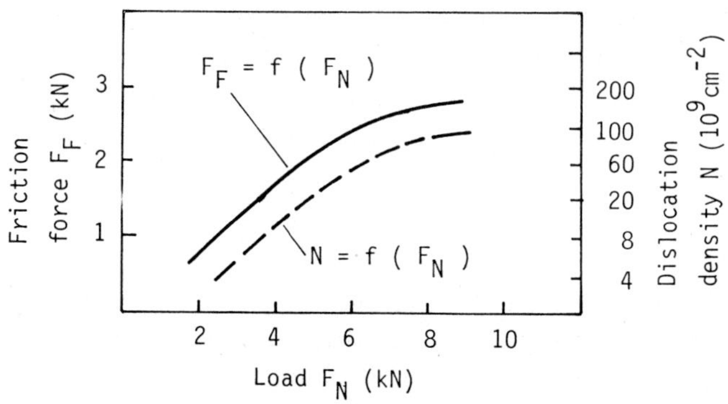

Figure 4.19 Friction force and dislocation density (Ref. 4.73).

From the theory of dislocations it follows that the work per unit length to create a dislocation in an isotropic undisturbed medium is given by (Ref. 4.75)

$$\Delta E_e = \frac{Gb^2}{4\pi(1-\nu)} \ln \frac{r_1}{r_o} \qquad \text{(edge dislocation)}$$

$$\Delta E_s = \frac{Gb^2}{4\pi} \ln \frac{r_1}{r_o} \qquad \text{(screw dislocation)}$$

where
- G : shear modulus
- b : Burgers vector
- r_1 : core radius
- r_o : radius of influence
- ν : Poisson number

As an example, with the data of copper

$(G = 4 \cdot 10^{10} N/m^2;\ \nu = 0.34;\ b = 2.5 \cdot 10^{-10} m;\ r_o = 10^{-9} m;\ r_1 = 10^{-2} m)$

it follows that

$$\Delta E_e = 4.8 \cdot 10^{-9}\ Jm^{-1}$$

If this value is compared with the frictional energy measured under experimental conditions which are comparable with the conditions of the estimation of ΔE_e, it was found that the stored elastic energy in a condition of incipient plasticity can only account for less than 1% of the energy expended in friction (Ref. 4.76). A similar result was also found in conditions of more severe plastic-deformation conditions of sliding (Ref. 4.77). This may be due to the fact that in the plastic-deformation mode the dislocations generated would not have remained in the surface after the passage of the sliding counterface and the energy would appear as heat (i.e., transformed to the thermal plane of the tribological system). These results indicate that some part of the frictional energy is indeed stored within the sliding partner via a dislocation mechanism. Since this stored energy is obviously only a small part of the whole frictional energy, in general other dissipation mechanisms must also be considered.

(b) EMISSION OF ENERGY (OTHER THAN THERMAL)

In the microscopic model of friction it has been explained that during sliding or rolling, micro-contacts are formed and destroyed. These processes are statistically distributed in time and location within the geometric area of contact. It is likely that in the process of junction formation some of the asperities or some part of them are deformed elastically. If then the junctions are destroyed and the adhesion bonds are broken, the elastic deformed parts of the asperities snap apart thus generating vibrations and other lattice stimulations which lead to the emission of energy in different forms. In solid state physics, different processes of the transformation of energy into apparently other forms are known (e.g., the photoelectric effect or thermoionic emission). In analogy, the main "tribo-induced" energy emission effects which have been observed experimentally are sketched in brief (the effect of heat generation is treated separately below).

(i) Emission of phonons (acoustic waves, sound)
The generation of friction-induced acoustic waves (consisting of phonons) and the emission of sound is a common feature of most tribological processes. Since noise is considered as an "acoustic pollutant" the reduction of noise emission is an important task in today's industry (Ref. 4.78). Physically, the generation of sound in tribo-mechanical systems is connected with processes of elastic deformation and release of asperities. A theory of natural normal micro-vibrations in sliding contact was put forward by Tolstoi (Ref. 4.79). He showed that the self-excited micro-vibrations during sliding are invariably accompanied by simultaneous upward jumps of asperities. The frequency of the natural micro-vibrations is determined by the contact stiffness and the mass of the sliding surface. These micro-vibrations vanish only when the normal vibrations are damped by external means or when the velocity is lower than a certain limit. The value of the critical velocity, v_{cr}, may be roughly estimated in terms of the creep viscosity, η, of the rubbing solids, their yield strength, σ_y, the friction coefficient, f, and both the height, y, and spacing l_a of the surface asperities:

$$v_{cr} = \frac{f\ \sigma_y\ l_a^2}{2\eta\ y}$$

For the sliding of conventional steel surfaces it follows that $v_{cr} = 5 \cdot 10^{-10 \pm 1}$ cm/s. This very low value indicates that in almost any sliding situation a certain amount of the mechanical energy is dissipated through the generation of vibration and the following emission of acoustic waves. The absolute amount of energy emitted is, however, usually only a very small part of the whole frictional energy.

(ii) Emission of photons (tribo-luminescence)
The effect of triboluminescence is well known in physics. It occurs if certain solid bodies, such as uran-nitrate or ZnSMn, are rubbed against each other. In this case, the mechanical work which separates adhesion bonds activates photons which are emitted as visible light. This effect is not rare: so far it has been observed in more than a thousand materials (Ref. 4.80). A complete theoretical explanation of this effect is still needed. Some observations support the assumption that in the surface cracks of the crystals, electrostatic double-layers exist which cause a kind of spark discharging if the surface cracks are cleaved during the friction process. Obviously some part of the frictional mechanical energy acts as activation energy and is - via a mechano-electro-optical transformation mechanism - emitted as optical radiation. Clearly, if it occurs, this effect also consumes only a small part of the whole frictional energy.

(iii) Emission of electrons (exo-electron emission, EEE, Kramer effect)
At the end of the 1940's, Kramer at the Physikalisch-Technische Bundesanstalt (PTB) observed that electrons are emitted from freshly abraded metallic surfaces. Kramer initially thought that these electrons were due to exothermic processes occurring at the surface. It is assumed that the friction process acts as an activation process for the emission of exo-electrons and that some analogies between EEE and other emission effects like luminescence, field emission and photoemission exist. However, owing to the complexity of the sequence of events involved in the whole process, a complete theory of the EEE effect is not yet available. Recent experimental work on the EEE of a clean, annealed magnesium single crystal surface, performed in UHV, verifies that electron emission occurs already from a strain free surface simply upon adsorption of oxygen (Ref. 4.81). On the other hand, measurements of EEE rates during friction experiments on aluminium surfaces show that there are close connections between the emission rate of exo-electrons, the composition of the topmost surface layers and the friction coefficient (Ref. 4.82). An example of the correlation between the EEE rate and the friction coefficient for the sliding of aluminium against aluminium is shown in Figure 4.20. Although in the whole energy balance of a tribo-mechanical system the energy "loss" due to the emission of exo-

electrons is only a very small part, it is obvious that if exo-electrons are generated they may play an important role for the tribo-chemical processes occurring in the system.

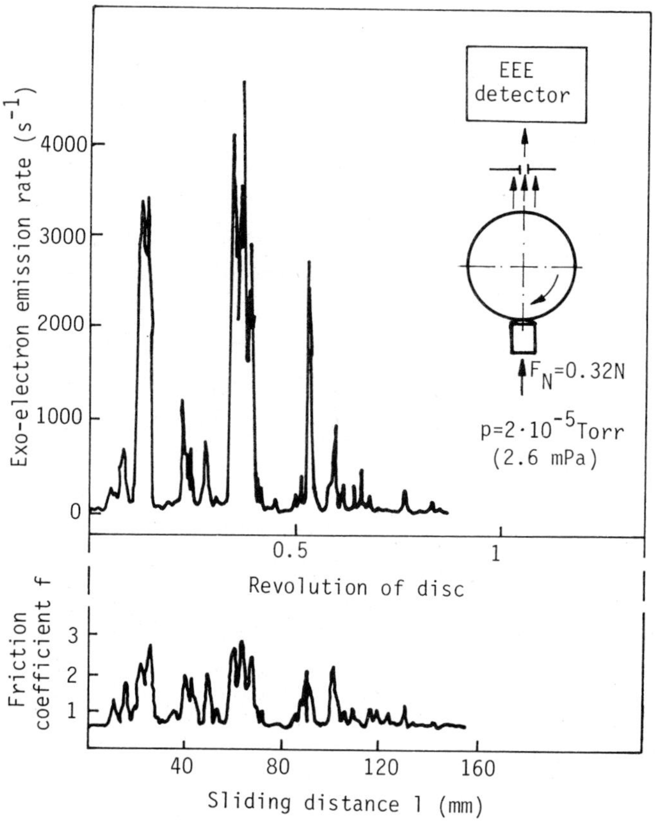

Figure 4.20 Friction coefficient and exo-electron emission rate (Ref. 4.82).

In summary, discussion of the effects (i), (ii), (iii) shows that there are some tribo-induced emission processes that must be included in the whole picture of the dissipation of energy in a tribological system. Although these tribo-induced processes are physically very interesting and may in certain cases be very important for the tribo-chemical processes occurring in the system, in general, they account for only a very small part of the whole energy dissipated.

(c) TRANSFORMATION TO THERMAL PLANE

The discussion of energy dissipation mechanisms hitherto has shown that there are some processes of storage and emission of energy (other than thermal) that contribute to the consumption of energy in friction. Since these contributions, in general, are very small, this rather detailed consideration of the various processes also leads to the conclusion that nearly all the energy dissipated by friction appears as heat. Indeed, this has been known from ancient times and is the basis of the mechanical equivalent of heat. From the microscopic model of friction it is obvious that all the processes involved in the formation and separation of asperity micro-contacts lead to lattice distorsions and vibrations which manifest as heat. In the general theory of tribology the generation of heat in friction is expressed as a transformation of work to the conceptual thermal plane of the system (see Section 3.4.3). Clearly, the heat generated in friction may either remain within the system (acting perhaps as activation energy for material changes or chemical reactions) or leave the system through convection, conduction or radiation.

An important feature of the generation of heat is the increase of temperature of the material elements of a tribological system. Since the increased temperature and the temperature gradients in turn influence the material properties of the friction partners and activate tribological processes, the knowledge of tribo-induced temperatures is of paramount importance in tribology. In the tribo-induced temperature rises we must distinguished between:

(i) a bulk temperature rise
(ii) temperature gradients
(iii) local temperature rises or "flash temperatures"

Theoretical treatments of the flash temperature problem were performed by Carslaw and Jaeger (Ref. 4.83), Holm (Ref. 4.84) and Blok (Ref. 4.85). These theories were combined and simplified by Archard (Ref. 4.86). Owing to the great importance of this theory to many tribological situations, the main results of this theory will be sketched in brief.

The model used in Archard's treatment is shown in Figure 4.21. A surface asperity of body (1) forms a circular area of contact $A = \pi a^2$, which moves with a velocity v over the flat surface of body (2). The temperatures are calculated on the assumption that the heat is generated at the true area of contact and that this heat is conducted away into the bulk of the rubbing members.

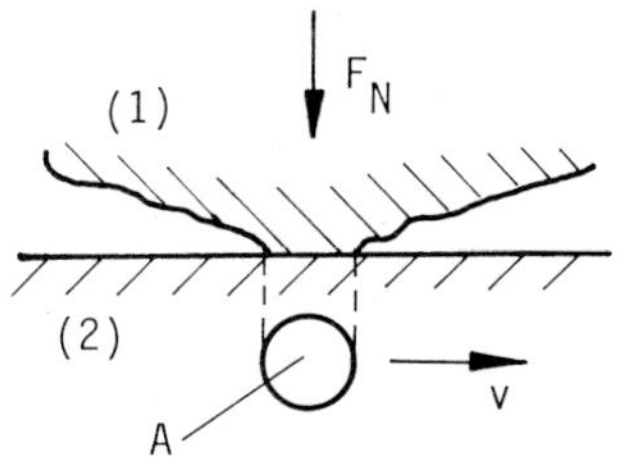

Figure 4.21 Model of a sliding asperity.

The theory firstly requires the solution of the equations for the flow of heat into each body. The derived surface temperatures are expressed in terms of the rate of supply of heat, the size and speed of the heat source, and the thermal properties of the material. Secondly, the proportion of the total heat flowing into each body is then determined by the criterion that the equation of heat flow for both bodies shall give the same average temperature over the contact area. Depending on the operating conditions of load F_N and speed v, the nature of materials and the deformation mode, different formulae have been derived. For the case of plastic asperity deformation of bodies of the same material using the abbrevations

$$N = \frac{\pi f g p_y}{J \rho c}$$

$$L = \frac{F_N^{1/2} v}{2\varkappa (\pi p_y)^{1/2}}$$

- f : friction coefficient
- g : acceleration due to gravity
- p_y : flow or yield pressure
- J : mechanical equivalent of heat
- ρ : density
- c : specific heat
- $\varkappa$: thermal diffusivity ($\varkappa = \frac{k}{\rho c}$)
- k : thermal conductivity

it follows that the maximum friction-induced temperature rise ΔT is given by the following expressions:

$\Delta T_{max} = 0.25\ NL$ for $L < 0.1$

$\Delta T_{max} = 0.25\ C_1 NL$ for $0.1 < L < 5$

($C_1 = 0.95$ at $L = 0.1$

$C_1 = 0.5$ at $L = 5$)

$$\Delta T_{max} = 0.435\ C_2 N L^{1/2} \qquad \text{for } 5 < L < 100$$
$$(C_2 = 1/\left[1 + 0.87\ L^{-1/2}\right])$$

$$\Delta T_{max} = 0.435\ N L^{1/2} \qquad \text{for } L > 100$$

The applications of these results to the example of a sliding pair of plain carbon steel with a hardness of V.P.N 150, 250 and 850 is shown in Figure 4.22.

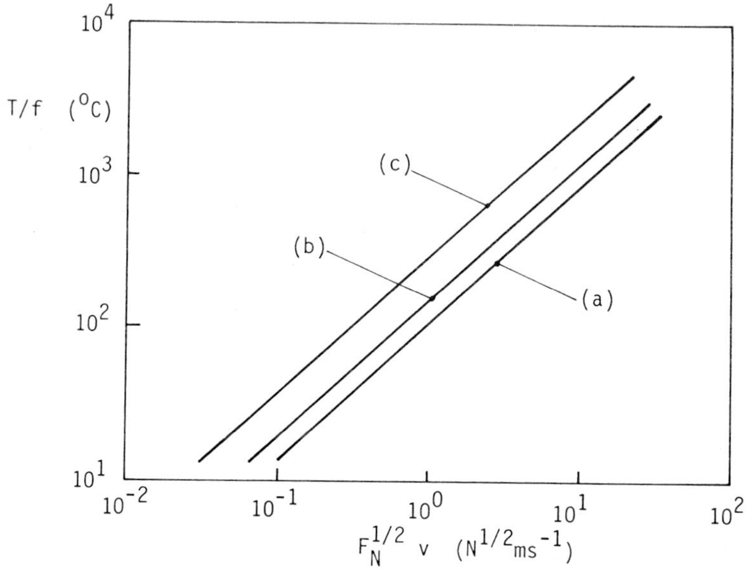

Figure 4.22 Maximum attainable flash temperatures for sliding contacts of steel.
(a) : V.P.N., 150. (b) : V.P.N., 250.
(c) : V.P.N., 850. (Ref. 4.86)

In addition to the temperature rise within a frictional interface, the distribution of heat to the two interacting bodies is important. Clearly, the distribution depends not only on the operating parameters and the materials properties but also on the geometry of both bodies (Ref. 4.87). For a system consisting of two bodies of equal kinematics and geometry it has been proposed that the frictional heat is distributed between the two bodies according to the factor (thermal conductivity x specific heat x density)$^{1/2}$ of both bodies (Ref. 4.88).

The discussion of the whole cycle of introduction-transformation-dissipation of work in a tribo-mechanical system shows the manifold of processes and effects involved in friction. Although with the contemporary techniques of experimentation it is possible to study a single isolated effect, in general there is an overlap and an interaction of the different processes. The detailed discussion as performed in the last sections has provided a basis for the compilation of the properties of materials relevant to the different steps of the transmission and dissipation of energy. These properties belong to different classes according to their relevance to the different interactions occurring at the interface in a tribological system:

(a) Properties "per se"
- volume properties: geometry, dimensions, chemical composition
- surface properties: surface topography, surface composition

(b) Properties relevant to tribo-mechanical interactions
- elasticity, plasticity, elastoviscosity

(c) Properties relevant to tribo-chemical interactions
- reactivity with the environment

(d) Properties relevant to tribo-physical interactions
- adhesion, transport properties

A data sheet to assist in the compilation of the relevant properties of friction pairs is given in Chapter 8.

4.4 Wear Processes

4.4.1 General description

In the foregoing section dealing with friction processes, the transactions on the work plane and the thermal plane of a tribo-mechanical system have been discussed according to the tribo-process diagram, Figure 3.4. In order to complete the conceptual framework of the tribology of mechanical systems as outlined in Figure 3.4 further, the transactions on the materials planes must be studied in connection with the relevant wear processes. Comparing friction with wear, depending on the technical function of the tribo-mechanical system under consideration, basic differences between friction and wear processes must be noticed. According to Table 3.2, the various tribo-mechanical systems can be classified broadly into two different groups:

(a) A first group in which the main inputs and outputs of the system are given by physical (i.e., non-material) quantities, headed by the terms motion, work and information.
(b) A second group in which the main inputs and outputs are given by a flow of materials through the system, as in the case of a wiredrawing system.

According to the definition of wear as "the progressive loss of substance from the operating surface of a body occurring as a result of relative motion at the surface" it follows that in case (b) wear may be seen as a process which leads to a loss of a material input quantity of the system. In these cases, material losses due to wear may be treated formally in analogy to energy losses due to friction. On the contrary, for systems of group (a), which will be considered mainly in this section, the technically relevant input and use-output quantities are essentially non-material. Therefore, in this case wear leads to property changes and losses from the elements of the system. To repeat this evident but important point: from a systems point of view, friction is a process which leads to losses from the input quantities of the system whereas wear is a process which leads to losses from the elements of the system. Therefore, friction influences primarily the input-output relations (i.e., the function of the system) whereas wear primarily influences or changes the properties of the elements of the system (i.e., the structure of the system). In the following, the main wear processes relevant to tribo-mechanical systems of group (a) will be discussed.

Consider, as a starting point, the dry wear of metals occurring in a simple tribo-mechanical system consisting of the partners

(1) One machine element
(2) A second machine element
(3) The interfacial volume (gaseous atmosphere)

For example, such a system may be given by the configuration shown schematically in Figure 4.23 in which the surfaces of (1) and (2) slide against each other in air (3) under a normal load F_N with a sliding velocity v.

Consider now the wear processes that take place on one of the two partners, say on element (1). According to Figure 3.8 the material losses due to the wear processes may be described as:

(i) material transfer from (1) to (2)
(ii) material loss from (1) to (3)
(iii) formation of reaction products

(There may be back-and-forth reactions between the elements (1), (2), (3); these will be considered below in Section 4.4.6).

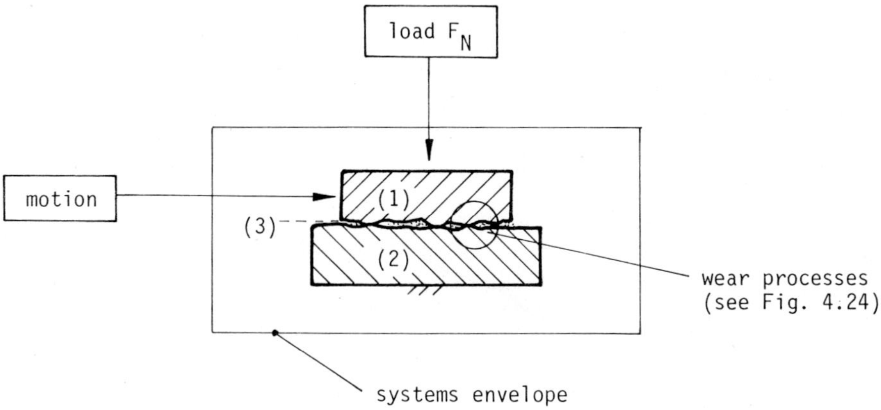

Figure 4.23 A simple tribo-mechanical system.

Depending on the amount of material which is lost from element (1), the wear may be broadly classified as "mild wear" or "severe wear". Obviously this classification can be related to the composition of the surface, (Figure 4.5). In mild wear, the wear processes occur at the "outer" surface layers. The surfaces remain relatively smooth and are usually protected by surface oxide layers generated in rubbing. The worn debris consists of small particles down to some nm. In severe wear, the contact is metallic, the surfaces are deeply torn, and the worn debris consists of metallic particles up to some 100 μm. Clearly, the classification of wear into "mild" and "severe" is primarily a distinction in the scale of size.

As in the case of sliding friction, there are some "macroscopic rules" of the wear of metals under dry sliding conditions which have been experimentally observed in steady-state situations, provided the "nature" of surfaces does not change:

I. The wear rate w, i.e. the volume V of material removed per unit sliding l is proportional to the normal load F_N

$$w = \frac{V}{l} \propto F_N$$

II. The wear rate w is independent of the apparent area of contact.

In order to explain these macroscopic rules, a microscopic model of the wearing process similar to that used in the discussion of the friction process (Figure 4.15) may be considered. Clearly, the processes occurring between interacting asperities - like elastic and plastic deformation, and so on - as discussed in the foregoing sections, must similarly be taken into account in any microscopic model of the wear process. In Figure 4.24 a schematic representation of a unit event of the wear process is shown.

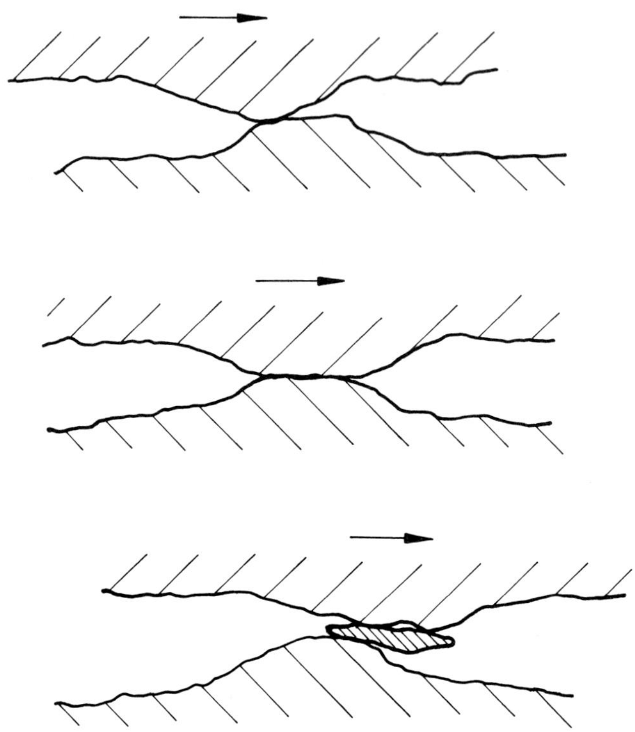

Figure 4.24 Schematic representation of a unit event in the wear process.

There have been several attempts (see Section 1.2.2) to derive formulae expressing the macroscopic observed wear rules mentioned above. The essential concept - as summarized by Archard (Ref. 4.89) - is that the worn volume V, produced in sliding a distance l, can be related to the true area of contact, A_r. It is usually assumed that the unit event which must be considered is the contact of two asperities on the opposing surfaces.

As idealized in Figure 4.24, the unit event is concerned with the establishment of an area of contact considered to be a circle of radius, a, and area $\Delta A = \pi a^2$. If it is assumed that in sliding a distance $\Delta l = 2a$ a hemispherical particle of radius a and volume $\Delta V = 2/3 \pi a^3$ is generated it follows that

$$\frac{\Delta V}{\Delta l} = \frac{\Delta A}{3}$$

Archard has suggested that not every unit event results in the formation of a wear particle. Introducing a factor K which represents the probability that any given event will produce a worn particle and summing for all micro-contacts, the total wear rate is

$$\frac{V}{l} = \frac{1}{3} K A_r$$

It has been shown above that, depending on the deformation mode, the following relations between the real area of contact, A_r, and the normal load F_N exist:

- elastic deformation (plasticity index $\psi < 0.6$)

$$A_r \propto \frac{F_N}{E} \qquad E \; : \; \text{composite elastic modulus}$$

- plastic deformation (plasticity index $\psi > 1$)

$$A_r \propto \frac{F_N}{p_y} \qquad p_y \; : \; \text{yield pressure}$$

From these relations it follows that the proportionality

$$V \propto K F_N l$$

exists, as observed experimentally in many situations of the dry wear of metals under steady-state conditions.

In trying to describe systematically the various wear processes that may occur in a tribo-mechanical system a couple of severe difficulties arise These are not only due to the great complexity of wear processes but also to the imprecise and ambiguous use of the term wear. For example, the type of wear of an unlubricated metal pair sliding in a dusty atmosphere may be termed

dry wear
or metallic wear
or sliding wear
or scratching wear
or abrasive wear

depending on the emphasis intended. Since most of the imprecise wear terms in common use are connected with practical wear problems, a wear classification scheme is needed which maintains the known terms but avoids ambiguity. In the review of the historical development of wear studies, Section 1.2.2, two wear classification schemes have been already mentioned:

(a) Classification in relation to the type of relative motion, suggested in 1938 by Siebel (Ref. 1.22).
(b) Classification in relation to the type of wear mechanism, suggested in 1957 by Burwell (Ref. 1.26).

Comparing these wear terminologies with the terminology used in the field of the bulk strength of materials, the classification (a) can be seen in analogy to the type of "external stress" (e.g., compression, tension, bending, etc.) and the classification (b) can be seen in analogy to the type of the mechanism of "internal damage" (e.g., ductile fracture, brittle fracture, etc.).

From a system's point of view, the wear loss-output of a given tribo-mechanical system should be related to the inputs of the system (e.g., type of motion, work) as well as to the structure of the system (system's elements, properties of the elements, interrelations between the elements). Combining these requirements with the existing classifications (a) and (b), for an unequivocal characterization of wear, the following characteristics should be specified:

(i) the type of relative motion,
(ii) the interacting elements,
(iii) the dominant wear mechanism.

Obviously, the combination of these three characteristics specifies the "type" of wear. For a more quantitative characterization, in addition the following characteristics should be specified:

(iv) the input work: normal load, friction coefficient, distance of motion (at stated velocity),

Interacting elements	Type of relative motion		Wear mechanisms: mainly stress interactions		stress + material interactions	
			surface fatigue	abrasion	adhesion	tribo-chemical
solid/solid (metals, polymers, minerals, etc.) without or with lubricants	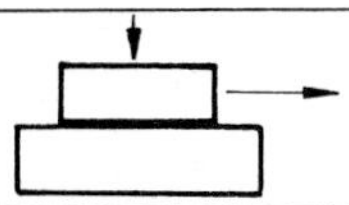	sliding	sliding wear			
	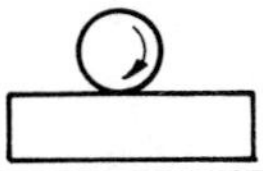	rolling	rolling wear			
	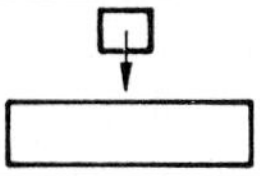	impact	impact wear			
	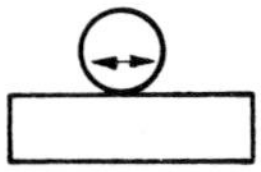	oscillation	fretting wear			
solid/liquid	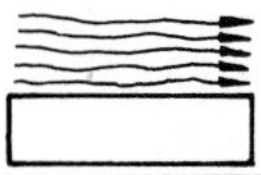	flow	cavitation wear			
solid/fluid + particles	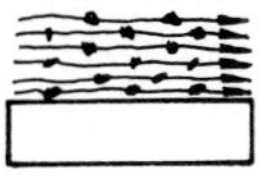	flow	fluid erosion			

Table 4.2 Characterization of wear.

Wear mechanisms	Appearance of the worn surfaces
Surface fatigue	cracks, pits
Abrasion	scratches, grooves, striations
Adhesion	cones, flakes, pits
Tribochemical	reaction products (films, particles)

Table 4.3 Appearances of worn surfaces.

(v) the material properties relevant to wear,
(vi) the wear rate
(vii) the appearance of the worn surfaces.

In Table 4.2 a classification scheme which tries to combine the above characteristics (i), (ii), (iii) is shown. In this table, the types of wear are termed according to the type of relative motion in analogy to the classification of friction modes. (This is the most evident way of distinguishing between different types of wear since in practical engineering situations the actual interfacial wear processes are seldom easy to detect.) In mechanical engineering, the wear of the components of tribo-mechanical systems is connected with four basic types of relative motion: sliding, rolling, impact, oscillation. It has been found that the interfacial wear mechanisms acting during these motions can be classified into four groups under the headings: surface fatigue, abrasion, adhesion, tribo-chemical. As indicated in Table 4.2 in any of the different types of relative motion one or more of the four basic wear mechanisms may act. In practice it is often difficult to determine what type of interfacial wear mechanism (or what combination) acts in a given situation. For example, in pure sliding of a metal/metal pair the abrasion or the adhesion wear mechanism may dominate, whereas during oscillation of a metal/metal couple fretting occurs essentially through the combination of all four basic wear mechanisms (see Section 4.4.6).

A customary (indirect) way of detecting the type of wear mechanism acting in a given wear situation is to study the appearance of the worn surfaces. In Table 4.3 some typical characteristics of surface damage due to the four basic wear mechanisms are listed. With the combination of the characteristics shown in Tables 4.2 and 4.3 the basis for an unambiguous distinction between wear of different types may be given.

After the general description of wear and its extremely complex nature, in the following the main aspects of the four basic types of wear mechanisms in relation to sliding, rolling, and oscillating will be studied from a physical point of view (Ref. 4.90). The discussion is restricted to the wear mechanisms between solid bodies. The effects of cavitation and erosion cannot be treated within the scope of this volume. (For a discussion of the mechanisms of impact wear the reader is referred to a recent book by Engel) (Ref. 4.91.) It should be emphasized again that the occurrence of one single wear mechanism is met only under specific conditions and that, in general, a combination of wear mechanisms occurs. This is discussed below in Section 4.4.6.

4.4.2 Surface fatigue wear mechanisms

At a first glance, surface fatigue seems to be a relatively simple wear mechanism since it appears to be based only on the action of stresses on the surfaces without needing a direct physical solid contact of the surfaces under consideration. This follows from the observation that surface fatigue effects are observed to occur in journal bearings where the interacting surfaces are fully separated by a thick lubricant film (Ref. 4.92). The effect of fatigue wear processes is normally associated with repeated stress cycling in rolling or sliding contact. In fatigue, failure in the material arises because of stress-reverse effects, i.e., fracture can develop under alternating stresses with a peak level which could be safe if imposed in only tension or compression. Scanning electron micrographs shown in Figure 4.25 exhibit the typical appearance of the surface fatigue damage after repeated unidirectional sliding of a pin over a disc.

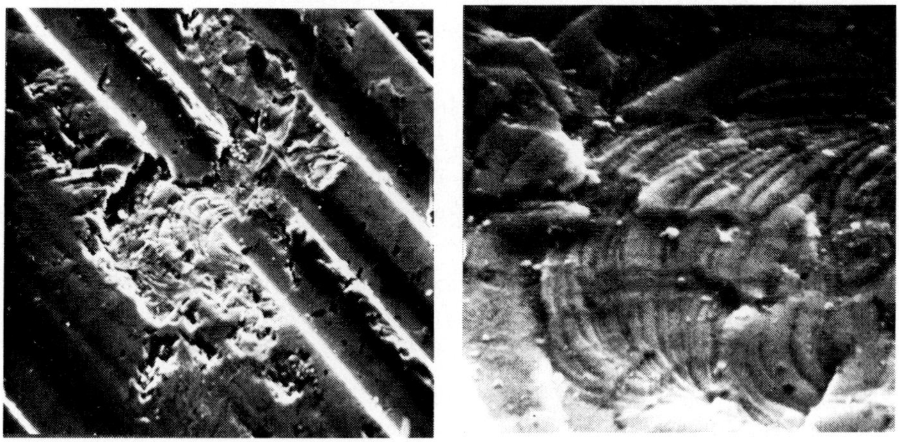

20 μm   4 μm

Figure 4.25 Appearance of surface fatigue damage.

Since surface fatigue is the classic failure mode of rolling contact, as in ball and roller bearings, the mechanisms of surface fatigue in rolling contact will be considered first.

Comparing the surface fatigue of rolling contacts with the ordinary fatigue of bulk material, there are two differences to be noted. First, the fluctuations in the time to failure, the "lifetime", are much more severe in rolling contact. Second, there is the important phenomena of fatigue limit stress which is observed in ordinary bulk testing, i.e., there exists a certain stress limit for a given material below which the material enjoys an infinite fatigue life. Such limit has not been detected in the surface fatigue case (Ref. 4.93). Empirically, it has been found in a series of experiments that the following relation for the time to failure exists:

$$t = \frac{\text{const}}{\sigma_m{}^9}$$

where σ_m is the maximum elastic stress.

To understand how and where surface-fatigue cracks form, it is necessary to consider the nature of the stress fields which exist beneath the surface. The Hertzian theory of static elastic contact shows that the maximum compressive stress occurs at the surface and the maximum unidirectional shear stress is some distance, y, below it (e.g., in the case of two contacting cylinders with a contact width of $2a_H$ at $y = 0.78a_H$). Under rolling contact conditions the important stress parameter is the maximum reversed shear stress and this can be shown to lie closer to the surface than the maximum unidirectional shear stress. If the contact is subjected to considerable surface traction forces then the positions of these shear stress maxima change and move towards the surface region. The two situations are illustrated in Figures 4.26 and 4.27 which represent combined stress distributions calculated for the case of a cylinder/flat contact by finite element methods (Ref. 4.94).

Although there has been considerable progress in the calculations of rolling stress distributions, a couple of additional influencing factors must be considered. Littmann (Ref. 4.95) has compiled the following list of those factors which modify the nominal contact stress distribution from that computed for homogeneous, isotropic, elastic, smooth-surface materials in dry rolling contact:

(a) Subsurface stress raisers
   Oxides and other hard, brittle inclusions,
   Sulfides, carbides, and other second phase particles,
   Grain boundaries, sub-boundaries, twins, and other dislocation arrays.

(b) Surface character
   Surface topography and texture,
   Residual stresses,
   Surface energy level,
   Microstructure,
   Contaminants.

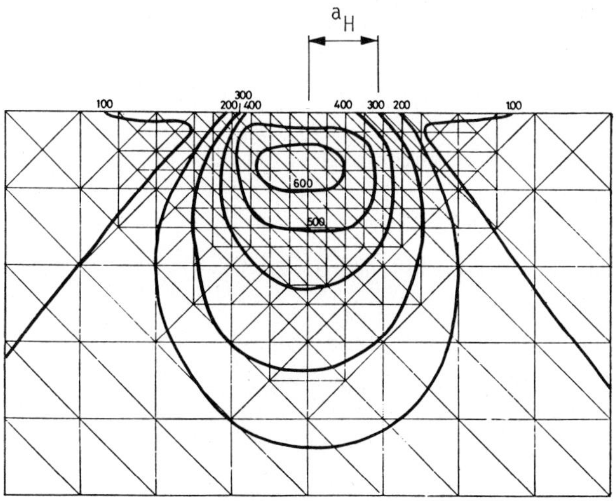

Figure 4.26 Map of maximum shear stresses (x $10^6 Nm^{-2}$) for cylinder/plane contact due to normal load alone (σ_{Hz} = $10^9 Nm^{-2}$) (Ref. 4.94).

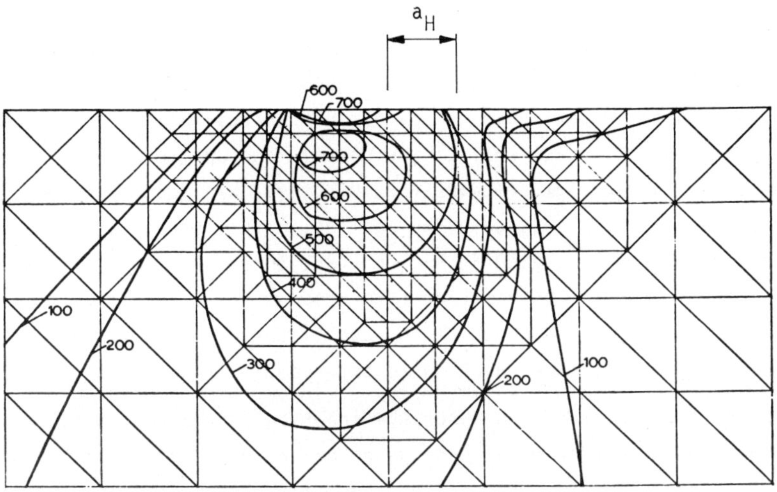

Figure 4.27 Map of maximum shear stresses (x $10^6 Nm^{-2}$) for cylinder/plane contact due to combined normal and tangential loads (F_T/F_N = 0.3; σ_{Hz} = $10^9 Nm^{-2}$) (Ref. 4.94).

(c) Surface flaws
 Inclusions and second phase particles,
 Nicks and dents including true and false brinelling,
 Grooves and scratches,
 Corrosion pits, rust, water etch,
 Fretting damage,
 Skidding damage.

(d) Discontinuities in contact geometry
 End of "line" contact geometry,
 Debris particles in the contact area.

(e) Load distribution within the bearing
 Elastic deflections,
 Misalignment of bearing parts,
 Internal clearance - bearing adjustment.

(f) Elastohydrodynamics

(g) Tangential forces
 Without gross sliding,
 Rolling plus sliding.

This compilation of influencing factors indicates the complexity of surface fatigue effects in rolling contact. An engineering model of fatigue failure in rolling contact which takes into account the influence of many factors has been proposed by Tallian and co-workers (Ref. 4.96).

A fundamental experimental study of the contact-fatigue processes under cyclic variation (26 kHz) of the normal load applied to the contact between a ball and a flat specimen (52,100 steel) was made by Tyler, Burton and Ku (Ref. 4.97). In this case the applied contact stress was beyond the elastic range. As illustrated in Figure 4.28, there are two distinct contact circles formed on both specimens corresponding to the minimum and maximum loads. As a consequence of the oscillatory loading three distinguishable types of surface damage were observed:

(a) In the toroidal region between the inner and the outer contact circles "fretting" occurred as a consequence of shear tractions induced during the loading and unloading cycles and influenced by the difference of the elastic moduli of the specimens. (The fretting mode of wear processes is discussed in Section 4.4.6.)

(b) The central part of the contact area underwent a depression due to plastic deformation which was found to accumulate with repeated loading.

(c) The ultimate failure of the contact progressed as the following succession of events generated in the toroidal area: crack formation, crack growth and interweaving and flaking out of the crack-surrounded areas. The fatigue cracks did not initiate in the most hardened material but in the material adjacent to it. It was found that a rapid temperature rise occurred in the initial stage of crack formation. The propagation of micro-cracks tended to follow the boundary of the hardened region along carbide stringers and influenced by the prior austenitic grain boundaries.

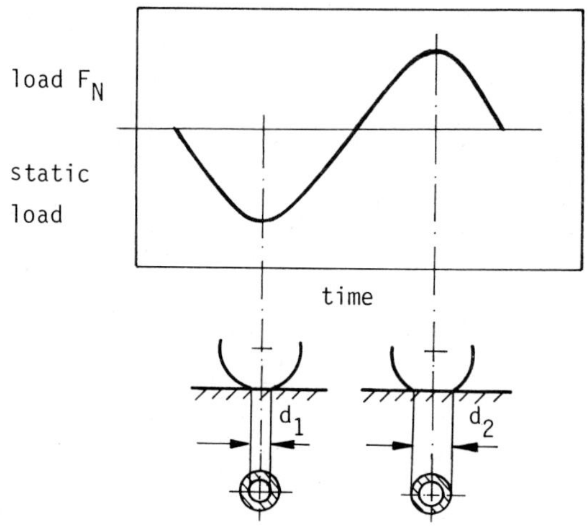

Figure 4.28 Variation of load and contact area on fatigue specimens (Ref. 4.97).

The experimental observations of surface fatigue indicate that - from a physical point of view - the study of contact fatigue is basically a study of stress concentration effects that govern fatigue crack initiation coupled with crack propagation behaviour. In the following, the physical mechanisms of (i) crack initiation and (ii) crack propagation will be discussed. The models described may be of significance not only for the (classic) surface fatigue mode of wear but also - in a broader sense - for the discussion of recently developed wear models related to the generation and propagation of surface cracks.

On the basis of dislocation theory, there are several possible mechanisms for crack initiation. These mechanisms were originally proposed in reference to bulk-material behaviour but they can be applied equally well to near-surface phenomena (Ref. 4.36). The main forms of crack initiation are (Ref. 4.98):

(a) Pile-up of dislocations in slip band against grain boundary (Cl. Zener).
(b) Coalescence of two slip dislocations to form a crack along a cleavage plane (A.H. Cottrell).
(c) Crack formed at tilt boundary (A.N. Stroh).

These mechanisms are presented schematically in Figure 4.29.

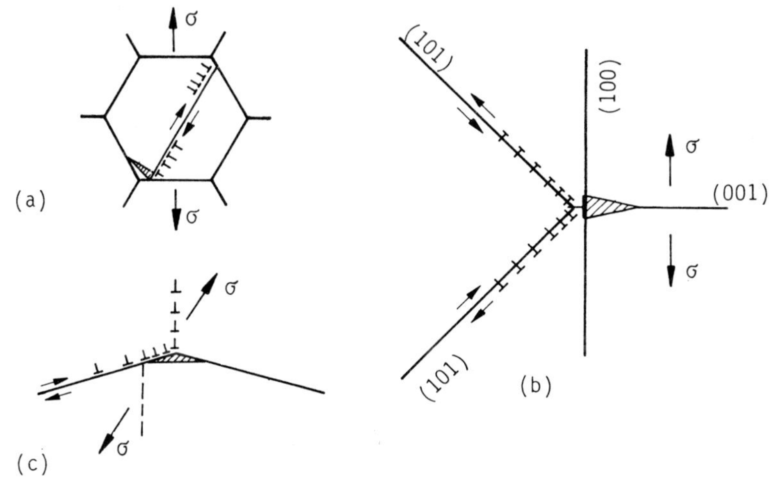

Figure 4.29 Mechanisms of crack initiation.

Another mechanism that can initiate subsurface cracks is based on the effect of inclusions. The mechanism of crack initiation can be started by such obstacles as oxide inclusions which are very prevalent in the near-surface layers of relatively soft metals such as copper. Further, with bearing and gear steels, carbide inclusions are always present. The presence of such obstacles can give rise to dislocation coalescence and the initiation of subsurface crack nuclei.

Although the described mechanisms of crack initiation are mostly used as starting points for a discussion of metal fatigue, there are some other details of dislocation behaviour that should be considered which are beyond the scope of this volume (Ref. 4.99).

Once an atomic crack has been formed by one of the mechanisms discussed above, the processes that govern the crack propagation must be considered. For brittle materials the well known classic continuum theory of Griffith may be applicable. Consider an elliptical crack of length 2a under a tensile stress σ. The surface energy of the crack is equal to $2\gamma_{cr}\pi a^2$. The distorsions in front of the crack tip are of the order of σ/E and extend over dimensions of the order of a. Thus an opening of the crack in the distored material leads to an energy change

$$\Delta U = 2\gamma_{cr}\pi a^2 - \frac{\sigma^2}{2E}\frac{4\pi}{3} a^3$$

When the energy put into the material by the applied stress exceeds the surface energy of the material, the crack surface area will increase. At that point

$$\frac{d\Delta U}{da} = 0$$

and

$$\sigma_{crack} = \left[\frac{2\gamma_{cr} E}{a}\right]^{1/2}$$

For example, for steel (σ = 700 N/mm^2, γ_{cr} = 1200 erg/cm^2) it follows that a $\approx$ 1 µm. The above equation shows the importance of the parameter

$$K_c = \sigma_{cr} a^{1/2}$$

for the ability of crack propagation. Recently, there have been attempts to relate the wear behaviour of metals with this parameter (Ref. 4.100).

These considerations apply to crack propagation in brittle materials. If the material is able to deform plastically, some of the energy input into the material will be absorbed in plastic flow in the regions of the crack tip.

In studying the plastic-elastic stress fields in the subsurface regions of sliding asperity contacts and the possible dislocation interaction, a "delamination theory of wear" has been put forward by Suh (Ref. 4.101) in which the generation of sheet-like particles is explained on the basis of the following chain of events:

(i) generation of subsurface dislocations,
(ii) dislocation pile-up,
(iii) formation of voids,
(iv) coalescence of voids leading to cracks parallel to the surface,
(v) production of sheet-like particles when the crack reaches a certain critical length.

In refining this theory, some implications of crystal plasticity effects have been discussed including a specific dislocation model for hexagonal close-packed (h.c.p.) metals under wear conditions and a correlation with stacking fault energy for face-centred cubic and h.c.p. metals (Ref. 4.102).

It appears that, by the use of dislocation models, it is now possible to explain the occurrence of sheet-like wear particles, observed as early as 1929 by Füchsel (Ref. 1.19) and interpreted through a process of material separation.

4.4.3 Abrasive wear mechanisms

As in surface fatigue processes, the attrition of materials particles in the abrasive wear mechanism is also caused mainly by contact deformation processes.

The effect of abrasion occurs in contact situations in which direct physical contact between two surfaces is given where one of the surfaces is considerably harder than the other. The harder surface asperities press into the softer surface with plastic flow of the softer surface occurring around the asperities from the harder surface. When a tangential motion is imposed the harder surface will move, ploughing and removing the softer material. The effect of abrasion is comparable with a "micro-cutting process", as can be seen in the scanning electron micrograph, Figure 4.30, showing an Al surface.

In industrial situations, abrasive wear processes are very widespread; in fact, it has been estimated that about 50% of wear encountered in industry is due to abrasion mechanisms (Ref. 4.103). As a consequence of the great varieties of abrasive wear situations, different types of abrasive wear are distinguished in practice.

As illustrated schematically in Figure 4.31, firstly we distinguish between "two-body abrasion" and "three-body abrasion". For example, the two-body situation is most frequently encountered in the transport of loose

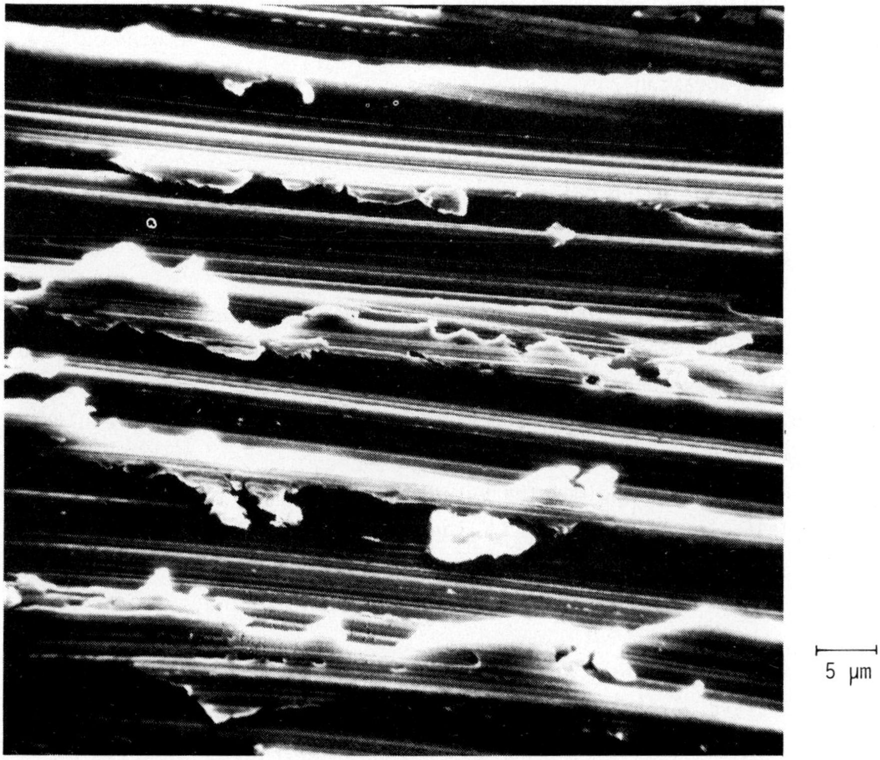

Figure 4.30 Appearance of surface damage due to abrasion.

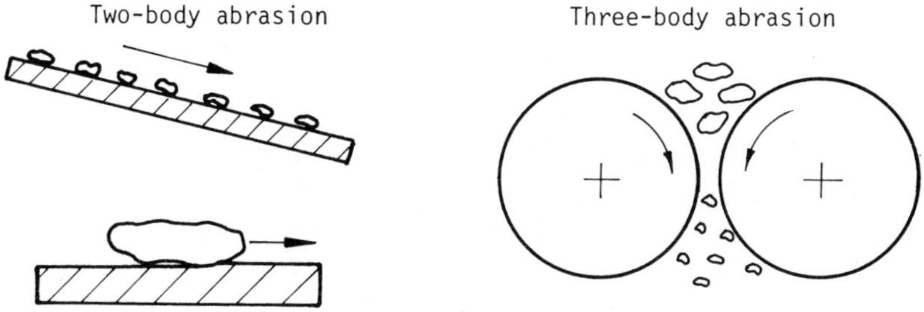

Figure 4.31 Two-body and three-body abrasion.

minerals and the three-body in mineral treatment and on the ingress of foreign particles into bearings. Secondly we distinguish between:

(a) Gouging abrasion
Examples: Shovel dipper teeth, Hammers in impact pulverizers, etc.
(b) Grinding abrasion
Examples: Grinding balls, etc.
(c) Erosion abrasion
Examples: Agitator impellers, Sandblast nozzles, etc.

Research into the mechanism of abrasion in the laboratory takes the form of examining the wear of solids sliding on abrasive paper. Extensive studies in this field have been performed by Wellinger and Uetz (Ref. 4.104, 4.105), Khruschov (Ref. 4.106, 4.107) and Richardson (Ref. 4.108). The results of these studies may be summarized as follows:

In a given wear regime, the abraded wear volume V of metals increases in most cases linearly with load F_N and sliding distance l

$$V \propto F_N \cdot l$$

If deviations occur, they are usually due to a reduction in particle size of the abrasive or clogging of the surface (Ref. 4.103). In order to compare the abrasion behaviour of different materials, the following measures have been defined:

(a) the wear volume V_w

(b) the wear resistance

$$\varepsilon = \frac{1}{\text{wear volume}} = \frac{1}{V_w}$$

(c) the relative wear resistance

$$\varepsilon_w = \frac{\varepsilon\,(\text{specimen})}{\varepsilon\,(\text{standard})}$$

It has been found that the abrasive wear depends on the correlation between the hardness of the abrasive, H_a, and the hardness of the metal, H_m, leading to three distinct wear regimes, as shown in Figure 4.32:

(I) a low-wear regime, if

$$H_a < H_m$$

(II) a transition regime, if

$$H_a \approx H_m$$

(III) a high-wear regime, if

$$H_a > H_m$$

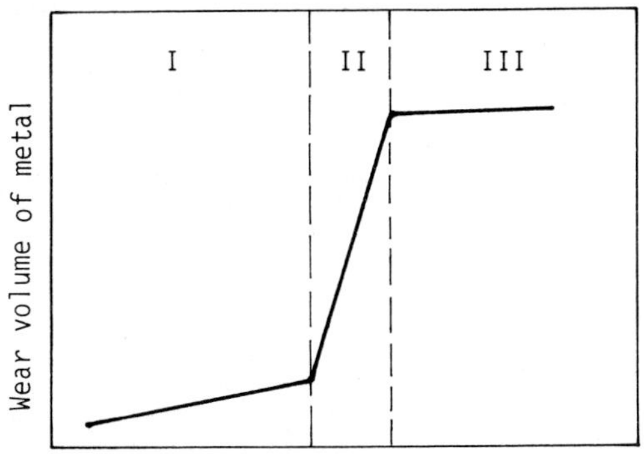

Figure 4.32 Influence of the hardness of the abrasive on the wear of metals (schematic).

This leads to the important conclusion that in order to reduce an abrasive wear component, the hardness of the material, H_m, should be higher than the hardness of the abrasive by a factor of about 1.3, i.e.,

$$H_m \approx 1.3\ H_a$$

may be used as a criterion for a low-wear abrasion rate. (It is not necessary to increase the hardness of the material beyond 1.3 times that of the abrasive because no further significant improvement is obtained, Ref. 4.103.)

In order to determine the relevant material properties that govern the abrasive wear behaviour in the high-wear regime (III) of Figure 4.32, a great variety of materials has been tested. In Figure 4.33 the wear resistance ε_W of various groups of materials as a function of the hardness of the materials, H_m, measured under conditions of regime (III) of Figure 4.32, is shown schematically (Ref. 4.107).

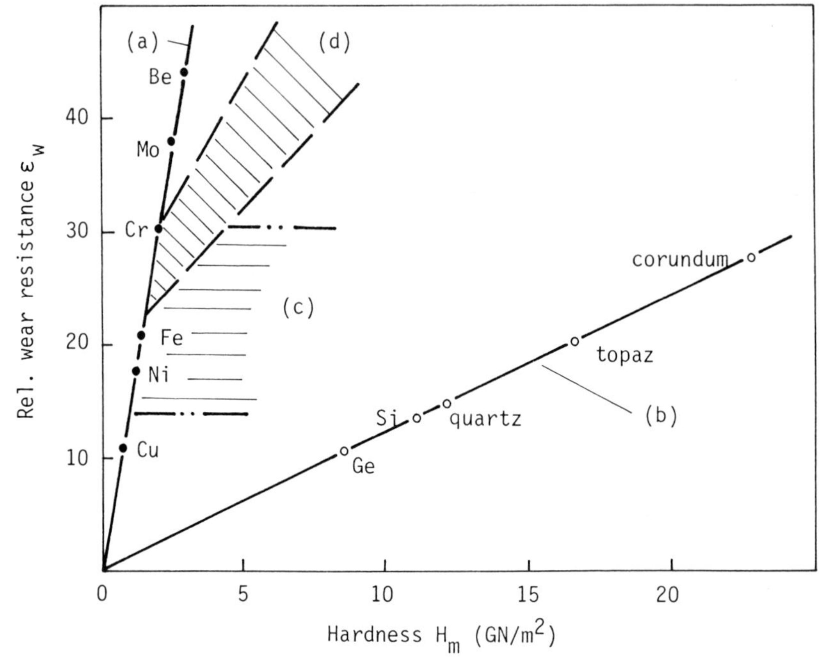

Figure 4.33 Khrushov diagram: wear resistance as a function of the hardness of materials.

The data were obtained using an electrocally produced corundum abrasive ($H = 22{,}900$ N/mm^2) and a Babbit-type lead-tin alloy (containing antimony) as the standard. The hardness of the test materials was determined before the test by diamond indentation. The basic conclusions that can be drawn from the Khrushov diagram, Figure 4.33, may be summarized as follows:

(a) Technically pure metals in an annealed state and annealed steels show a direct proportionality between the relative wear resistance, ε_W, and the pyramid hardness, H_m:

$$\varepsilon_W = C_{metal} \cdot H_m$$

where $C_{metal} = 13.8 \cdot 10^{-3}\ N^{-1}\ mm^2$

(b) For non-metallic hard materials and minerals, a linear relationship between wear resistance and hardness is similarly found:

$$\varepsilon_w = C_{minerals} \cdot H_m$$

where $C_{minerals} = 1.3 \cdot 10^{-3}\ N^{-1}\ mm^2$

(c) For metallic materials cold work-hardened by plastic deformation, the relative wear-resistance does not depend on the hardness resulting from cold work hardening.

(d) A heat-treatment of structural steels (normal hardening and tempering) improves the abrasive wear resistance.

These results indicate the great importance of the hardness of materials and of other metallurgical factors on the abrasive wear resistance of materials (Ref. 4.109).

The linear relationship between abrasive wear resistance and the material hardness can be verified using the simple model of ploughing shown in Figure 4.16. In this model an abrasive asperity is approximated by a cone that ploughs out and removes material from the counterface.

If the load, ΔF_N, on the indenter is only supported over the leading half of the contact, ΔF_N is balanced by the yield pressure, p_y, of the counterface acting via the contact area A:

$$\Delta F_N = A\ p_y$$

$$\Delta F_N = \frac{1}{8} \pi\ d^2 p_y$$

The volume of material ΔV removed in sliding a distance Δl is given by the cross-sectional area of the indenter times Δl

$$\Delta V = \frac{d^2}{4} \Delta l \cdot \cot \varphi$$

Thus, substituting for d from above

$$\frac{\Delta V}{\Delta l} = \frac{2 \cot \varphi}{\pi\ p_y} \Delta F_N$$

and assuming that only a proportion, K, of all the contacts produce worn particles, it follows that

$$\frac{V}{l} = K \frac{2 \cot \varphi}{p_y} F_N$$

This expression shows that the wear volume is proportional to the load and the sliding distance, and that the only material property involved is the yield pressure. If the yield pressure is assumed to be equal to the indentation hardness, H, for given values of F_N and l, the last equation can be simplified to

$$V = \text{const} \frac{1}{H}$$

Introducing the wear resistance defined above as the reciprocal wear volume it follows that

$$\text{Abrasive wear resistance} = \text{const. x hardness}$$

$$\varepsilon_w = C \cdot H$$

This simple derivation and the experimental results shown in Figure 4.33 indicate that the abrasive wear resistance is only dependent on the material hardness. Therefore, the abrasive wear resistance may be interpreted via the material hardness, as a kind of "tribological material property", i.e. as an apparently system-independent property. It should be borne in mind, however, that the abrasive wear resistance, ε_w, is defined as a relative measure and that the linear ε - H relationship is valid only for conditions of regime (III) of Figure 4.32, i.e., for the cases in which the abrasion process acts mainly in the bulk of the material under study. If, on the other hand, an abrasive wear process acts in the "outer" surface layers, then the influence of surface contaminants and the action of the environmental atmosphere must additionally be taken into account (Ref. 4.110).

4.4.4 Adhesive wear mechanisms

Whereas the wear mechanisms of surface fatigue and abrasion can be explained mainly in terms of forces, stresses and deformation processes, in the adhesive wear mechanism, material interactions play an important role. It was described in Section 4.2.2 that if two solid bodies are brought into contact they will first experience long-range van der Waals forces. At distances of about 1nm strong short-range surfaces forces come into action at the real areas of contact, the type of which depends on the nature of the surfaces in contact. Thus, strong adhesive junctions may be formed with the assistance of the dispersal of surface contaminants and junction growth (see Section 4.3.1). In order that adhesive wear should take place, fracture must occur in the subsurface of one or both materials. The situation is illustrated schematically in Figure 4.34.

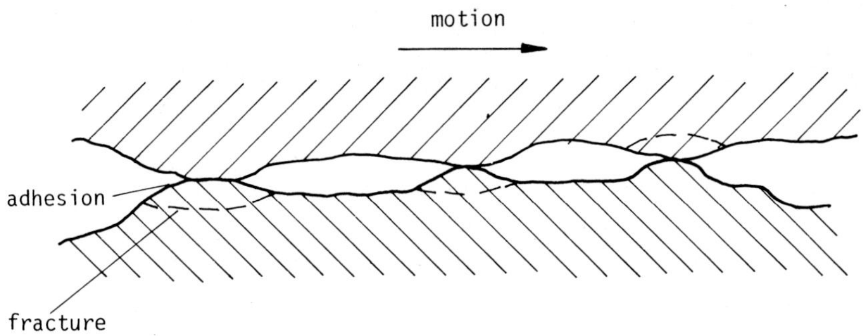

Figure 4.34 Adhesive wear situation.

In contrast to the other wear mechanisms which generally take some time to develop or to reach a critical destructive magnitude, adhesive wear effects occur quite precipitously leading in their severe forms to dangerous failure in the form of "scuffing" or "seizure" of moving parts in mechanical engineering systems. In Figure 4.35 the typical appearance of the surfaces of a sliding steel/steel pair after failure due to adhesive wear processes is shown. It can be seen that large material lumps were transferred from the counterpartner and adhere strongly on one surface in "cold-welded" junctions.

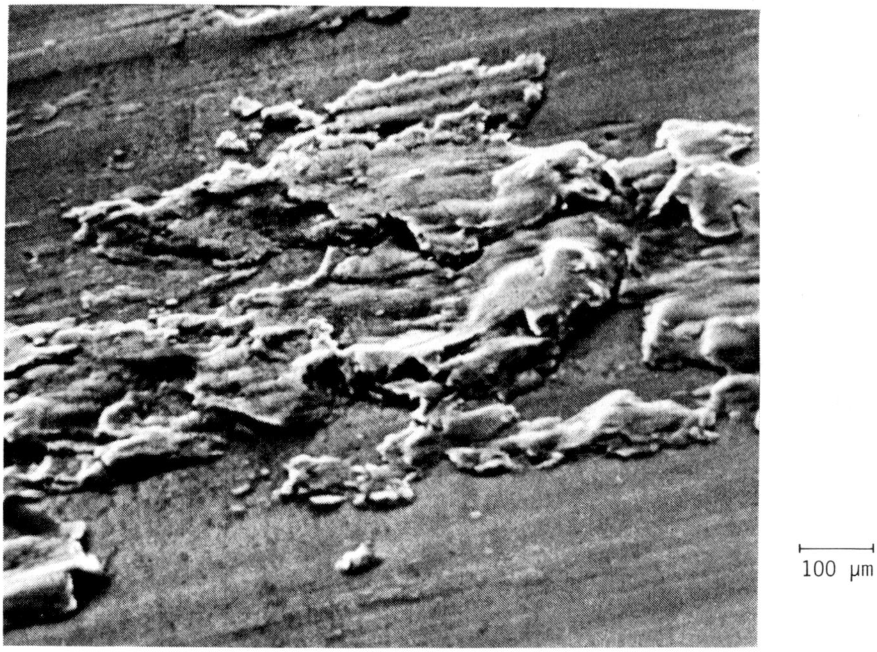

Figure 4.35 Appearance of surface damage due to adhesive wear processes.

Concerning the physical mechanisms of adhesive wear, obviously the processes and parameters of adhesion (Section 4.2.2) as well as those of fracture (Section 4.4.2) must be taken into account. Since both adhesion and fracture are influenced by surface contaminants and the effect of the environment, it is quite difficult to relate adhesion wear rates with elementary bulk properties of materials. In vacuum where these influences are eliminated, the following experimental results have been observed to be generally valid for the adhesive wear processes of metal/metal pairs (Ref. 4.111):

(i) Interfacial metallic adhesion bonding occurs between any pair of metals; there is no direct relation between the (volume) solubility of metal pairs and their (surface) adhesion bonding.

(ii) Crystal structure exerts an influence on adhesive wear. Hexagonal metals in general exhibit lower adhesive wear characteristics than either body-centred cubic or face-centred cubic metals. This difference is assumed to be related to different plastic asperity contact deformation modes and the number of operable slip systems in the crystal systems.

(iii) Crystal orientation influences the adhesive wear behaviour. In general high atomic density, low surface energy grain orientations exhibit lower adhesion and less adhesive wear than do other orientations.

(iv) When dissimilar metals are in contact the adhesive wear process will generally result in the transfer of particles of the cohesively weaker of the two materials to the cohesively stronger.

(v) Small amounts of alloying elements (a few parts per million) such as carbon and sulfur are sufficient to inhibit appreciably the adhesion of metallic alloys thereby minimizing their adhesive wear. (Sulfur and carbon will, activated by frictional heating, diffuse to a surface and occupy a concentration at the surface far in excess of their concentration in the bulk material.)

These results clearly show the influences of the various characteristics of contact adhesion and contact deformation processes, as dicussed above, on adhesive wear. The combined effects of adhesion and fracture also act in the formation of loose adhesive wear particles.

In studying the details of the adhesive wear mechanism, Cocks (Ref. 4.112) and Antler (Ref. 4.113) considered the process of "prow" formation by which contact is maintained for a longer period while the junction is growing and the junction base is moving along one or both of the running surfaces. In refining this model, Landheer and Zaat (Ref. 4.114) emphasized the following phenomena in the severe adhesive wear process:

(a) a junction development, where under the influence of shear resistance in the interface, a field of iso-strain lines (plastic strain) moves through the metal of the wearing surface in a direction opposite to the sliding direction, so that metal accumulates in the junctions and the distance between the running surfaces is enlarged,

(b) cracking of the metal at the backside of the junction, by which material detaches for transfer.

An instructive experimental example of the generation of an adhesive wear particle has been given by Buckley (Ref. 4.115). He performed sliding friction experiments with a polycrystalline copper rider sliding across the surface of a bicrystal of copper. One grain of the bicrystal was of the (111) orientation while the second was a (210) plane. With a single pass of the rider across the surface, cracks developed in the surface of both grains of the bicrystal. These surface cracks are shown in the scanning electron micrograph of Figure 4.36. The fracture cracks developed at room tempera-

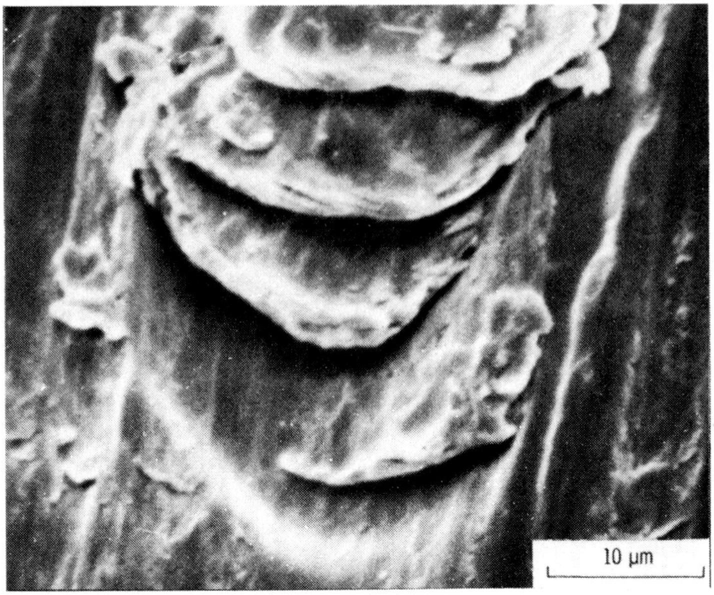

Figure 4.36 Areas of wear on (210) copper grain (Ref. 4.115).

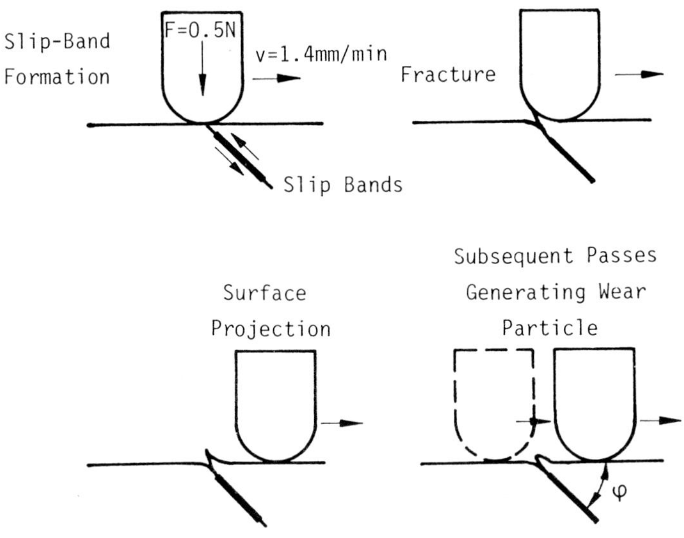

Figure 4.37 Origin of surface fracture and formation of wear particle (Ref. 4.115).

ture within the copper. Close examination of these cracks indicates that the wall of the crack is very smooth. Further, the material at the leading edge of the crack has been curled up above the plane of the bicrystal surface. The measured angle between the crack and the orientation of the surface plane indicates that the fracture crack occurs along slip bands.

The mechanism responsible for the cracks shown in Figure 4.36 is presented in Figure 4.37. Adhesion of the rider to the bicrystal occurs. With tangential motion fracture occurs along slip bands as the tangential force parts atomic planes. At some applied tangential force, the applied force exceeds the adhesive bonding force and the rider breaks away from the bicrystal surface leaving a curl of metal projecting above the surface. A second pass shears off the curl and an adhesive wear particle has been generated.

4.4.5 Tribo-chemical wear mechanisms

Whereas the mechanism of surface fatigue wear, abrasive wear, and adhesive wear can be understood in terms of deformation and adhesion interactions between two contacting surfaces, (1) and (2), in tribo-chemical wear as third partner the environment (3) and the dynamic interactions between (1), (2) and (3) determine the wear process (Ref. 4.116). If the two surfaces react actively with the environment, the rubbing of surfaces together in such an environment results in the continuous formation and removal of reaction products. Since the material of the contacting surfaces are contained in the reaction products, material is being removed from the surfaces. The typical appearance of tribo-chemically formed wear particles is shown in Figure 4.38 in scanning electron micrographs.

The differences in the mechanisms of surface fatigue, abrasion and adhesion on one hand and tribo-chemical wear effects on the other can be seen in the conceptual diagram of the wear processes within a tribological system, Figure 3.8 and Figure 3.4. The first three mechanisms can essentially be described - in the simplest case - as interactions between only two partners, namely the two solid surfaces (1) and (2), leading directly to the generation of debris $\{Z^M\}$ from (1) and (2). In the case of tribo-chemical wear, the whole mechanism clearly consists of the interactions of three partners, namely the solid surfaces (1) and (2) and the environment (3). These interactions may be expressed as a cyclic stepwise process:

(i) At the first stage, the surfaces (1) and (2) react with the environment (3). In this process, reaction products (1)⟷(3) and (2)⟷(3) are formed on the surfaces of (1) and (2), shown in Figure 3.8 below.

(ii) The second step consists of the attrition of the reaction products as a result of crack formation and abrasion in the contact process interactions of (1) and (2). When this occurs "fresh", i.e., reactive surface parts, of (1) and (2) are formed and stage (i) continues.

Clearly, in attempts to study the mechanisms of tribo-chemical wear, besides the effects of deformation and adhesion, the chemistry of the reaction product formation must be considered additionally, taking into account the contribution of frictional energy in these processes.

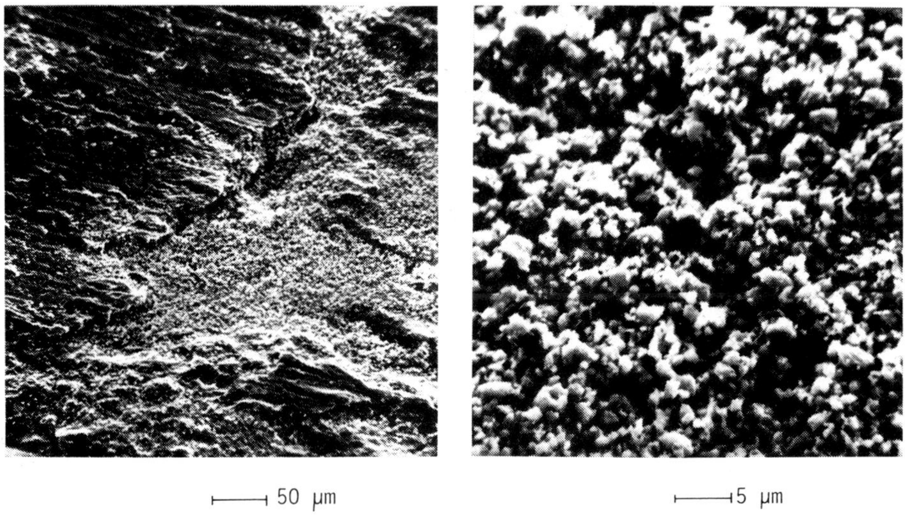

Figure 4.38 Appearance of tribo-chemical wear particles.

As a consequence of thermal and mechanical activation, the asperities undergo the following changes:

(a) the reactivity is increased due to the increased asperity temperature; therefore the formation of surface layers is accelerated,

(b) the mechanical properties of the surface asperity layers are changed: in general they have a tendency to brittle fracture.

Starting with the assumption that tribo-chemically formed surface asperity layers are detached at a certain critical thickness, Quinn proposed an oxidation hypothesis of the wear of steels (Ref. 4.117). Since this theory indicates the main influencing factors of the tribo-chemical wear process, it will be reproduced here. It was assumed that, at a given asperity contact, on average, 1/K encounters are necessary (K: statistical wear coefficient, see Section 4:4.1) for a critical oxide film thickness ξ to build up before it becomes detached to form a wear particle. If t_c is the duration of a single wearing contact, then the total time t to produce a wear particle of thickness ξ is given by

$$t = t_c/K$$

But $t_c = d/v$, where d is the distance along which a wearing contact is made and v is the speed of sliding.
Hence

$$t = d/vK$$

Now the mass per unit area of oxide growth Δm depends on the time of oxidation t through the parabolic relation

$$\Delta m^2 = k_p \cdot t$$

where k_p is the parabolic oxidation rate constant. It is reasonable to assume $\Delta m = \xi\rho$, where ρ is the density of the oxide, so that we get

$$\xi^2 = k_p \cdot t/\rho^2$$

Eliminating t we obtain an expression for K, namely

$$K = \frac{d \cdot k_p}{v\rho^2 \xi^2}$$

It is now generally accepted that k_p is exponentially dependent on the temperature of oxidation T_o namely

$$k_p = A_c \exp(-Q/R_c T_o)$$

where A_c is the Arrhenius constant, Q is the activation energy, and R_c is the gas constant. Combining equations and writing T_c (the contact temperature) for T_o, we get

$$K = \frac{d \cdot A_c \cdot \exp(-Q/R_c T_c)}{\xi^2 \cdot \rho^2 \cdot v}$$

From the results of Section 4.4.1 it follows that under the assumption of a plastic asperity deformation mode (plasticity index $\psi > 1$, yield pressure $p_y \approx$ hardness H) the wear coefficient K can be related with the wear volume V via the expression

$$V = \frac{K}{3H} F_N \cdot l$$

Combining this relation with the expression for K, calculated on the basis of Quinn's theory, it follows

$$V = \frac{d \cdot A_c \cdot \exp(-Q/R_c T_c)}{3 \cdot \xi^2 \cdot \rho^2 \cdot v \cdot H} F_N \cdot l$$

This expression indicates the dependence of the wear volume V on the following groups of parameters of the tribological system:

(a) operating variables (F_N, v, l, T_c)
(b) general constants and material parameters (A_c, Q, R_c, ρ, H)
(c) interaction characteristics (d, T_c, ξ)

Although it is quite difficult to perform reliable calculations of the wear rates due to the uncertainties of some of these quantities, the above equations may serve as a basis for a physical picture of the tribo-chemical wear mechanism.

4.4.6 The complexity of wear processes

In general, the different wear mechanisms as discussed separately in the foregoing sections, are superimposed. The processes of tribo-chemical reactions and adhesion lead to changes in the properties of the surfaces. Thereby reaction products are formed or material is transferred to the other partner. The processes are activated by the mechanical and thermal energy transferred and dissipated in the surface asperities and in the bulk material. All these processes promote the formation of loose wear particles which, in any case, are generated through the mechanisms of surface fatigue and abrasion.

A well known wear mode which results through the superposition of all four basic wear mechanisms described above is fretting wear (Ref. 4.118, 4.119). Fretting wear occurs if two contacting solid bodies move relatively to each other in an oscillatory motion with an amplitude of movement of less than 100 µm. In industry, fretting occurs, for example, in the following mechanical systems and components: lap joints, splines, push-fitted shaft-wheels, metal static seals, engine mounts, clutches, etc. The surface damage due to fretting is characterized by the initiation of fatigue cracks and the formation of fretting debris. As a consequence of fretting, fatigue-strength-reduction factors of 3 to 6 are quite normal (Ref. 4.120). The chain of events that leads to fretting is compiled schematically in Table 4.4 (Ref. 4.121).

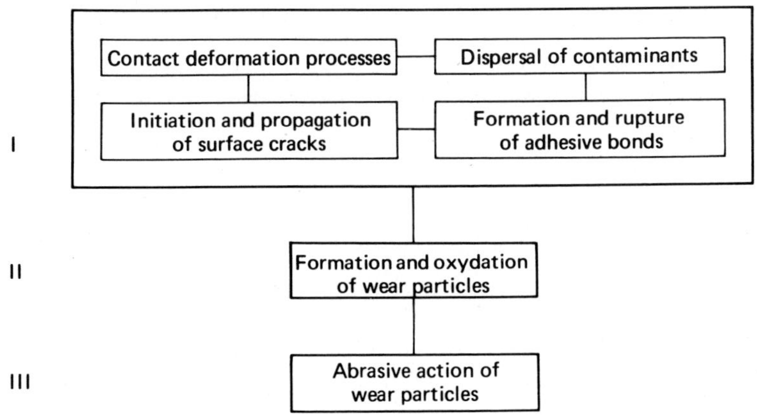

Table 4.4 Schematic representation of the different stages of fretting.

The first stage (I) is initiated by the oscillatory contact deformation leading to the generation and propagation of surface cracks as well as to the dispersal of surface contaminants and the formation and rupture of adhesive bonds. The second stage (II) is the formation of wear particles either by the surface fatigue mechanism or the adhesion/deformation process. This is followed by the oxidation of wear particles in tribo-chemical reactions. The third stage (III) consists of the abrasive action of formed wear particles. This in turn assists the processes of stage (I), and so on.

The fretting mode of wear which occurs under oscillatory motion is a very pronounced example of the superposition of the different wear mechanisms described separately above. From various wear studies it is obvious that under sliding and rolling conditions in most cases a superposition of wear mechanisms also occurs. Consider, for example a simple tribo-mechanical system consisting of two solid bodies (1), (2) sliding in a reactive gaseous atmosphere (3) as illustrated schematically in Figure 4.39 above. As indicated by the schematic diagram in the centre of Figure 4.39, the following processes occur in general simultaneously at the interface:

(i) formation of surface layers by adsorption or tribo-chemical reactions of the sliding surfaces (1) and (2) with the gaseous environment (3),

(ii) adhesive transfer of material and reaction products from (1) $\longrightarrow$ (2) and (2) $\longrightarrow$ (1),

(iii) generation of wear particles from the bulk material, the reaction products or the adhesive transferred material of both (1) and (2) through the effects of surface fatigue and abrasion.

The changes of the material components of a tribo-mechanical system due to the combined action of the wear mechanisms have been studied systematically by Mølgaard (Ref. 4.122) and are shown schematically in Figure 4.39. The possible material components of a wearing system, consisting initially of the three partners (1), (2), (3), are:

the given materials	M^1		M^2
the gaseous atmosphere		G	
deformed material parts	D^1		D^2
transferred deformed material	D^1_2		D^2_1
back-transferred material	D^1_1		D^2_2
reaction products	R^1		R^2
transferred reaction products	R^1_2		R^2_1
debris of given material		Z^M	
debris of reaction products		Z^R	
adsorbed gas	G^1_a		G^2_a

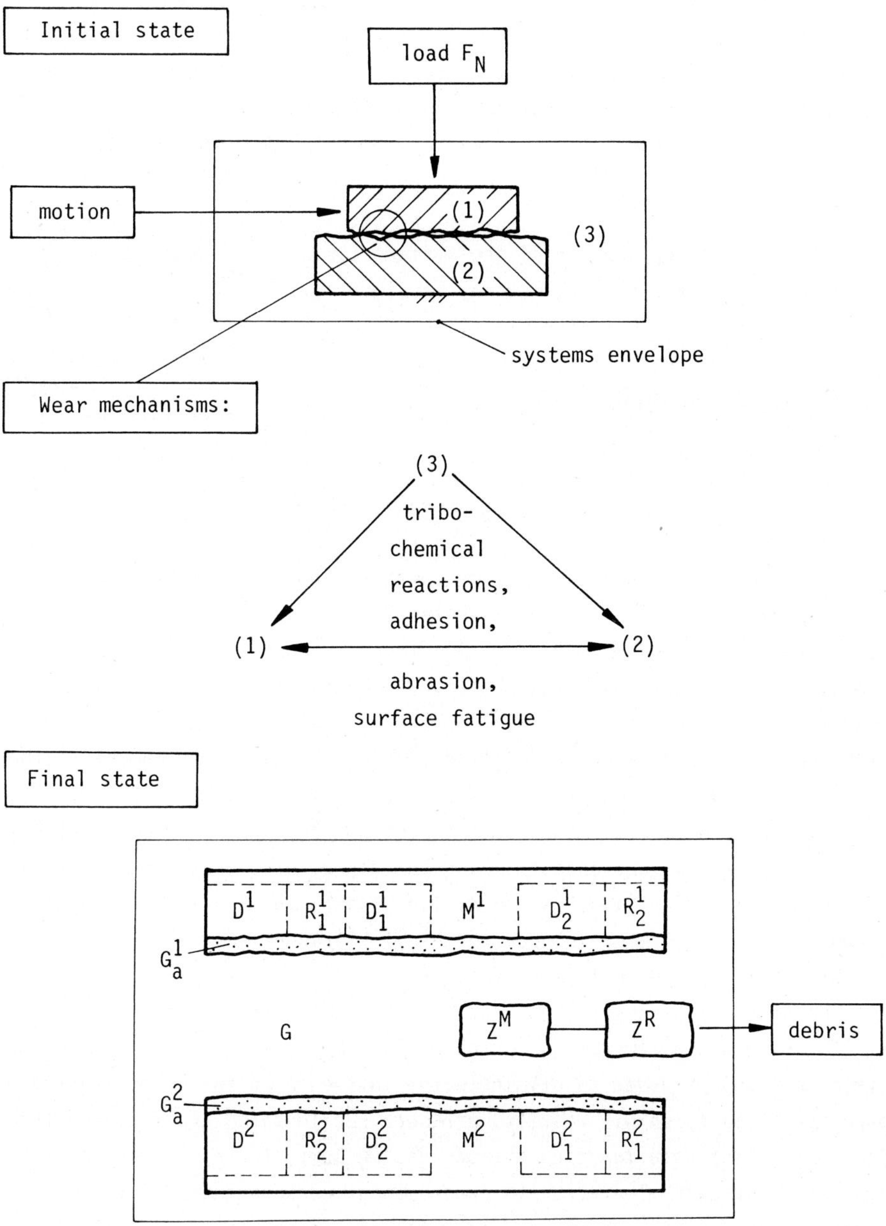

Figure 4.39 Superposition of wear mechanisms in a tribo-mechanical system and the changes of the system's elements due to wear processes.

In summary, we see from Figure 4.39 that the study of wear in a tribomechanical system is one involving a complex interdependence of many processes and physical parameters. It is most difficult to isolate any one process or parameter for investigation, and hence we are forced into an empirical approach to the subject. In order to give this empirical approach a definite direction likely to prove successful, one must consider, systematically, all the processes and parameters relevant to a given wear situation. For the practical application of the systems approach to this subject the data-sheet described in Chapter 8 may be used.

4.5 Lubrication Modes

4.5.1 Stribeck curve and lubrication modes

The tribological processes, i.e., the contact, friction and wear processes discussed hitherto are, in general, related to direct physical interactions between relatively moving surfaces. All these processes can be influenced or modified by the process of lubrication. The purpose of lubrication is to separate the surfaces moving relative to each other with a film of a material which can be sheared with low resistance without causing any damage to the surfaces. Depending on the thickness of the lubricant film (which may range from some tenth of a millimeter down to the nm range), the interfacial height distribution of the lubricant film and the degree of geometric conformity, different lubrication modes can be distinguished. The lubrication modes can be conveniently discussed in considering the Stribeck curve.

Stribeck in 1900 to 1902 performed comprehensive experiments on the friction of sliding and rolling bearings, measuring the friction coefficient as a function of the operating variables load F_N, velocity v and temperature T (Ref. 4.123). In order to eliminate the influence of the temperature dependence of viscosity on his results, Stribeck recalculated the measured friction coefficient as a function of load and velocity for a constant bulk-oil temperature of 25°C. The accurate experimental measurements of Stribeck served as a basis for the theoretical work of Sommerfeld, Gümbel and successors in establishing the theory of hydrodynamic lubricated bearings, see Section 1.2.3.

It is now generally accepted that the Stribeck curve represents the general characteristic of lubricated moving surfaces as a function of the lubricant viscosity η , the velocity v and the normal load F_N (or pressure p). In Figure 4.40 the shape of the Stribeck curve is plotted in a simplified manner as a function of the parameter $\eta \cdot v \cdot F_N^{-1}$. (This parameter is related to the reciprocal of the Sommerfeld number So, see Section 4.5.2.)

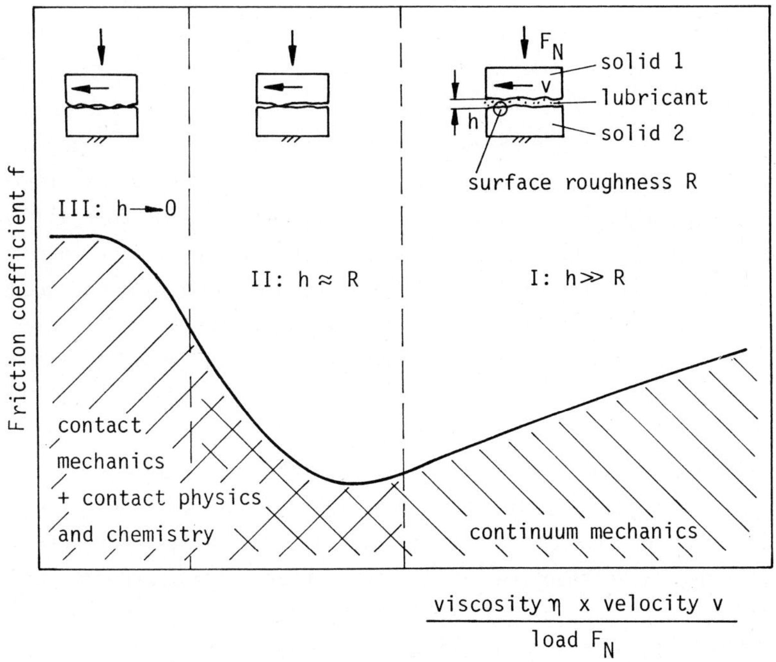

Figure 4.40 Stribeck curve and lubrication regimes (schematic).

As illustrated in Figure 4.40, depending on the geometry, the materials, the operating conditions and the separation h of the surfaces, three main lubrication regimes may be distinguished:

I : Hydrodynamic lubrication (and Elastohydrodynamic, EHD, lubrication),
II : Partial EHD lubrication or Mixed lubrication,
III : Boundary lubrication.

In regime I, the rigid surfaces are separated by a continuous lubricant film, whose thickness is much larger than the combined surface roughness measure R of the surfaces. The friction resistance is due to the internal friction of the lubricant. In this regime, the tribological behaviour of the system is determined by the rheology of the lubricant and can be calculated or estimated by the methods of fluid mechanics. If the lubricated system under consideration consists of nonconformal concentrated contacts, the elastic (Hertzian) deformation of the surfaces and the pressure dependence of the lubricants viscosity must also taken into account. This leads to the regime of elastohydrodynamic lubrication. Since in regime I no direct physical contact interactions between the surfaces occur, wear processes cannot take place (except surface fatigue wear, cavitation wear or fluid erosion).

If under conditions of hydrodynamic or EHD lubrication, the lubricant viscosity or the velocity decreases or the load increases, the lubricant film gets "thinner" and the separation of the surfaces decreases. If then the first asperity contact interactions occur, region II of partial EHD lubrication or mixed lubrication is reached. In this regime the load is carried partly by the fluid film and partly by the contacting surface asperities. Consequently, in this regime the friction resistance is due partly to the shearing of the lubricant film and partly to the asperity interactions. In regime II, in principle, all wear mechanisms may take place as discussed in Section 4.4. The processes explained above are, of course, influenced and modified by the action of the lubricating film.

If in a lubricated system which runs in regime II of Figure 4.40 the operating conditions change further to the left of the Stribeck curve, the amount of asperity interactions within the contact area increases, and the film thickness decreases down to some monolayers or below. In regime III of boundary lubrication, bulk rheological properties of the lubricant are of less importance and the load is carried almost entirely through the deformation of the asperities. Clearly, in regime III the physico-chemical interactions at the solid/lubricant/solid interface determine the friction and wear behaviour of the system.

After this simplified survey of the different lubrication modes with the use of the generalized Stribeck diagram, the different lubrication modes will be discussed in some detail in the following.

4.5.2 Hydrodynamic lubrication

In the process of hydrodynamic lubrication of a tribological system, a continuous fluid film is brought between the surfaces in relative motion in order to minimize friction and to eliminate wear. The tribological behaviour of a hydrodynamically lubricated system is then mainly determined by the two following features:

(i) The resistance to motion is given by the "internal friction" of the fluid, i.e., the shear resistance or "viscosity" of the fluid film.

(ii) The effects of wear are eliminated if the geometry of the surfaces is such that a load-carrying pressure is set up in the lubricant film during the motion of the surfaces leading to a complete separation of the surfaces.

In (i), the resistance to motion of a hydrodynamically lubricated system is determined by the shear forces which exist in a moving fluid. To get a measure of the shear resistance or viscosity of the fluid, consider a system of two solid plane surfaces, as sketched in Figure 4.41.

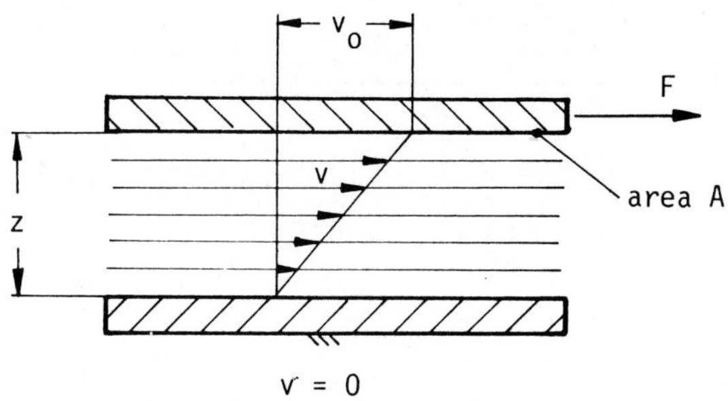

Figure 4.41 Viscous drag between two parallel plates.

In Figure 4.41 the lower surface is kept stationary while the upper moves parallel to it at a velocity v_o. Since the molecules of the fluid adhere to the surfaces, at velocities which are not too high, a laminar flow profile results within the fluid with $v = v_o$ at the upper surface and $v = o$ at the lower surface. The force, F, required to keep the upper surface moving is then proportional to the area of the surfaces, A, and to v_o/z, where z is the distance between the plates. Thus the shear stress F/A is proportional to the change of shear strain v_o/z:

$$\frac{F}{A} = \eta \frac{v_o}{z}$$

or more generally:

$$\tau = \eta \frac{\partial v}{\partial z}$$

The constant of proportionality η is called coefficient of viscosity, and fluids which obey proportionality between the viscous shear stress and the rate of change of the shear strain $\frac{\partial v}{\partial z}$ are termed Newtonian fluids.

As regards point (ii), for the generation of a load-carrying pressure the formation of a converging, wedge-shaped film is necessary, as shown by Reynolds in 1886 (Ref. 4.124). A physical picture of the processes leading to a generation of a pressure in a converging oil film can be gained by considering a simple tilting pad bearing, shown schematically in Figure 4.42.

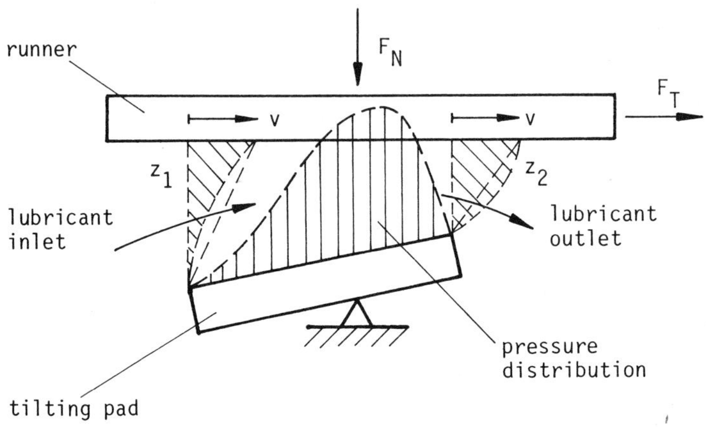

Figure 4.42 A tilting pad bearing.

The tilting pad is stationary with an inlet gap z_1 larger than the outlet gap z_2. The whole gap is flooded with a viscous incompressible lubricant. Assume that initially the inlet velocity distribution may be linear (as in Figure 4.41). Since $z_1 > z_2$ there would be an excess of inlet flow over the outlet flow in contrast to the condition of continuity of flow. This automatically leads to the generation of a pressure distribution against which the inlet flow has to work. This in turn modifies the velocity profiles: at the inlet the velocity profile is concave and at the outlet it is convex as indicated schematically in Figure 4.42. The area under each velocity-profile curve along the pad is the same, thus the condition of continuity of flow is fulfilled. If the surfaces are fully separated, relative motion occurs entirely within the lubricant film.

For the tribological behaviour of a hydrodynamically lubricated mechanical system, i.e., the transmission and dissipation of mechanical energy, the following aspects are important:

(a) the influences of the operating variables like load, velocity, temperature on the pressure and film thickness generated in the lubricant film,

(b) the relations between (a) and the system geometry and the materials and lubricant parameters,

(c) the amount of mechanical energy dissipated and the value of friction coefficient, depending on (a) and (b),

(d) the limits of hydrodynamic lubrication in relation to (a), (b) and (c).

The main features of these points will be discussed in the following, starting from a consideration of the general equation of motion for a viscous fluid, the Navier-Stokes equation. The main physical ideas involved in the derivation of this equation are the following (Ref. 4.125):

Consider an element of volume ΔV within a fluid of viscosity η and density ρ which moves under the action of a specific pressure p and an external conservative force field of the potential ϕ with a mean velocity v, as in Figure 4.43.

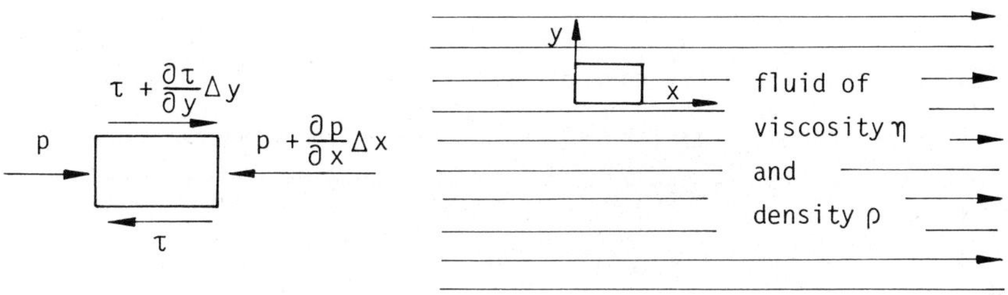

Figure 4.43 A volume element within a viscous fluid.

According to Newton's law of motion, the total force per unit volume is equal to the time rate of change of linear momentum

$$\text{force} = \frac{d}{dt}\ (\text{mass x velocity})$$

$$F = \rho \ \text{x} \ (\text{acceleration})$$

The force per unit volume F is given as sum of three terms:

$$F = F_1 + F_2 + F_3$$

(i) F_1: the pressure force per unit volume

The pressure on the face at x gives the force $p\Delta y \Delta z$, and the pressure on the face at $x + \Delta x$ gives the force $-\left[p + (\frac{\partial p}{\partial x})\Delta x\right]\Delta y \Delta z$, so that the resultant force is $-(\frac{\partial p}{\partial x})\Delta x \Delta y \Delta z$. If we take the remaining pairs of the faces of the cube, we see that the pressure force per unit volume is

$$F_1 = -(\frac{\partial p}{\partial x} + \frac{\partial p}{\partial y} + \frac{\partial p}{\partial z})$$

or with the use of the Nabla operator

$$\vec{\nabla} = (\frac{\partial}{\partial x}, \frac{\partial}{\partial y}, \frac{\partial}{\partial z})$$

$$F_1 = -\vec{\nabla} p$$

(ii) F_2: the gravity forces per unit volume

It is well known that forces due to gravity are conservative forces. With a potential per unit mass, Φ, the force density results as

$$F_2 = -\rho \vec{\nabla} \Phi$$

(iii) F_3: the viscous forces per unit volume

At the bottom face of the volume element, the stress $\eta \frac{\partial v}{\partial y}$ acts according to the definition of the viscosity given above. At $y + \Delta y$ the stress is

$$\eta \frac{\partial v}{\partial y} + \frac{\partial}{\partial y}(\eta \frac{\partial v}{\partial y})$$

so that the resultant stress is

$$\frac{\partial}{\partial y}(\eta \frac{\partial v}{\partial y})$$

Summing, it follows that

$$F_3 = \eta \nabla^2 v$$

with $$\nabla^2 = \frac{\partial^2}{\partial x^2} + \frac{\partial^2}{\partial y^2} + \frac{\partial^2}{\partial z^2} \qquad \text{(Laplace operator)}$$

This expression is valid for incompressible fluids. In the general case of compressible fluids there is another term in stress which depends on other derivatives of the velocity. The general expression for the viscous forces is given by

$$F_3 = \eta \nabla^2 v + (\eta + \eta')\vec{\nabla}(\vec{\nabla} v)$$

where η' is the so-called second coefficient of viscosity (Ref. 1.7). This quantity is usually said to be related to the normal or shear coefficient of viscosity η by the expression

$$3\eta' + 2\eta = 0$$

After the determination of the forces acting on a volume element of a viscous fluid, we need an expression for the acceleration of a fluid particle. In trying to get an expression for the acceleration of a fluid particle, it must be taken into account that if v(x, y, z, t) is the velocity of a fluid particle which is at the location (x, y, z) at time t, then the velocity of the same particle at time t + Δ t is given by

$$v\,(x + v_x \Delta t,\ y + v_y \Delta t,\ z + v_z \Delta t,\ t + \Delta t).$$

The acceleration $\frac{\Delta v}{\Delta t}$ is then

$$v_x \frac{\partial v}{\partial x} + v_y \frac{\partial v}{\partial y} + v_z \frac{\partial v}{\partial z} + \frac{\partial v}{\partial t}$$

or symbolically

$$(v\,\vec{\nabla})\,v + \frac{\partial v}{\partial t}$$

Now, inserting in Newton's equation

$$F = \rho \times (\text{acceleration})$$

the expressions obtained for the force $F = F_1 + F_2 + F_3$ and the acceleration $\Delta v/\Delta t$ and combining everything, the general Navier-Stokes equation for the motion of a viscous fluid is

$$\rho\left\{\frac{\partial v}{\partial t} + (v\,\vec{\nabla})\,v\right\} = -\vec{\nabla}p - \rho\vec{\nabla}\Phi + \eta\nabla^2 v + (\eta + \eta')\,\vec{\nabla}\,(\vec{\nabla}\,v)$$

This equation is the basis for all of hydrodynamics (similar to the Maxwell equations which are the basis for all of electrodynamics). The equation and its diverse application in tribology have been studied at length by different authors. For a discussion of this aspect, see (Ref. 4.126).

Starting from the Navier-Stokes equation of fluid mechanics, Reynolds in his fundamental paper of 1886 laid the basis of hydrodynamic lubrication theory and practice. The derivation of the Reynolds equation obtained either from the Navier-Stokes equation or otherwise, can be found in any standard textbook of hydrodynamics, see for instance the books by Cameron (Ref. 1.7) or Hersey (Ref. 4.126). The Reynolds equation is a differential equation for the pressure distribution, taking into account the velocities of the solid surfaces and the profile or contour of the lubricant film. Since, from a consideration of Reynolds equation, a physical picture of the hydrodynamic lubrication process as well as basic tribological relations for the load-carrying capacity and friction resistance can be obtained, a brief discussion of Reynolds equation and its consequences is given in the following.

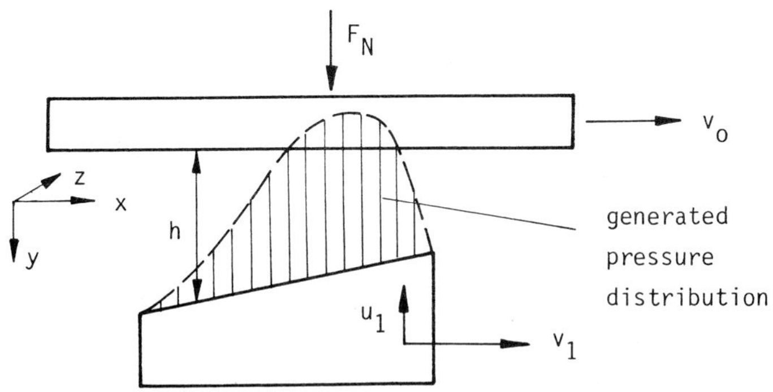

Figure 4.44 Two-dimensional fluid-film geometry.

Consider again a simple tilting pad bearing, shown schematically in Figure 4.44, with the following assumptions:

(a) The fluid is Newtonian
(b) The fluid flow is laminar
(c) The fluid adheres to the bearing surfaces
(d) The fluid inertia can be neglected
(e) The fluid weight can be neglected
(f) The fluid pressure is constant across the thickness of the fluid film
(g) The fluid is incompressible.

For this case, the Reynolds differential equation can be written as:

$$\frac{1}{6}\left\{\frac{\partial}{\partial x}\left(\frac{h^3}{\eta}\cdot\frac{\partial p}{\partial x}\right)+\frac{\partial}{\partial z}\left(\frac{h^3}{\eta}\cdot\frac{\partial p}{\partial z}\right)\right\}=$$

$$= (v_o - v_1)\frac{\partial h}{\partial x} \quad \longleftarrow \quad \text{"wedge" term}$$

$$+ \quad h\frac{\partial(v_o + v_1)}{\partial x} \quad \longleftarrow \quad \text{"stretch" term}$$

$$+ \quad 2\frac{\partial h}{\partial t} \quad \longleftarrow \quad \text{"squeeze" term}$$

The right-hand side of this equation contains three terms of different physical significance.

(i) "Wedge" term

This contribution is caused by the shape of the fluid film, $\frac{\partial h}{\partial x}$, and the relative velocity $(v_o - v_1)$ of the surfaces.

(ii) "Stretch" term

This term requires that the sum of the velocities in the x-direction change as a function of x-location. This term never exists in rigid plane-slider bearings but is applicable to special cases of dynamically loaded journal bearings.

(iii) "Squeeze" term

This contribution is caused by relative motion normal to the bearing fluid.

Solution of the Reynolds equations leads to the pressure distribution which, when integrated in the x- and z-directions, yields the load-carrying capacity:

$$F_N = \int_0^{z_1}\int_0^{x_1} p\cdot(x, z)\, dxdz$$

The flow of fluid per unit width in the direction of velocity $v = (v_o - v_1)$ is

$$q_x = \frac{vh}{2} - \frac{h^3}{12\eta}\cdot\frac{\partial p}{\partial x}$$

and the z-direction is

$$q_z = -\frac{h^3}{12}\cdot\frac{\partial p}{\partial z}$$

The friction force to be overcome in a fluid-film-lubricated system is that caused by velocity-induced and pressure-induced fluid shear stresses, namely

$$F_F = \pm\int_0^{z_1}\int_0^{x_1} \frac{\eta v}{h}\, dxdz - \frac{1}{2}\int_0^{z_1}\int_0^{x_1} h\frac{\partial p}{\partial x}\, dxdz$$

where the plus sign is used for the force acting on the slower moving surface. Clearly, in integrations of the Reynolds equation the film thickness $h = h(x)$ must be known. A further complication is the variation of the viscosity as a function of both temperature and pressure.

In general, the viscosity of lubricant fluids decreases with increasing temperature and increases with increasing pressure. The influence of these effects on the behaviour of lubricated systems will be discussed in Section 4.5.3 and 4.5.6.

Solutions of Reynolds equation, as outlined above, have been performed for various types of bearings, such as thrust bearings and journal bearings, and the theoretical suppositions and practical consequences of these solutions have been discussed in detail (see for instance Ref. 4.127). These results will not be reproduced here. In the following, the solution of the Reynolds equation for the simple case of lubrication of a cylinder near a plane will be considered in brief, in order to get an impression of the implications of the Reynolds equation and to obtain a starting point for the discussion of the elastohydrodynamic lubrication mode. Consider the geometry of a rigid cylinder near a plane, as in Figure 4.45.

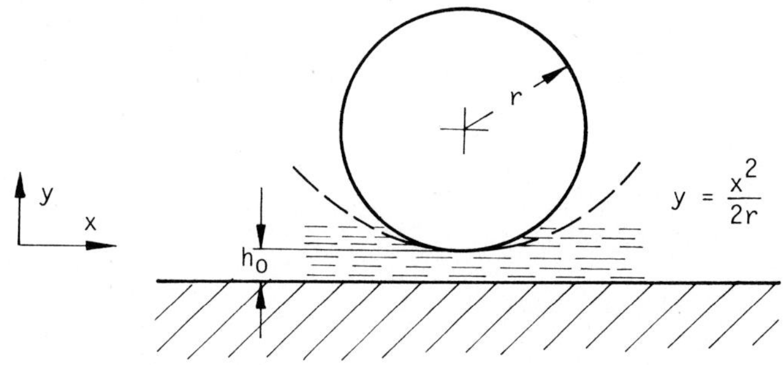

Figure 4.45 A lubricated cylinder-plane system.

Reynolds equation in one dimension for a steadily running bearing is

$$\frac{dp}{dx} = 6\eta v \frac{h - \bar{h}}{h^3}$$

v : surface velocity
η : viscosity
h : film thickness
$\bar{h}$: film thickness when $dp/dx = 0$

In the surrounding of the minimum of the gap the circle of radius r can be approximated by the parabolic equation

$$y = \frac{x^2}{2r}$$

Using this film-thickness equation and a boundary condition for p(x), the Reynolds equations can be integrated. In the literature, different boundary conditions for p(x) are known:

(a) Sommerfeld condition, $p = 0$ at $x = +\infty$

(b) Half Sommerfeld condition, $p = 0$ at $x \geqslant 0$

(c) Reynolds condition, $p = 0$ when $dp/dx = 0$ at $x > 0$

The integration of the Reynolds equation in using the Sommerfeld condition leads to the pressure distribution

$$p = -2\eta v \frac{x}{h^2}$$

Now using the Half Sommerfeld condition, it is possible to calculate easily the total load carried per unit length of the cylinder F_N/L:

$$F_N/L = \int_{\infty}^{0} p\,dx$$

$$F_N/L = 2\frac{\eta v r}{h_o}$$

Use of the Reynolds condition leads to

$$F_N/L = 2.45\frac{\eta v r}{h_o}$$

In rewriting these equations for the minimum film thickness as a function of the operating parameters load F_N and velocity v, and the viscosity η and the cylinder radius r and length L, it follows that

$$\frac{h_o}{r} = 2\frac{\eta\ v}{F_N/L} \qquad \text{(Half Sommerfeld)}$$

$$\frac{h_o}{r} = 2.45\frac{\eta\ v}{F_N/L} \qquad \text{(Reynolds)}$$

These results show that for the cylinder/plane system the separating thickness of the lubricant film decreases with decreasing viscosity η, decreasing velocity v, or increasing normal load F_N.

After the study of the relation between the lubricant film thickness and the relevant system parameters, the question of friction must be considered. As described above, the friction force can be found by integrating the viscous shear forces on the cylinder surface and adding to them any hydrodynamic force acting in the direction of movement. The results of this calculation, applying the Reynolds condition leads to the friction coefficient (Ref. 1.7):

$$f = \frac{F_F}{F_N} = 0.76\left[\frac{\eta\ v}{F_N/L}\right]^{1/2}$$

It is interesting to compare the results for the cylinder/plane system with that for a journal bearing - practically, the most important hydrodynamically lubricated mechanical system (Ref. 4.128). In Figure 4.46 the geometry of an operating journal bearing is shown schematically, the clearance being grossly exaggerated.

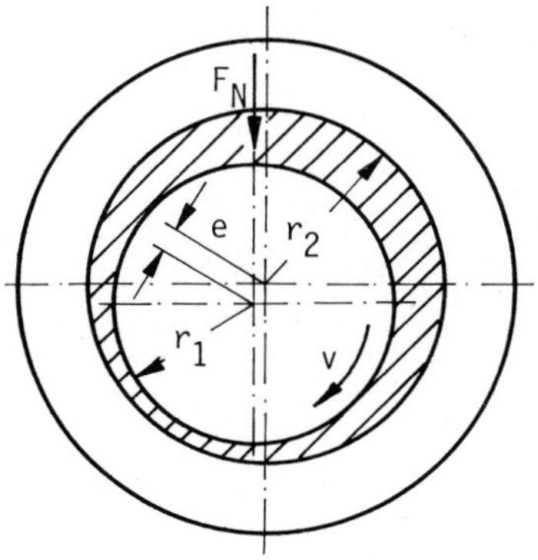

F_N : load
v : velocity
e : eccentricity
r_1 : shaft radius
r_2 : bearing radius
$c_r = r_2 - r_1$: radial clearance

Figure 4.46 Schematic representation of the geometry of a journal bearing.

Since in any journal bearing a natural lubricant wedge exists as a consequence of the clearance, at sufficiently high velocities the pressure generated in the wedge of lubricant supports the shaft and separates it completely from the bearing bushing. The frictional behaviour can be described conveniently as a function of operating variables and the journal bearing data in terms of the "Sommerfeld number" (sometimes also called "Gümbel number"). The Sommerfeld number is defined as

$$So = \frac{p\,\psi_r^2}{\eta\,\omega}$$

where

$$p = \frac{F_N}{2rL} \quad : \text{load/bearing projected area}$$

$$\psi_r = \frac{c_r}{r} \quad : \text{radial clearance/shaft radius}$$

$$\eta \quad : \text{viscosity}$$

$$\omega = \frac{v}{r} \quad : \text{angular velocity}$$

According to Vogelpohl (Ref. 4.129), the friction coefficient can be expressed approximately as

$$f = \frac{3\psi_r}{So} \qquad \text{for} \qquad So < 1$$

$$f = \frac{3\psi_r}{So^{1/2}} \qquad \text{for} \qquad So > 1$$

it follows that

$$f \propto \frac{\eta v}{F_N/L} \qquad \text{for} \qquad So < 1$$

$$f \propto \left[\frac{\eta v}{F_N/L}\right]^{1/2} \qquad \text{for} \qquad So > 1$$

In this connection it should be remembered that the parameter $\eta v F_N^{-1}$ was used as abscissa of the Stribeck curve Figure 4.40, the general characteristic of a lubricated system.

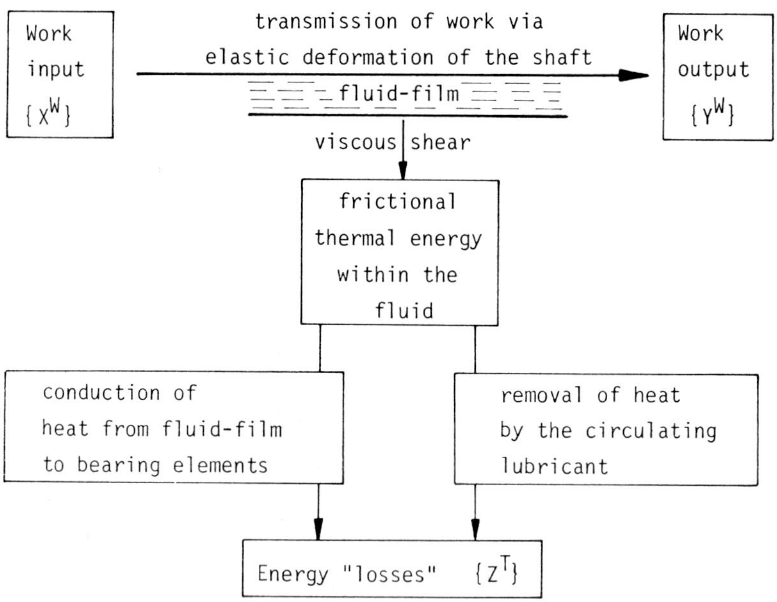

Figure 4.47 Schematic diagram of the transmission and dissipation of work in a hydrodynamically lubricated system.

Supplementing the study of the friction resistance in hydrodynamic lubrication, the aspect of energy dissipation should be studied. As explained earlier, in a journal bearing the transmission of the mechanical input work into the technically used output work is performed via the torsional elastic deformation of the bearing shaft. The frictional energy losses are due to "parasitic" guidance losses of the shaft. In a full-fluid journal bearing the friction resistance against motion is given by the viscous shear resistance leading to the generation of heat within the lubricant which is either removed by the circulating lubricant or transferred to the elements of the bearing. Referring to the conceptual Figures 3.6 and 3.7, in a simplified manner, the processes of mechanical energy transmission and dissipation are depicted schematically in Figure 4.47.

If the whole frictional power can be expressed as $f\,F_N\,v$, then in steady-state the dissipation process is described by

$$f\,F_N\,v = \rho\,c_1\,q\,\Delta T_1 + C_h\,A\,\Delta T_2$$

where

- f : friction coefficient
- F_N : normal load
- v : velocity
- ρ : density of lubricant
- c_1 : specific heat of lubricant
- q : volume flow rate of lubricant
- ΔT_1 : mean temperature rise of the lubricant in passing through the bearing
- C_h : convective heat-transfer coefficient at the oil bearing interface
- $A = 2\pi rL$: area of the oil bearing interface
- ΔT_2 : mean temperature difference between the oil in the bearing and the bearing surface

If $\Delta T_1 = \Delta T_2 = \Delta T$

it follows that

$$f\,F_N\,v = \Delta T\,(\rho\,c_1\,q + 2\pi\,C_h rL)$$

$$\Delta T = \frac{f\,F_N\,v}{\rho\,c_1\,q + 2\pi\,C_h rL}$$

where, as shown above, the friction coefficient can be written as

$$f = \frac{3\,\psi_r}{So} \qquad \text{for} \qquad So < 1$$

$$f = \frac{3\,\psi_r}{So^{1/2}} \qquad \text{for} \qquad So > 1$$

From these equations it follows that the determination of the frictional temperature rise in a full-fluid journal bearing is extremely complicated. It follows further that in a practical journal bearing design problem an iterative procedure must be adopted (Ref. 4.130). This is due to the interdependence of the operating parameter, load, velocity, the Sommerfeld number which includes the viscosity and the temperature ΔT within the bearing. The consequences of these effects on the practical design of bearings are described in books on practical lubrication (see for instance the Standard Handbook of Lubrication Engineering, Ref. 4.131).

4.5.3 ELASTOHYDRODYNAMIC LUBRICATION

The discussion of the hydrodynamic lubrication mode has been performed mainly for systems of high "geometrical conformity", like tilted pad bearings. In these cases the surfaces are assumed to remain completely rigid under the operation. In the cases of poor geometrical conformity, as in the cases of Hertzian concentrated contact situations, the effect of the elastic deformation of the surfaces must be taken into account. Owing to the high pressures to be found in lubricated Hertzian contacts, the effect of pressure on the lubricant's viscosity must also be considered. All these leads to the concept of elastohydrodynamic (EHD) lubrication (Ref. 4.132).

In the following, the main features of the elastohydrodynamic (EHD) lubrication mode are to be discussed in two steps: Firstly, the isothermal smooth surface EHD theory is studied and secondly, the effects of temperature and surface roughness are included. Also in this section the emphasis is on the tribological fundamentals and on the systems aspects of the subject matter.

As a starting point, consider a lubricated cylinder/plane system deformed elastically under the action of the load, as illustrated in Figure 4.48.

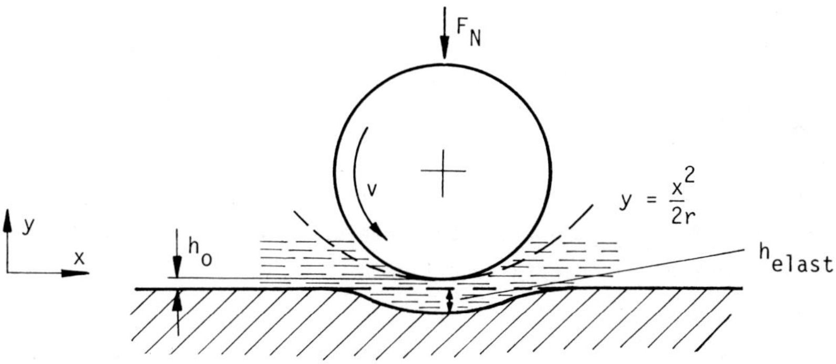

Figure 4.48 An EHD contact of a cylinder near a plane.

The situation is determined by the following equations:

(i) Hydrodynamic equation
As shown above, the one-dimensional Reynolds equation in its integrated form is

$$\frac{dp}{dx} = 6\eta v \frac{h - \bar{h}}{h^3}$$

The solution of this equation for the hydrodynamic case has been given above. This solution is not applicable, however, to EHD situations owing to the dependence of the film profile h and the viscosity η on the pressure p, i.e., $h = h(p)$ and $\eta = \eta(p)$.

(ii) Elasticity equation
The film profile h(x) is given by

$$h = h_o + \frac{x^2}{2r} + h_{elast}$$

In applying Boussinesq's equation of elastostatics (see Section 4.2.1) for h_{elast} it follows that

$$h(x) = h_o + \frac{x^2}{2r} - \frac{2}{\pi E} \int_{s_1}^{s_2} p(s) \ln(x-s)^2 \, ds + \text{const}$$

where $$\frac{1}{E} = \frac{1}{2}\left[\frac{1-\nu_1^2}{E_1} + \frac{1-\nu_2^2}{E_2}\right]$$

(iii) Viscosity-pressure relation

Assuming that the fluid is incompressible and that an isothermal situation is given, the viscosity-pressure relation is usually expressed by an exponential equation

$$\eta(p) = \eta_o e^{\alpha p}$$

η_o : viscosity at atmospheric pressure and T_0

α : pressure-viscosity coefficient

The validity of an exponential law indicates an enormous increase of viscosity with pressure. For instance, at a Hertzian pressure of about 1 GN/m^2 the viscosity of a mineral oil may increase by a factor of 10^6 compared with its value under atmospheric pressure.

The solution of the EHD problem involves an iterative procedure to establish a compatibility between equations (i), (ii) and (iii). In a simplified description the iterative procedure runs as follows. Assuming an initial film-thickness distribution which is inserted in the Reynolds equation, a pressure distribution is obtained. This in turn is inserted in the elastic equation giving an estimation of the elastic displacements which is compared with the initial film-thickness distribution. The iteration continues until a sufficient degree of conformity is obtained.

Before considering the main solutions of the EHD problem, a physical picture of the EHD pressure and film profile will be gained in considering the drawings shown schematically in Figure 4.49 for the contact of two cylinders (Ref. 4.133).

Figure 4.49 (a) shows the well known Hertzian-contact-pressure distribution of dry surfaces studied in Section 4.2.1. Various studies have also indicated that under lubricated conditions the central part of the gap is approximately parallel. This requires that at the inlet region the hydrodynamic pressure is lower than the Hertzian pressure distribution as sketched in Figure 4.49 (b). In the parallel region of the gap, the flow volume per unit gap width is approximately constant. Now, at the outlet the Hertzian pressure rapidly decreases and the viscosity falls several orders of magni-

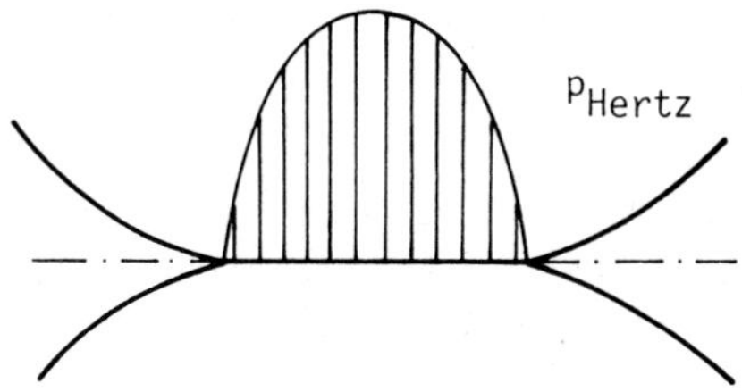

(a) dry Hertzian contact

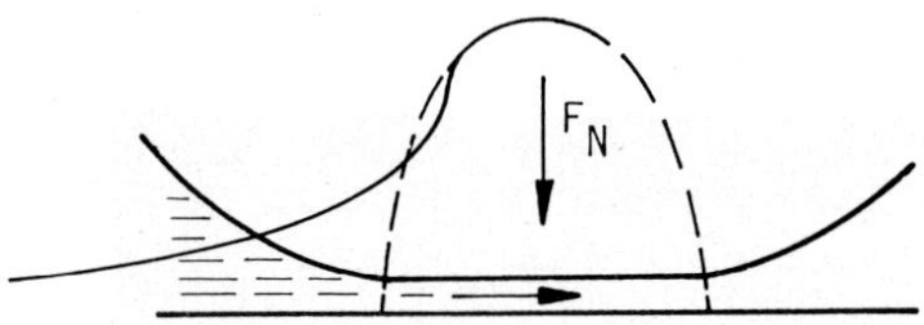

(b) pressure profile at the inlet

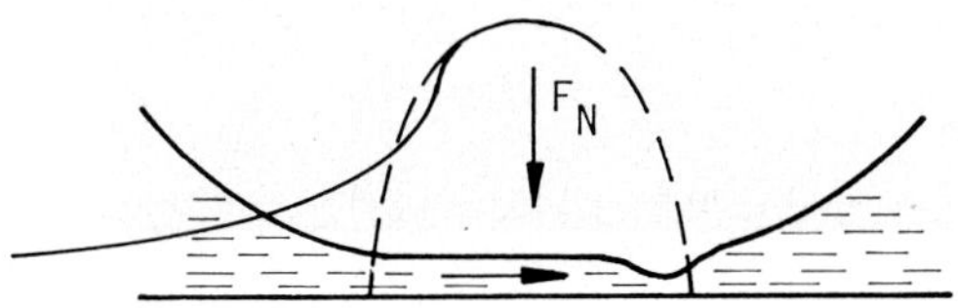

(c) film profile belonging to the pressure profile

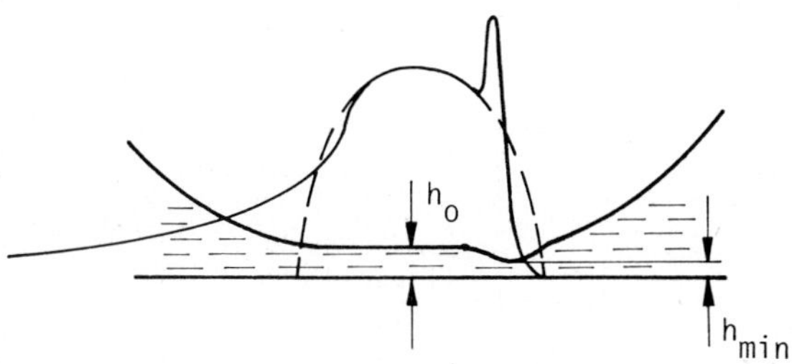

(d) pressure profile and film profile of isothermal smooth-surface EHD contact

Figure 4.49 Schematic representation of an EHD contact.

tude to its value under atmospheric pressure. Under these conditions, the requirement of the continuity of flow leads to a constriction in the film profile as indicated in Figure 4.49 (c). This is connected, in accordance with the elasticity equation, with a pressure spike at the outlet. All these considerations lead to a film profile and a pressure-distribution shape as sketched in Figure 4.49 (d). Recently, these distributions have been determined experimentally using special pressure transducers (Ref. 4.134). A typical result is shown in Figure 4.50 (Ref. 4.94). The features of EHD lubrication as described in Figures 4.49 and 4.50 are now generally accepted.

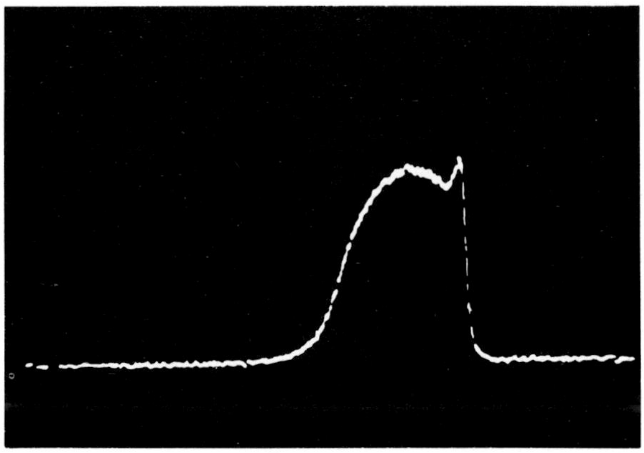

Figure 4.50 Measured pressure distribution in an EHD cylinder/cylinder contact (Ref. 4.94) ($\sigma_{Hz} = 10^4$ bar, $n = 20s^{-1}$, $r_1 = r_2 = 5$cm).

Now, turning our attention from the physical picture of EHD to the influences of the various operating variables and systems parameters, a relation describing the film thickness was first obtained by Grubin (Ref. 4.135) in the following form

$$\frac{h_o}{r} = 1.95 \frac{(\eta_o \alpha \, v)^{8/11} \, E^{1/11}}{r^{7/11} \, (F_N/L)^{1/11}}$$

Later, Dowson and Higginson (Ref. 4.136) were able to show that the various solutions of the EHD problem could be summarized to the formula:

$$\frac{h_{min}}{r} = 1.6\left(\frac{\eta_o v}{Er}\right)^{0.7} (\alpha E)^{0.6} \left(\frac{F_N/L}{Er}\right)^{-0.13}$$

where h_{min} : minimum film thickness

r : radius of relative curvature

η_o : viscosity at atmospheric pressure and at the temperature of the surfaces as they enter the conjunction

E : composite elastic modulus

α : pressure viscosity coefficient, defined by $\eta = \eta_o \exp \alpha p$

$v = \frac{1}{2}(v_1 + v_2)$: velocity

F_N/L : load per unit width of contact

The results were derived for pure rolling but they are also applicable to moderate degrees of sliding. The Dowson-Higginson formula has been extended in various directions (Ref. 4.137, 4.138). Further, refinements of this formula have been obtained by the experimental methods of optical elastohydrodynamic fluid testing (Ref. 4.139, 4.140). In addition, a hypothetical constitutive relation for EHD oil films has been proposed which combines a linear elastic response with a non-linear shear thinning viscous response (Ref. 4.141).

The analyses of elastohydrodynamic lubrication show that the major variables encountered in isothermal smooth-surface EHD problems can be represented by the following dimensionless groups:

Film-thickness parameter : $H = \frac{h}{r}$

Load parameter : $W = \frac{F_N/L}{Er}$

Velocity parameter : $U = \frac{\eta_o v}{Er}$

Materials parameter : $G = \alpha E$

The set of the four dimensionless groups can be reduced to three without any loss of generality (Ref. 4.133, 4.142, 4.143). Two of the possible sets, for instance are:

$$HU^{-1/2},\quad WU^{-1/2},\quad GU^{1/4}$$

$$HG^2,\ WG^2,\ UG^4$$

The results of isothermal, smooth-surface EHD theory have been confirmed experimentally and are now generally accepted as an idealized basis of this field. In more realistic situations, however, two further aspects must be considered (Ref. 4.144):

(i) the dissipation of energy within the lubricant film, i.e., the viscous heating and the generation of interfacial temperatures and its influence on the lubricant viscosity and the film thickness,

(ii) the surface roughness and its influence on the film shape and the effect of asperity interactions.

Another operating condition which may frequently be important but rarely considered is the supply and distribution of lubricant in the neighbourhood of the conjunction region. It has been shown that lubricant "starvation" alters the EHD behaviour of the system in several important ways (Ref. 4.145). Moreover, the influence of starvation can be an overriding factor on the performance of machine elements (Ref. 4.146).

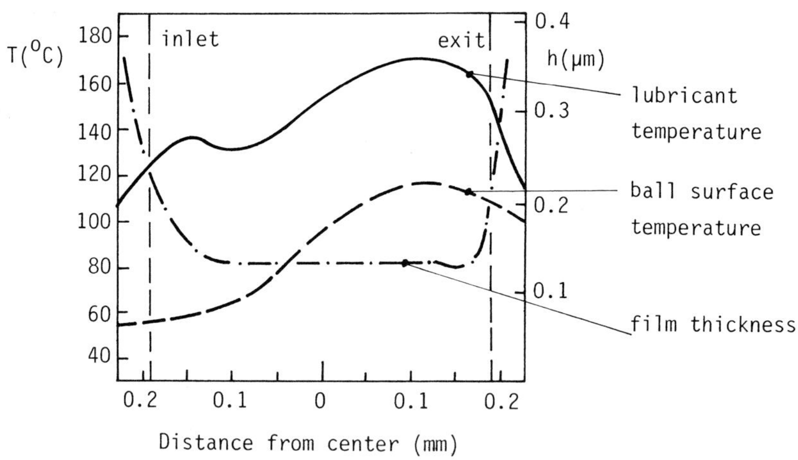

Figure 4.51 Temperatures in sliding EHD contact along contact center line (Ref. 4.147) (ball-plane contact, $r = 15.9\text{mm}$, $\sigma_{Hz} = 1\text{GN/m}^2$, $v = 1.4\text{ms}^{-1}$, $T_V = 40^{\circ}\text{C}$).

In applying the IR temperature-detecting technique, Winer and co-workers were able to measure both lubricant temperatures and surface temperatures in sliding EHD point-contact situations (Ref. 4.147). Some representative results are reproduced in Figure 4.51.

The friction-induced temperature data measured under various operating conditions clearly indicate the importance of the effect of the operating variables sliding velocity and contact pressure, on both ball-surface and fluid temperatures. They also show significant inlet zone heating which in turn will reduce the lubricant viscosity and pressure viscosity coefficient, the dominant mechanical properties affecting the EHD film thickness.

The influence of friction-induced temperatures on lubricant viscosity lowers the film thickness. Therefore the effect of surface roughness and the possibility of asperity contacts cannot be neglected in EHD lubrication. The inclusion of these effects has led to the concept of partial EHD lubrication.

A characteristic factor of partial EHD is the ratio, λ , of the theoretical (isothermal) smooth-surface EHD film thickness h_o to the combined roughness, $\bar{R}$, of the surfaces:

$$\lambda = \frac{h_o}{\bar{R}}, \quad \text{where } \bar{R} = \left[R_{a_1}^2 + R_{a_2}^2 \right]^{1/2}$$

The value of $\bar{R}$ is calculated either on the basis of readily measured r.m.s. or c.l.a. (R_a) values, or in more sophisticated way, by using the standard deviation of the peak-height distribution of the surface asperities as described in Section 4.2.1. The λ-ratio not only is important in the limits of lubrication and film failure in rolling and sliding situations (see Section 4.5.6), but it can also be used in describing the change-over from full EHD to partial EHD lubrication or "mixed" lubrication and to boundary lubrication. If $\lambda > 3$ to 5 we expect a full-fluid EHD film to separate the surfaces. At $\lambda \approx 3$ the first asperity "contacts" occur. At this point, the full EHD regime of the Stribeck's curve is left and the region of partial EHD (at the minimum of Stribeck's curve) is reached.

4.5.4 MIXED LUBRICATION

A simple and elegant explanation of the regime of mixed lubrication is to assume that the total applied load is carried partly by the hydrodynamic action of the lubricant film and partly by asperity contacts. Consequently, the total friction force is supposed to be due partly to viscous friction and partly to the asperity contact. Based on this assumption a couple of models of mixed lubrication have been proposed which can be broadly classified in two groups:

(a) Models of mixed lubrication, in which the (classical) hydrodynamic lubrication theory is used as a starting point. Theories on this basis are due to Gümbel (Ref. 4.148), Vogelpohl (Ref. 4.149), Christensen (Ref. 4.150) and Berthe and Godet (Ref. 4.151).

(b) Models of mixed lubrication, which start from the EHD lubrication theory taking into account the asperity contact component, suggested by Johnson, Greenwood and Poon (Ref. 4.152) and by Tallian (Ref. 4.153).

Vogelpohl in 1954 (Ref. 4.149) extended the expression for the friction coefficient of a hydrodynamically lubricated journal bearing, described in section 4.5.2

$$f = \frac{3\,\psi_r}{So^{1/2}} \qquad (So > 1)$$

to the following formula of mixed lubrication

$$f = f_o\left[1 - \frac{\eta\,\omega}{\bar{p}\,\psi_r^{\,2}}\left(\frac{c_r}{h_{min}} - 1\right)\right] + 3\left[\frac{\eta\,\omega}{\bar{p}}\right]^{1/2}$$

where f_o is the static coefficient of friction and c_r and h_{min} are the radial clearance and the film thickness at the minimum of the Stribeck curve, respectively.

Christensen in his theory of mixed lubrication (Ref. 4.150) used a modification of Reynold's equation for surfaces with roughnesses in the form of continuous longitudinal ridges and with no large-scale side leakage. In its integrated form, this equation becomes

$$\frac{dp}{dx} = 6\eta(v_1 + v_2)\frac{E^*(h') - E^*(h'_o)}{E^*(h')^3}$$

where p is the hydrodynamic pressure,
x is a Cartesian co-ordinate in the direction of motion,
η is the local dynamic viscosity of the lubricant,
v_1 and v_2 are the velocities of the surfaces relative to their conjunction,
h' is the local film thickness,
and E^* is the expectation operator, defined by

$$E^*(x^n) = \int_0^\infty x^n f(x)\, dx$$

A modification of the classical Reynolds equation for application to the contact of rough surfaces has also been proposed by Berthe and Godet (Ref. 4.151). Their approach also emphasized the pronounced influence of sliding velocities in the regime of mixed lubrication.

Johnson, Greenwood and Poon (Ref. 4.152) combined the smooth-surface EHD theory of Dowson and Higginson with the theory of the contact of dry rough surfaces developed by Greenwood and Williamson (see Section 4.2.1) in order to apply to conditions of mixed or partial contact.

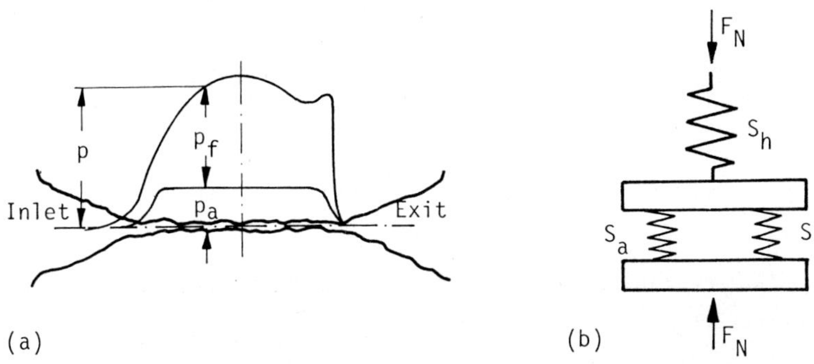

Figure 4.52 An EHD contact with rough surfaces: (a) pressure distributions, (b) diagrammatic representation of the flexible elements in an EHD contact.

The situation is illustrated in Figure 4.52 (a) which represent either a line or a point contact. The total pressure p is made up of the fluid pressure p_f and the asperity contact pressure p_a. As shown diagrammatically in

Figure 4.52 (b), the bulk (Hertzian) deformation of the two solids under the applied load F_N, which gives rise to the nominal contact area A_o, is modelled by the spring S_h. The oil film and the asperities are represented by the parallel springs S_f and S_a, respectively. All springs are non-linear; the stiffness of each increases with compression. From this model it follows that the asperity pressure is determined primarily by the film-thickness-to-roughness ratio λ, defined above.

Likewise, in Tallian's theory of partial elastohydrodynamic contacts (Ref. 4.153), the λ-ratio is considered as an important parameter. From the theories of partial EHD, as outlined by Tallian and by Johnson, Greenwood and Poon, it may be concluded that if λ is lower than a certain limit, less than $\lambda = 0.4$ say, the lubrication mode changes to boundary lubrication.

4.5.5 Boundary lubrication

The lubrication regime of the utmost left part of the Stribeck curve is termed boundary lubrication. This lubrication mode is characterized by the following features:

(a) the solid surfaces are so close together that appreciable contact takes place between the asperities,
(b) hydrodynamic effects and influences of the bulk rheological properties of the lubricants are of little or no importance,
(c) the tribological behaviour is determined by surface interactions between thin layers of boundary lubricants and the solid surfaces.

Due to these features, the processes which determine the tribological behaviour of solid surfaces (see Sections 4.2, 4.3, 4.4) are valid likewise under boundary lubrication conditions, namely:

(i) the contact mechanics and the elastic and plastic asperity deformation processes,
(ii) the contact physics and chemistry and the action of surface forces,
(iii) the friction processes, especially the shearing of adhesive junctions and the deformation of asperities,
(iv) the wear processes which are given in the general case by the superposition of effects of surface fatigue, abrasion, adhesion and tribochemical reactions.

All these solid/solid interactions are modified through the action of the boundary lubricant, so that the tribological behaviour of a boundary lubricated system is determined by the processes at the solid/lubricant/ solid interface influenced by the environmental atmosphere. Compared with the other lubrication regimes of the Stribeck curve and the dry state, the tribology of a boundary lubricated system is most complex due to the variety of interactions and parameters involved.

Concerning the lubricating action of a boundary lubricant, its main purpose is to interpose between the moving surfaces a film that is able to reduce the amount of direct solid/solid interaction and that is itself easily sheared. This is provided best by an interfacial film consisting of long chain molecules possessing the following properties:

(a) strong attraction between the chains to resist penetration by surface asperities (thus mitigating wear processes),

(b) low shear strength to give a low friction,

(c) high melting point so that it provides solid-film protection up to a high temperature.

Since the mechanisms of boundary lubrication are determined mainly by the physics and chemistry of the solid/lubricant/solid interface, a brief review of the chemical composition of lubricants is first given followed by a discussion of the processes of boundary film formation and its tribological behaviour.

The various kinds of lubricants may be classified into three general groups:

(I) "Nonreactive" or "indifferent" lubricants

Hydrocarbons and most other base oils fall in this class. Usually the most important property of these materials is viscosity, which depends on chemical structure in a fairly regular fashion. The hydrocarbons or base oils from petroleum may be classified according to their structure into saturated (paraffins and naphthenes) and unsaturated ones (olefins and aromatics). The four classes can be typified by the following structures:

(a) Paraffin

$$\begin{array}{c} CH_3 - CH_2 - \underset{\displaystyle CH_3}{\underset{|}{CH}} - CH_2 - CH_3 \end{array}$$

(b) Naphthene

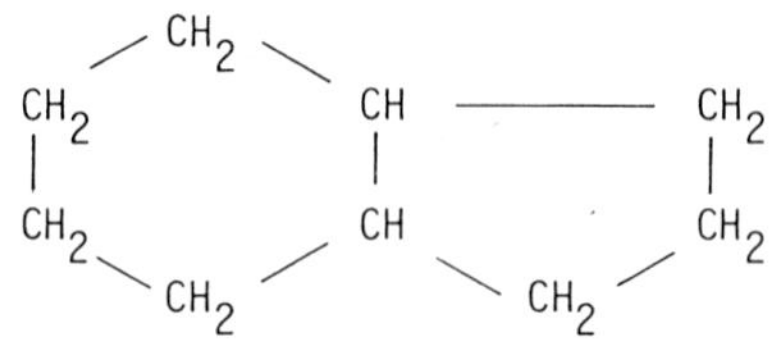

(c) Olefin

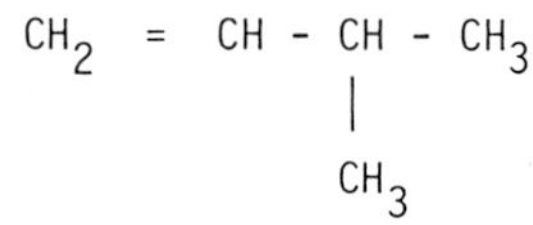

(d) Aromatic

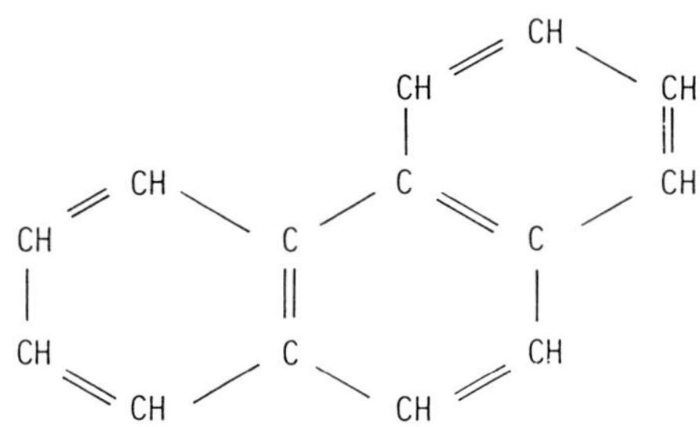

All mineral oils consist of molecules which generally contain more than one type of hydrocarbon moiety. Classification of oils as paraffinic, naphthenic (i.e., cycloparaffin) or aromatic indicates the relative concentrations of the moieties.

(II) "Oiliness" agents

Lubricant additives containing a polar group attached to long chains of methylene groups (i.e., $(-CH_2-)_n$ where $n > 10$) are often referred to as "oiliness" or "lubricity" additives; oxygen is the most usual hetero-atom in the commonly used material. Typical examples are:

Stearic acid:

$$CH_3\,(-CH_2-)_{15}\,\overset{\overset{O}{/\!/}}{C}-OH$$

Oleic acid:

$$CH_3\,(-CH_2-)_7\,CH = CH\,(-CH_2-)_7\,\overset{\overset{O}{/\!/}}{C}-OH$$

Each of these additives possesses an unbranched hydrocarbon chain with a polar group at its extremity. Besides the long-chain fatty acids certain alcohols and esters of similar structure are also used. These materials can be dissolved in small quantities (about 1 per cent) in a base oil and will then attach themselves to the metal surfaces.

(III) " Extreme pressure" (EP) and "anti-wear" (AW) additives

These lubricant additives usually contain the heteroatoms phosphorus, sulfur or chlorine and may additionally or even principally contain oxygen, lead, zinc, boron, selenium, etc.

Examples are:

Dibenzyldisulfide:

Zinc dioctyl-dithio-phosphate:

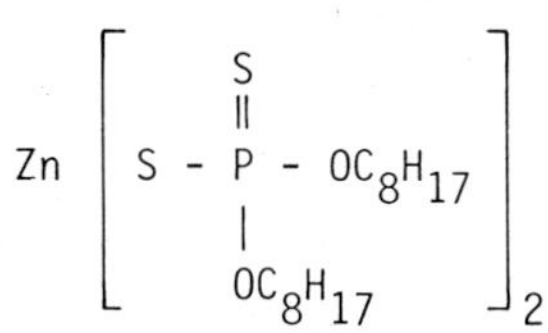

These additives function under boundary lubrication conditions by reacting with the contacting metal asperities to form films which aid in the reduction of intermetallic contact and wear and to prevent catastrophic surface failures (Ref. 4.154, 4.155). The major determined operating factor for these actions is the operating temperature in the interface.

In order to understand the mechanisms of boundary lubrication, the physico-chemical processes of interfacial film formation must be considered.

The lubricant/solid interactions which lead to the generation of a protective boundary film can be classified into three mechanisms (Ref. 4.156, 4.157):

(a) PHYSICAL ADSORPTION (PHYSISORPTION)

In a boundary-lubrication system, physical adsorption occurs when the molecules of the lubricant are held to the surface by van der Waals surface forces (i.e., dipole-dipole interactions, see Section 4.2.2). The molecules are weakly bonded, and the film formation is characterized by reversibility and monomolecular or multimolecular layers. Polar molecules, particularly long chain hydrocarbons, adsorb with preferred vertical orientation, see Figure 4.53. (For instance: adsorbed stearic acid molecules are 19 Å high, Ref. 4.158.)

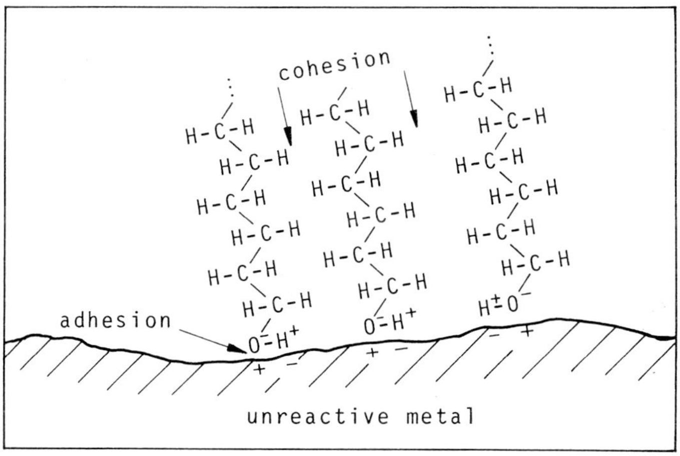

Figure 4.53 Schematic representation of physisorption.

The polar additives condense on the surface to form a solid film. Many molecules pack in as closely as possible and strengthen the film with lateral cohesive forces. This solid film, adhering to the surface and with the molecules cohering to each other, then has the ability to resist penetration of asperities and thus inhibit metal-to-metal contact. In addition, a zone of low shear strength is formed between the outermost surfaces of two monolayers adsorbed on opposing metal surfaces. A boundary lubricated system with physical adsorption of the film is temperature sensitive because heat causes desorption, disorientation, or melting of the film. Thus, boundary lubrication dependent upon physical adsorption is limited to low bulk temperatures and conditions of low frictional heat generation, i.e., low loads and low sliding velocities.

(b) CHEMICAL ADSORPTION (CHEMISORPTION)

In a boundary-lubrication system, chemisorption occurs when the molecules of the lubricant are held to the surface by chemical bonds, i.e., short-range surface forces. Compared with physisorption, it is generally not completely reversible and shows higher heats of adsorption (Physisorption: 2,000 to 10,000 cal/mole; Chemisorption: 10,000 to 100,000 cal/mole). A well known example of chemisorption in boundary lubrication is the reaction of stearic acid with iron oxide in the presence of water to form a "metal soap" film of iron stearate on the surface, as in Figure 4.54.

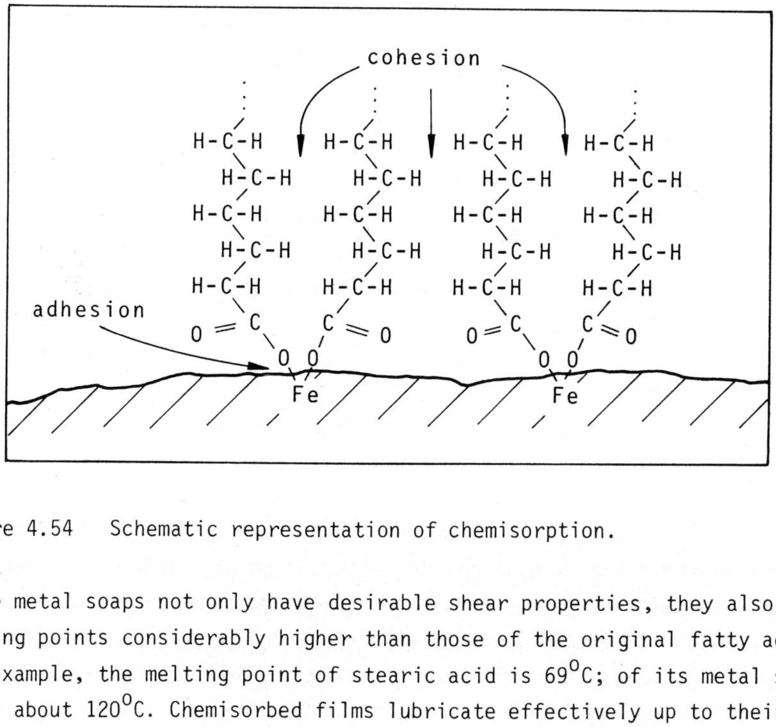

Figure 4.54 Schematic representation of chemisorption.

These metal soaps not only have desirable shear properties, they also have melting points considerably higher than those of the original fatty acid. For example, the melting point of stearic acid is 69^{0}C; of its metal soaps, it is about 120^{0}C. Chemisorbed films lubricate effectively up to their melting point, whether the film is formed "in situ" on a reactive surface or spread on an unreactive surface. Boundary chemisorbed films provide lubrication at moderate loads, temperatures and sliding velocities and fail under severe operating conditions. The failure to lubricate is a result of the disorientation, softening or melting of the film under the combined action of the operating variables and the energy transactions.

(c) CHEMICAL REACTION

In a boundary-lubrication system, chemical reaction between the solid surfaces and the lubricant molecules occurs when there is an exchange of valence electrons and a new chemical compound is formed as illustrated in Figure 4.55.

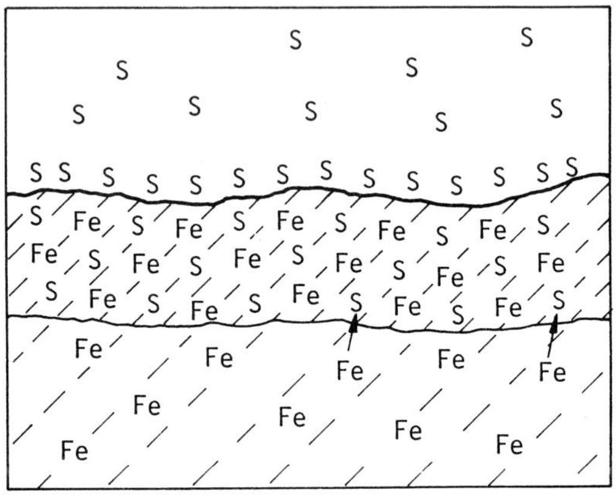

Figure 4.55 Schematic representation of a boundary film formed by chemical reaction.

The boundary films are unlimited in thickness (governed by diffusion processes through the crystalline lattices) and characterized by high activation and bonding energies and irreversibility. Most of the chemically reactive boundary lubricants contain sulphur, chlorine and phosphorus atoms in the molecule. They are believed to function by forming metal-salt films of low shear strength but of high melting point, such as sulfide, chloride or phosphide, at the interface. These films are more stable than any physically or chemically adsorbed film. Boundary lubricants dependent upon chemical reaction are suitable for high load, high temperatures, and high sliding speeds and limited to reactive metals. These conditions are commonly referred to as "extreme-pressure" conditions, although the greatest severity arises from high temperatures. Although "EP" additives function by reacting with the surfaces they must not be too reactive, otherwise tribo-chemical wear may reach very high rates.

Concerning the tribological behaviour of a boundary-lubrication system, since the boundary films may be considered to be solid and to behave "as a rigid continuation of the solid body" (Ref. 4.159), the contact, friction and wear processes as discussed in Sections 4.2, 4.3 and 4.4 are valid in principle. In order to describe the frictional behaviour of the various boundary films, stylized diagrams have been used (Ref. 4.160, 4.161) as in Figure 4.56.

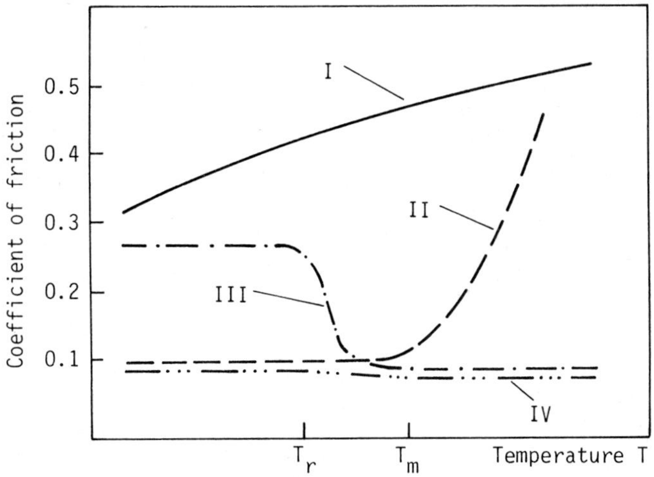

Figure 4.56 Friction behaviour of boundary-lubrication systems (schematic).

In this diagram, the coefficient of friction f is plotted against the temperature T. Curve I is for a system lubricated with a non-polar base oil. It shows that the friction is initially relatively high and increases with increasing temperature as the weak physical adsorption bonds are loosened. Curve II is for a fatty acid dissolved in the base oil. This lubricant reacts with the metal surface forming a metallic soap which is easily sheared at the area of real contact. Up to T_m, the melting point of the metal soap, the friction is low and constant but above this point rises steeply. Curve III is representative of an EP additive dissolved in the base oil. The additive reacts very slowly below a certain reaction temperature T_r. When T_r is reached, the chemical reaction starts and the friction coefficient decreases as the rate of film formation permits a solid film to carry the load. In this case effective lubrication is provided up to high temperature. Curve IV is a hypothetical curve for an effective combination of II and III.

Good lubrication is provided by the fatty acid below T_r, while above this temperature the greater part of the lubrication is due to the EP additive.

If the wear behaviour of the different boundary-lubrication systems is to be described in the form of stylized diagrams, the characteristics shown in Figure 4.57 result.

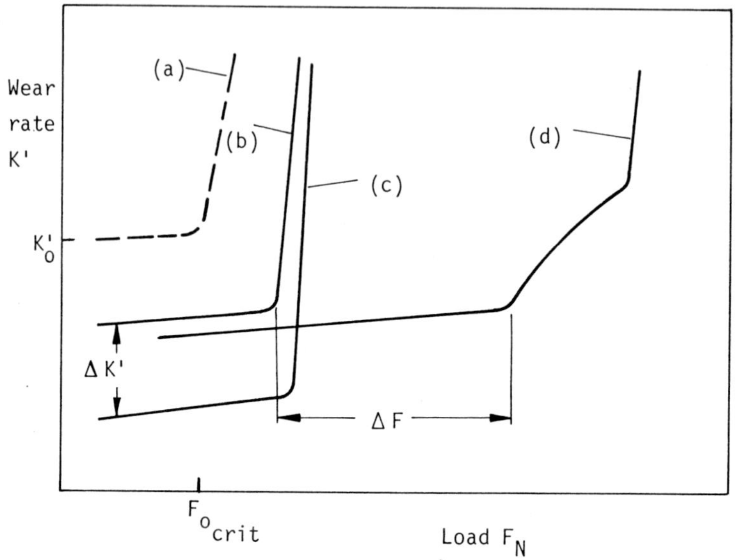

Figure 4.57 Wear behaviour of boundary-lubrication systems (schematic).

In this diagram, the wear coefficient K' (wear volume x load^{-1} x sliding distance^{-1}) on a log scale is plotted versus the load F_N.

Curve (a) represents the wear behaviour of a dry metal/metal sliding system. As a result of the wear processes described in Section 4.4 occurring initially at the "outer" surface layers (see Figure 4.5) a certain wear rate K'_0 results. If the load is relatively low, the surface layers worn off may be replaced under the action of the gaseous environment ("mild wear" regime). If however, the load reaches a certain limit, F_{0crit}, the wear process may penetrate to the "inner" surface layers leading to "severe wear" and to an increase of the wear rate over several orders of magnitude.

Curve (b) is for a system lubricated with a base oil. The influence of chemical additives in a base oil may lead to two different effects:

(i) a reduction of the wear rate K' by a certain amount $\Delta K'$,

(ii) an increase of the load-carrying capacity F_{crit} by a certain amount ΔF.

If the action of an additive leads primarily to the effect (i), as shown in curve (c), it is called "anti-wear" (AW) additive. The most common types of AW additives are those containing phosphorus, like metal-dialkyldithiophosphate, tricresyl phosphate or dialkyl phosphite. Organic carboxylic acids, like stearic acids and oleic acids and their esters have also good AW properties. If the action of an additive leads primarily to the effect (ii), as shown in curve (d) of Figure 4.57, it is termed "extreme pressure (EP)" additive. As mentioned above, the most common type of EP additives are those containing sulfur and/or chlorine. Typical sulfur-containing additives are sulfurized fatty esters, terpenes and olefins or dibenzyl disulfide. The most common chlorine-containing EP additives are chlorinated paraffins, containing up to 70% Cl.

Although it is clear that the additives under boundary conditions form protective surface layers by the physico-chemical processes explained above, the chemical details and the kinetics of the film formation and its dispersal are still under discussion. Clearly, in the discussion of additive performance besides the properties of the lubricant also the properties of the metal substrate must be considered (Ref. 4.162). The literature on boundary lubrication properties of additives and their various combinations is enormous and cannot be discussed in detail in the scope of this volume. For recent reviews, see the articles by Godfrey (Ref. 4.156, 4.157), Campbell (Ref. 4.161), Fein (Ref. 4.163).

The treatise on the boundary-lubrication mode of a tribological system should not be closed without a compilation of the main influencing parameters of this most complex field. Some of these parameters are contained in a formula proposed by C.N. Rowe (Ref. 4.164, 4.165) for the rate of adhesive wear of a boundary lubricated system. The formula contains two dimensionless groups, one of which is descriptive of the sliding metal couple and the other is of the effectiveness of the lubricant film at the surface, considering that the heat of adsorption may be the controlling factor in the effectiveness of the boundary lubricant. The model leads to the expression:

$$\frac{V}{l} = K_m \frac{(1 + 3f^2)^{1/2}}{p_m} \cdot \frac{F_N}{v} \cdot \frac{d_a}{t_o} e^{-\frac{Q}{R_c T_s}}$$

The parameters involved in this relation may be classified under the following headings:

Operating variables:

F_N : load
v : velocity
l : sliding distance
T_s : absolute temperature of the sliding surface

Materials parameters:

p_m : flow pressure of the softer metal
Q : heat of adsorption
R_c : molar gas constant

Interaction characteristics:

d_a : diameter of the area associated with an adsorbed molecule
t_o : fundamental oscillation time of an adsorbed molecule
f : friction coefficient
K_m : wear coefficient of the sliding metal surfaces
V : wear volume of the boundary lubrication system

The great number of influencing factors of a boundary-lubrication system urges the need for a systematic parameter compilation of tribological systems as attempted in Chapter 8.

4.5.6 THE LIMITS OF LUBRICATION

The surveys presented in the foregoing sections show that lubrication is the most important means of influencing the transactions of energy and materials in tribological systems and reducing friction and mitigating wear. Consequently, any disturbance or failure of lubrication has detrimental consequences for the tribological behaviour of the system. Besides an understanding of the mechanisms of proper lubrication, it is therefore necessary to gain some insight into the processes and factors which determine a failure of lubrication, i.e., the possible "limits" of lubrication.

Starting from conditions of full-fluid-film lubrication and following the course of the Stribeck curve either to the right or to the left, two different types of the limits of lubrication are met:

(a) To the right, as for instance with increasing velocity, the conditions of laminar fluid flow may change to turbulent flow.

(b) To the left, for instance with decreasing velocity, running through the various non-hydrodynamic lubrication regimes, the lubricant film may eventually break down leading to catastrophic failure in the form of "scuffing" or "seizure" (Ref. 4.166).

In fluid mechanics, considering a fluid of viscosity η and density ρ flowing with a velocity v past a long cylinder of diameter d, the change from laminar to turbulent flow is discussed in terms of the Reynolds number

$$Re = \frac{\rho}{\eta} v d$$

For Re < 1 a laminar fluid flow mode is found. For Re > 1 various effects of disturbances and turbulences occur, like the "Kārmān vortex street" appearing for Re > 40 (Ref. 4.167).

In journal bearings of the typical clearance ratios $0.001 < c_r/r < 0.01$, "Taylor vortex flow" will develop in the lubricant film before the onset of turbulence. In studying the flow between concentric cylinders in a classical investigation (Ref. 4.168), Taylor found that the transition to vortex flow occurs if the Taylor number

$$\frac{\rho v c_r}{\eta} \left[\frac{c_r}{r}\right]^{1/2} > 41.3$$

For non-concentric cylinders, i.e., bearings, the situation is less clear. A theory for predicting the transition to vortex flow for non-concentric cylinders has been developed by di Prima (Ref. 4.169). As the rotational velocity of a journal bearing is increased above the onset of vortices, full turbulence may occur at a Reynolds number of about Re $\approx$ 2500. This regime has been analyzed by Constantinescu (Ref. 4.170). For a review of further studies see (Ref. 4.171).

Turning now to the limits of full-fluid lubrication on the left-hand side of the Stribeck curve, the regions of the various lubrication modes can be specified if the Stribeck curve is expressed in terms of the film thickness to roughness ratio λ and combined with the no-contact time fraction ζ determined with the electrical-contact-resistance measuring technique (Ref. 4.172, 4.173). (The no-contact time fraction ζ is defined as the percentage of time during which the electrical contact resistance R_c of a contact is higher than a certain R_c-level indicating complete separation of the surfaces, see Section 7.2.3.) The variations of the friction coefficient f and the no-contact time fraction as function of the film-thickness to roughness ratio λ are shown schematically in Figure 4.58.

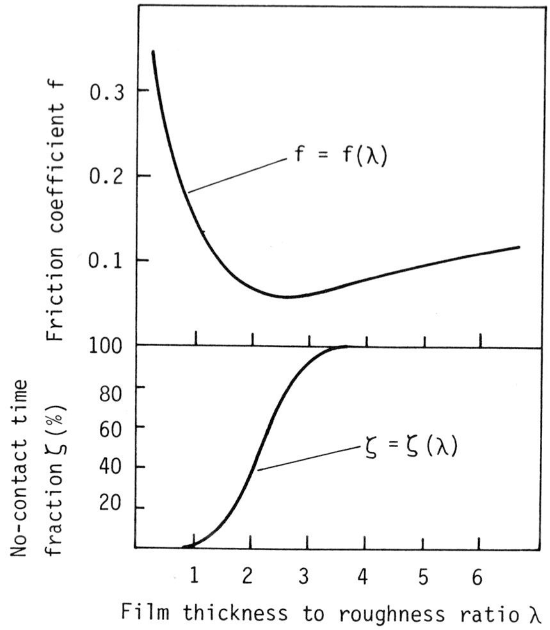

Figure 4.58 Stribeck curve and no-contact time fraction (schematic).

With the diagram shown in Figure 4.58, two transitions can be specified:

(a) The transition from full fluid-film lubrication to mixed lubrication or partial EHD lubrication, indicated by

$$\zeta < 100\% \qquad \text{and} \qquad \lambda < \approx 3$$

(b) The onset of continuous asperity contact within the interface, indicated by

$$\zeta = 0\% \qquad \text{and} \qquad \lambda < \approx 1$$

The description of the transitions (a) and (b) has been given in terms of the film-thickness-to-roughness ratio λ . The value of λ in turn depends on the operating variables and the material and lubricant properties as described in Section 4.5.3.

Another way of describing the transitions between the various lubrication modes is to study the change of λ as function of the operating variables for a given sliding system. In the following, a simplified physical picture of the different transitions from a full-fluid EHD lubrication to the breakdown of boundary-lubrication films is outlined (Ref. 4.174). The situation is illustrated in the schematic diagram of Figure 4.59. On the left the variation of λ with increasing contact pressure, p, is shown for a given sliding system of lubricated point or line contact, specified by the materials parameters (E, r, R, η_0, α) and the operating variables (F_N, v, T). On the right, the different portions of the contact pressure are described schematically.

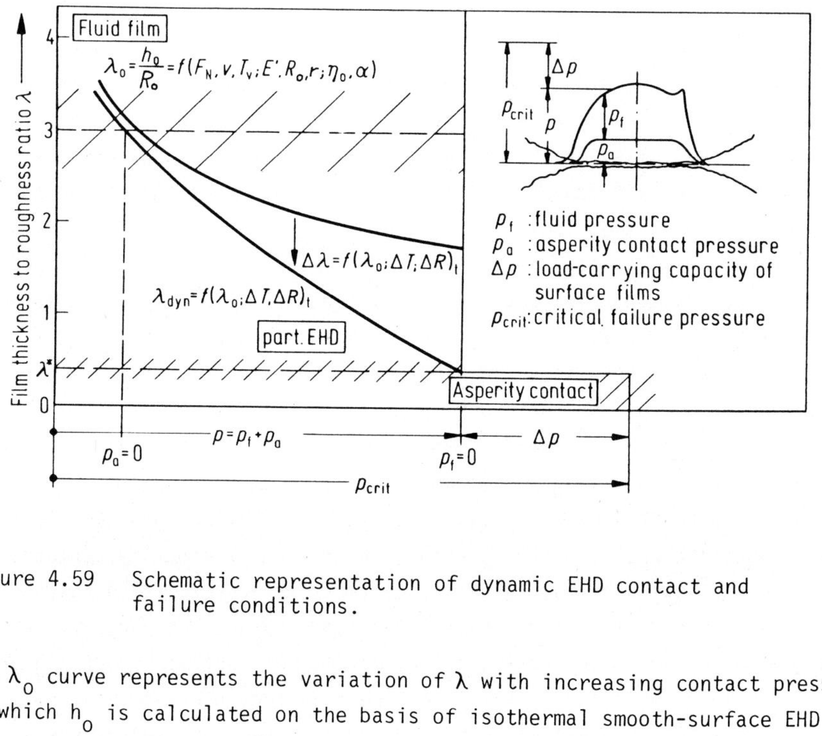

Figure 4.59 Schematic representation of dynamic EHD contact and failure conditions.

The λ_0 curve represents the variation of λ with increasing contact pressure in which h_0 is calculated on the basis of isothermal smooth-surface EHD theory and in which R_0 represents the initial surface roughness. In the lower λ_{dyn} curve, two dynamic influences are taken into account:

(i) The increase of the interfacial temperature ΔT, estimated on the basis of Blok's theory, see Section 4.3.4. (The assumption that the interfacial temperature rise, ΔT, in sliding EHD contacts can be estimated on the basis of Blok's theory is supported by the direct IR-temperature measurements performed by Winer and co-workers, Ref. 4.147, see also Figure 4.51.)

(ii) The increase in surface roughness during the sliding process. (The actual "dynamic" surface topography has been measured in separating the sliding surfaces rapidly at different stages of the test.)

From the values of (i), the viscosity $\eta = \eta\ (T + \Delta T)$ and the viscosity-pressure coefficient $\alpha = \alpha\ (T + \Delta T)$ are estimated and used in re-calculating the "dynamic" EHD film thickness h. If this (reduced) film thickness is divided by the actual dynamic surface roughness values (ii), the λ_{dyn} curve of Figure 4.59 is obtained. The whole chain of events which leads to lubricant failure can now be discussed on the basis of the schematic diagram of Figure 4.59:

If in sliding EHD contacts the experimental conditions lead to an initial film-thickness-to-roughness ratio of $\lambda > 3$ to 4, calculated on the basis of conventional smooth-surface isothermal EHD theory, it could be expected that the system operates under full fluid-film conditions. However, depending on the values of the operating variables (F_N, v, T_V) and the material data, a certain amount of interfacial viscous heating must be taken into account, which then lowers the viscosity η_0 and the pressure-viscosity coefficient α, leading to a decrease in the nominal values of λ. If then λ falls below a certain value ($\lambda < 3$, say), the first asperity contacts occur and the region of partial EHD is reached. In this region, the λ_{dyn} curve rather than the λ_0 curve is valid. The contact pressure is then made up of the fluid pressure p_f and the asperity pressure p_a. Through the action of interfacial viscous heating and asperity collision heating, as well as contact topography changes, λ_{dyn} decreases steeply with increasing p until a value of λ is reached at which the action of an EHD film ends, i.e., $p_f = 0$. Now, the whole contact pressure is carried by asperity contact, i.e., $p = p_a$. In the region of full asperity contact, the onset of "incipient scuffing" is determined, of course, by the stability of the asperity surface films, which in turn are affected by the metal/lubricant/atmosphere interactions and the generation of frictional heat in the interface. Failure occurs, if, from the point at which $p_f = 0$ and onwards, the contact pressure is increased by more than an amount Δp, the load-carrying capacity of the surface films (see Section 4.5.5). The breakthrough of the surface layers in itself may also occur as another stepwise process in which first the "outer" surface films (physi- or chemisorbed layers) and then the "inner" surface layers (oxide layers and the cold worked layers) are broken or thermally desorbed. It follows that the mechanisms of failure of thin-film lubrication of sliding concentrated contacts are determined by the physical and chemical nature of protecting surface layers as well as the EHD and asperity contact conditions as a function of the operating variables and the dynamic interfacial changes.

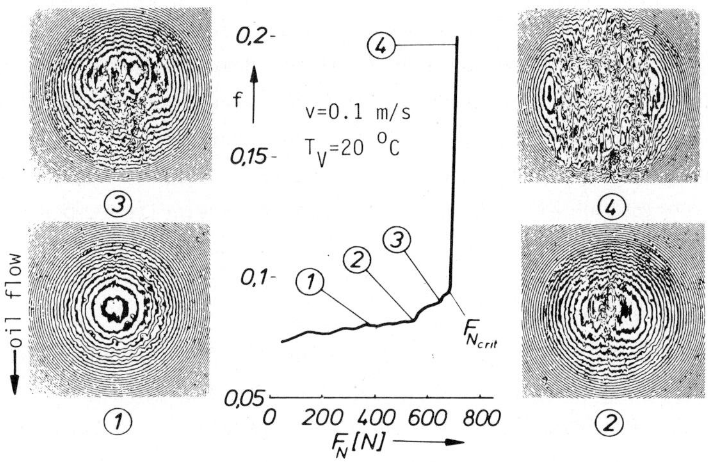

Figure 4.60 Friction coefficient and contact topographies at different stages of failure (stationary ball of four-ball system).

An empirical characterization of the critical values of the operating variables at the failure transition can now be obtained as follows:

If during a given change of operating variables the friction coefficient f and the wear coefficient K' (wear volume x load^{-1} x sliding distance^{-1}) are recorded, the failure transition indicating "incipient scuffing" manifests in sharp jumps of both the friction coefficient and the wear coefficient (Ref. 4.175, 4.176, 4.177). For the case of lubricated sliding concentrated contacts, the onset of incipient scuffing is characterized by the transitions

$$f < 0.1 \longrightarrow f > 0.3$$

$$K' < 10^{-6} \left(\frac{mm^3}{Nm}\right) \longrightarrow K' > 5 \times 10^{-6} \left(\frac{mm^3}{Nm}\right)$$

An experimental example of the friction transition, in connection with interferograms of the sliding surfaces at the different stages of film failure as described above, is shown in Figure 4.60 (Ref. 4.174).

The above transitions of f and K' can then be used as a criterion to specify the critical load F_N (or pressure p) determined for a given constant sliding velocity v and a given bulk oil temperature T_V during a given test duration t. If next the critical failure load is determined for various values of v and T_V, within a given time t, it has been found experimentally that the critical load F_N decreases with both increasing sliding velocity v and bulk oil temperature T_V as illustrated in Figure 4.61, (Ref. 4.178, 4.179). In combining these diagrams, it is then possible to describe the limits of lubrication in a three-dimensional load-velocity-temperature diagram, representing the critical values of all the operating variables (F_N, v, T_V) at the failure transition. This diagram has been termed "failure surface" (Ref. 4.180) and is shown in Figure 4.62.

The failure surface separates the region of effective lubrication (below the surface) from that of ineffective lubrication (above the surface) and shows the interrelation of the values of operating variables at the failure transition. (It should be mentioned that beyond the failure transition shown in Figures 4.60 to 4.62, in the region of ineffective lubrication, other friction and wear transitions may occur which are associated with metallurgical transformations in the sliding surfaces, Ref. 4.179. The different friction and wear transitions have been discussed in detail by G. Salomon, Ref. 4.181.)

Besides the dependence of failure on the operating variables, the influence of the lubricant properties is of great interest. From the schematic diagram of Figure 4.59, it is clear that both viscosity and the surface reactivity of lubricant additives influence the failure transition and the location and shape of the failure surface. Both influences shift the critical failure load to higher values. It has been found experimentally that the influence of viscosity on failure is most pronounced at low sliding velocities and low temperatures, whereas the influence of surface reactive agents is most pronounced at high sliding velocities and high temperatures (Ref. 4.182). The influence of lubricant viscosity on the location and the

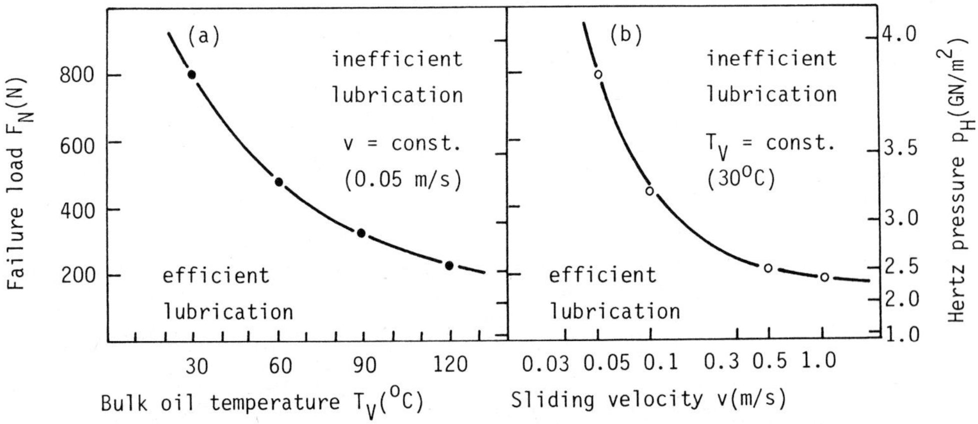

Figure 4.61 Failure load or nominal initial mean Hertz pressure as a function of (a) bulk oil temperature and (b) sliding velocity.

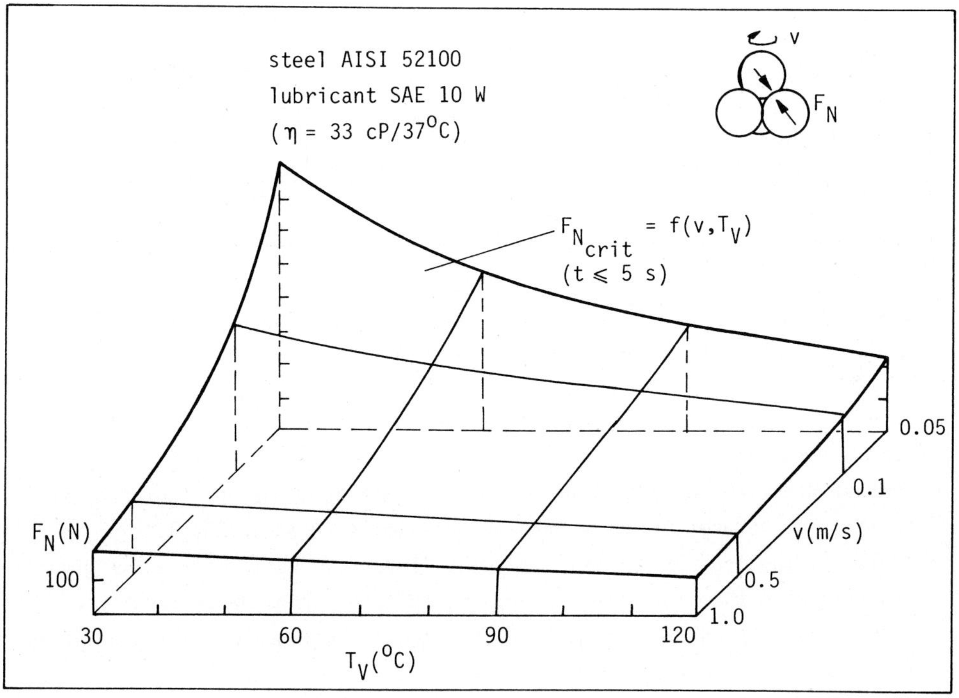

Figure 4.62 The failure surface: dependence of failure on load F_N, velocity v, and temperature T_V.

shape of the failure surface can be seen from Figure 4.63. The influence of lubricant additives on the failure transitions has been studied extensively by Sakurai and co-workers, (Ref. 4.183, 4.184, 4.185).

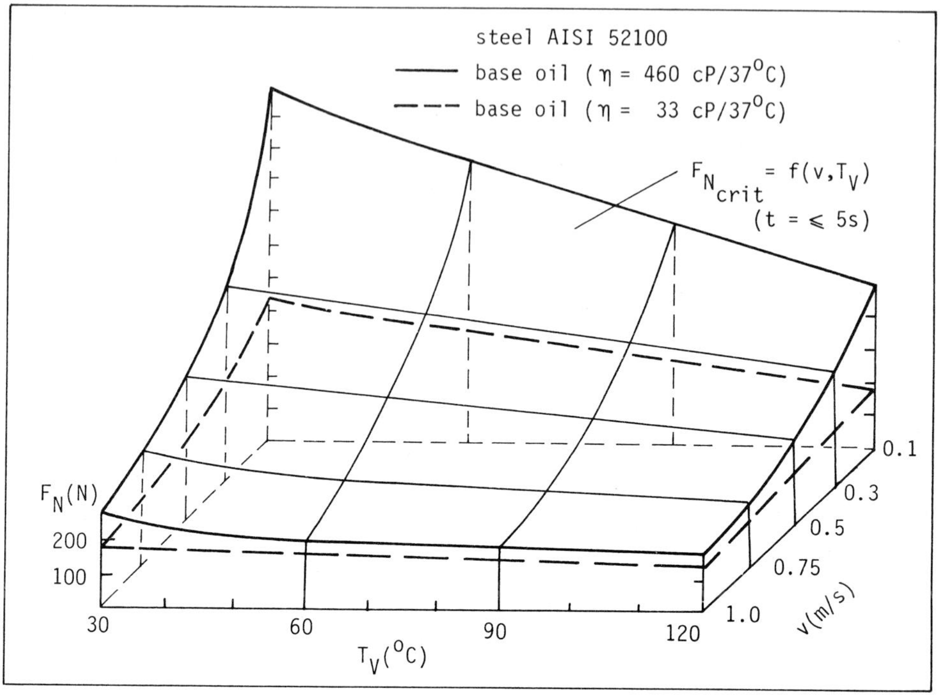

Figure 4.63 Influence of lubricant viscosity on the shape and location of the failure surface.

Finally, it should be mentioned that besides the influence of the lubricant's viscosity and chemical reactivity, the contact geometry, i.e., the type of Hertzian contact, like "point" contact or "line" contact and the radii of curvature influence the failure limits (Ref. 4.186, 4.187). In addition, the type of motion, e.g., the slide/roll ratio is also of great importance in that for pure sliding the failure phenomena are most pronounced whereas under pure rolling conditions scuffing failure is unlikely to occur.

In conclusion, it appears that three conditions must be met for (scuffing) failure to occur for a tribo-mechanical system which operates initially in the regimes of full hydrodynamic or elastohydrodynamic lubrication:

(A) The breakdown of the full-fluid hydrodynamic or elastohydrodynamic film.

(B) The breakthrough of (protective) surface layers.

(C) The occurrence of strong metal-metal adhesion bonds.

Depending on the type of tribo-mechanical system and operating conditions, the sequence of events (A), (B), (C) leading to failure may occur as a "stepwise" process, as explained in Figure 4.59 for point contact conditions and relatively low sliding velocities. On the other hand, for other experimental conditions, e.g., higher sliding velocities, the three conditions (A), (B), (C), may apparantly coincide in a "collapse" type of lubricant failure.

It follows that from a systems point of view the failure limits depend on the operating variables (load F_N, velocity v, temperature T, time t) as well as on the structure of the system (elements of the system A, properties of elements P, interrelations between the elements R) characterized symbolically as:

failure limits = f(operating variables; system's structure)

$$\boxed{\text{failure limits} = f\,(X;\,S)}$$

where $X = \{F_N, v, T, t\}$

$S = \{A, P, R\}$

With the diagrams presented in this section, a physical picture of the failure of lubricant film as well as a basis for a compilation of the important system's parameters that govern the limits of lubrication is obtained. Together with the results of the foregoing sections describing the various tribological processes, these data will be used in studying the influence of the tribological processes on the structure and function of mechanical systems.

5 Influence of tribological processes on the structure of mechanical systems

5.1 General Considerations

Having studied in some detail the various tribological processes, i.e., the contact, friction, wear and lubrication processes, we are now in a position to discuss the influences of these processes on the behaviour of whole tribo-mechanical systems. According to the systems concept outlined in Chapters 2 and 3, the description of a tribo-mechanical system must include:

(I) The structure of the system, characterized formally by the set

$$S = \{A, P, R\}, \text{ where}$$

$\{A\}$: the elements, i.e., the material components of the system,
$\{P\}$: the relevant properties of the elements of the system,
$\{R\}$: the interrelations between the elements.

(II) The function of the system, consisting of:
$\{X\}$: the inputs of the system,
$\{Y\}$: the outputs of the system,
T : the functional transformation of the inputs into the outputs.

In engineering applications, the interest is often mainly concentrated on the functional behaviour of the system under consideration. In this connection attention should be drawn to the great influence of the system's structure on the functional performance of a system. It was explained in Chapters 2 and 3 that it is only permitted to concentrate exclusively on the functional aspect of a given system if this system possesses a constant invariable structure. Only in these cases - which, for instance, are often given for electronic systems - is it possible to describe the behaviour of the whole system in terms of {X} - {Y} input-output relations and to model the behaviour of the system by mathematical or symbolic means, like the network techniques of system modelling. Most tribo-mechanical systems, however, possess structures which may change during the functional performance of the system due to the tribological processes described in the foregoing chapter. Therefore, in the description of tribo-mechanical systems it must be taken into account that these systems generally possess a dynamic system's structure as well as a dynamic system's function.

According to the definition of the structure of a system, the changes in a tribo-mechanical system's structure may involve:

(i) The creation or annihilation of elements of the system.

Examples are the generation of "third bodies" in the tribological contact of dry surfaces in the form of "friction polymers" or reaction products or, on the contrary, the destruction of lubricant films.

(ii) Changes in the interrelations of the elements.

Examples are the variation of lubrication modes in following the conditions of the Stribeck curve, or changes in wear mechanisms.

(iii) Changes in the properties of the system's elements.
Examples are the tribo-induced changes in surface topography or in the hardness and strength properties of sliding or rolling surfaces.

These aspects are not only important in the analysis of existing tribo-mechanical systems, they should also be considered carefully in the design of new systems (see Sections 5.4 and 6.6). Clearly, the details of the dynamic changes of the system's structure depend on the type of tribo-mechanical system under consideration as well as its functional purpose and its operational variables. Although tribo-mechanical systems differ very broadly in design and function (as can be seen from the compilation of tribo-mechanical systems given in the Appendix A), it should be attempted to study the basic phenomena of the tribo-induced changes of the structure of tribo-mechanical systems which are of general importance. The main features of the above aspects (i) and (ii) have already been discussed in Sections 4.4 and 4.5.6 under the headings of "wear processes" and "the limits of lubrication". In the following the basic aspects of class (iii) of the tribo-induced changes of the structure of mechanical systems are studied.

5.2 Changes in System's Elements Properties

In considering the possible tribo-induced changes of the properties of the elements of a tribo-mechanical system it should be borne in mind that all interacting components of a system may change their relevant properties to some extent during the functional performance of the system. In Chapters 2 and 3 it has been described that the central part of tribo-mechanical systems consists, in general, of four components as indicated schematically in Figure 5.1 (see also Appendix B):

(1) The first machine element } (friction and wear couple)
(2) A second machine element }
(3) The lubricant
(4) The atmospheric environment

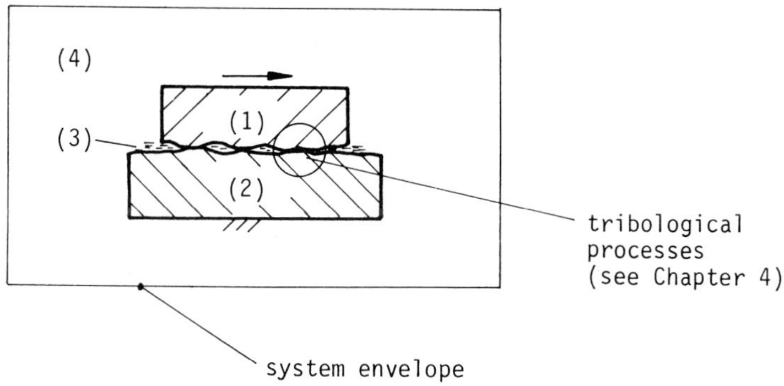

Figure 5.1 Schematic representation of the main components of tribo-mechanical systems.

The possible changes of the lubricant (3) and its influences on the structure and function of tribo-mechanical systems were discussed in some detail in the section on the limits of lubrication (Section 4.5.6). The influences of the element (4), the atmospheric environment and its changes, have in the past often not been paid the attention they deserve. These changes, which may include variations of the chemical composition of the environment and its humidity, have a strong influence on all tribo-chemical processes (see Sections 4.4.5 and 4.5.5). In these processes also the interactions between the environmental atmosphere (4) and the lubricant (3) must be taken into consideration.

The tribo-induced changes of the properties of the elements (1) and (2) - the main elements involved in the transmission of the functional system's inputs and outputs - may be classified, according to the results of Chapter 4, broadly into the following groups:

(a) changes in surface topography,
(b) changes in surface composition,
(c) changes in surface strength properties.

In the following, some of the main features of these tribo-induced changes of the properties of elements (1) and (2) of mechanical systems will be discussed. As in Chapter 4, the discussion is restricted to the basic physical phenomena of the subject matter.

5.2.1 Surface topography changes

In all tribo-mechanical systems in which solid-solid contact of surfaces in relative motion occurs, it must be expected that the surface topography may undergo some changes during the functional operation of the system. These changes may be due to two effects:

(i) plastic contact deformation,
(ii) wear processes.

Consider first the surface topography changes due to plastic contact deformation. According to Section 4.2.1 the type of contact deformation can be distinguished by the plasticity index ψ:

$$\psi = \left[\frac{E}{H}\right]\left[\frac{\sigma^*}{\beta}\right]^{1/2}$$

E : Combined elastic modulus
H : Hardness
σ^* : Standard deviation of peak height distribution
β : Mean radius of asperities

If for a given system, the calculation of ψ leads to a value of $\psi < 0.6$ the probability of plastic contact deformation is very small. If on the other hand, $\psi > 1$, permanent surface topography changes due to plastic asperity flow must be expected. The plastic contact deformation changes the surface roughness data, such as R_a or R_z (see Section 4.2.1) and modifies the shape and the characteristics of the asperity height distribution curve and the auto-correlation function of the surface profile.

It is well established that many engineering surfaces have a Gaussian height distribution. If the initial surface preparation technique involves a large number of repeated events which occur randomly over the surface, it follows immediately from the Central Limit Theorem of statistics (Ref. 5.1) that the topography of the surface will tend towards a Gaussian height distribution. Consequently, any plastic contact deformation changes the initial height distribution curve. The effect of plastic contact deformation on the changes of the surface contact topography is illustrated in Figure 5.2.

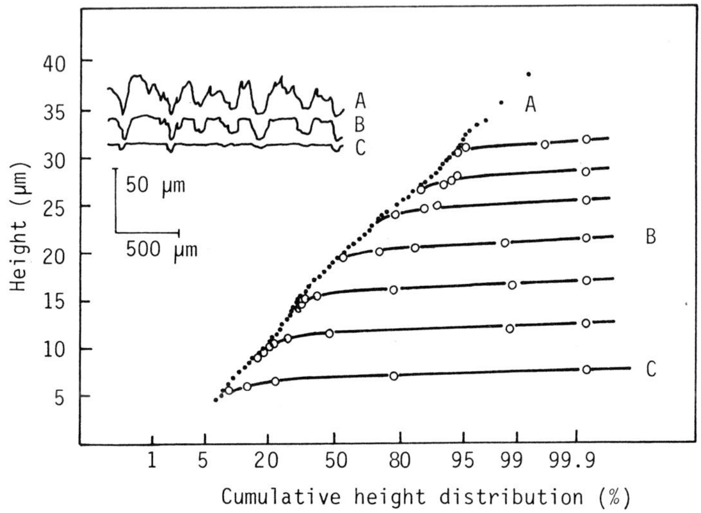

Figure 5.2 Cumulative height distributions showing effect of pressing an initially rough surface (Ref. 5.1).

In this diagram, the cumulative height distribution curves of the initial and the plastically deformed surfaces are plotted on probability paper (Ref. 5.1). (In this diagram paper, a Gaussian distribution is represented as a straight line.) It can be seen that the initial Gaussian height distribution (A) is increasingly disturbed with increasing severity of the contact deformation process (B), (C). Since most engineering surfaces initially possess a Gaussian surface height distribution, the curves shown in Figure 5.2 appear to represent basic features of the surface topography changes due to plastic deformation.

The other group of tribo-induced surface changes are the changes due to wear processes. The appearance of worn surfaces has been described in

Section 4.4 (see Table 4.3). For the changes of the surface topography due to wear, depending on the initial surface finish there are two possibilities:

(i) the wear processes may roughen smooth surfaces,
(ii) the wear processes may smooth out rough surfaces.

There is a general consensus that the surfaces of moving machine parts should not be too "smooth". For example, for the design of pistons, a suitable surface finish for the cylinder bores of R_a = 0.4 to 0.6 μm has been suggested (Ref. 5.2). (These figures should not be generalized because they depend on the system under consideration).

As in the case of plastic contact deformation, the wear-induced changes of surface topography can be conveniently illustrated by considering the cumulative height distribution of surface asperities (Ref. 5.3). In Figure 5.3 the effect of wear on the initially Gaussian height distribution is shown (Ref. 5.1).

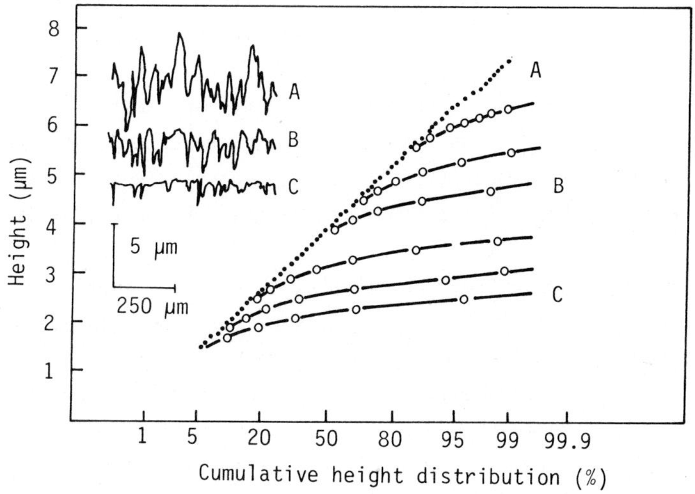

Figure 5.3 Cumulative height distributions showing the effect of wear on the initially Gaussian height distribution of a bead-blasted surface (Ref. 5.1).

The six non-Gaussian distributions represent, from (A) to (C), progressive stages in the wearing process. It can be seen that almost immediately the top 20% of the surface was altered, the rest remained completely unaffected. This bent distribution, and the corresponding changes in bearing curve are typically associated with lightly worn surfaces. As wearing progressed, the transition percentile between the new topography created by the wear process and the initial surface topography moved steadily lower until the original texture had been almost totally replaced.

In mechanical engineering, the surface finish and the wear-induced changes of surface topography are considered to be of great importance for the functional performance of moving mechanical components (Ref. 5.4, 5.5, 5.6). Generally speaking, the wear-induced changes of surface topography may have an indifferent, beneficial, detrimental or catastrophic influence on the functional behaviour of the whole tribo-mechanical system. Some aspects of the detrimental influence of surface topography changes have been discussed in connection with the limits of lubrication (Section 4.5.6). The beneficial aspects of surface topography changes lead to the well known running-in effect of bearing surfaces (Ref. 5.7, 5.8, 5.9).

Consider, as an example of the running-in effect, a simple tribo-mechanical disc-on-plane sliding system. The influence of a running-in process on the tribological behaviour has been studied in two steps. Firstly, the Stribeck curve of the sliding system lubricated with a pure paraffinic mineral oil has been measured together with the no-contact time fraction determined by electrical contact-resistance measurement techniques. These curves, valid for the initial surface roughness, were determined for constant values of load and temperature by varying the sliding velocity from high to low values within a few seconds. (With this procedure, Stribeck's curve is run through from right to left with neglible surface-roughness changes.) Secondly, after a "running-in" process has been performed under conditions of boundary lubrication, the Stribeck curve and the no-contact time fraction have been determined again. The results of these measurements are shown in Figure 5.4. In addition, the surface-roughness characteristics before and after the run-in procedure are compiled in Figure 5.5.

In comparing the curves for the system with the initial surface roughness data (a), with those of the run-in system (b), the following main consequences of the running-in process can be noted:

(i) A reduction of the height of the highest asperity peaks indicating a greater conformity of the sliding surfaces and consequently, an increase of the lubricant gap between the sliding surfaces.

(ii) A shift of the minimum of the Stribeck curve to the left, i.e., an extension of the regime of hydrodynamic lubrication.

(iii) A lowering of the friction value at the minimum of the Stribeck curve.

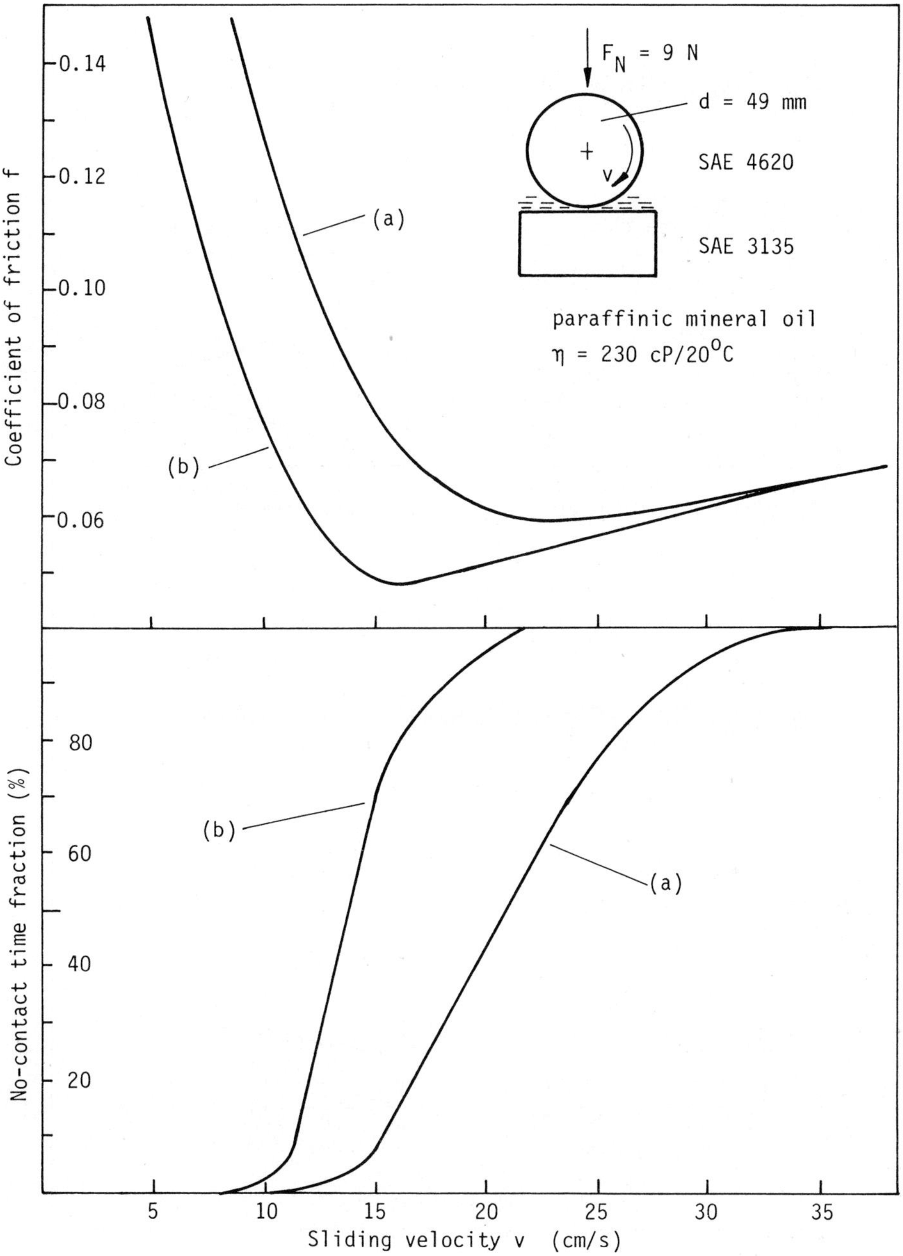

Figure 5.4 Stribeck curve and no-contact time fraction before (a) and after (b) running-in.

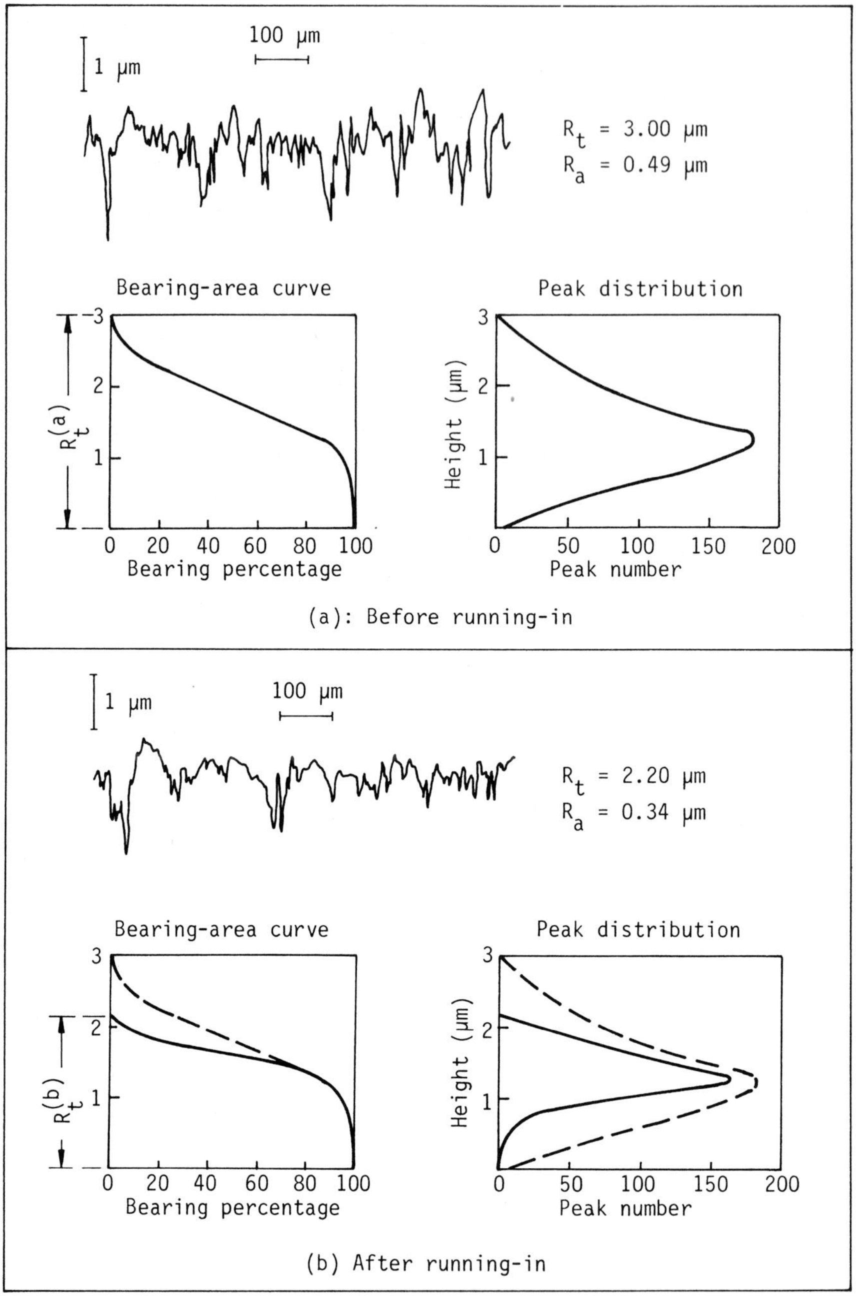

Figure 5.5 Surface roughness data before (a) and after (b) running-in.

Clearly, the extension of the hydrodynamic regime as a consequence of the running-in process enhances the "operating security" of the system.

In mechanical engineering, owing to the great importance of running-in procedures for the functional behaviour of moving machine elements, it is attempted to optimize the running-in procedure in order to obtain properly run-in surfaces within a minimum of running-in duration. For example, in the determination of running-in schedules for Diesel engines, the schedules have been optimized with respect to both reliability and economy through measuring the amount of wear during the running-in procedure by means of sensitive radionuclide measuring methods (Ref. 5.10). Detailed studies were made on the influences of operating conditions, like load and speed, on criteria like engine power, specific fuel and oil consumption and surface state of piston ring and liner. As a typical example, in Figure 5.6 the bearing area curves of the cylinder liner of a Diesel engine at different stages of running-in are shown.

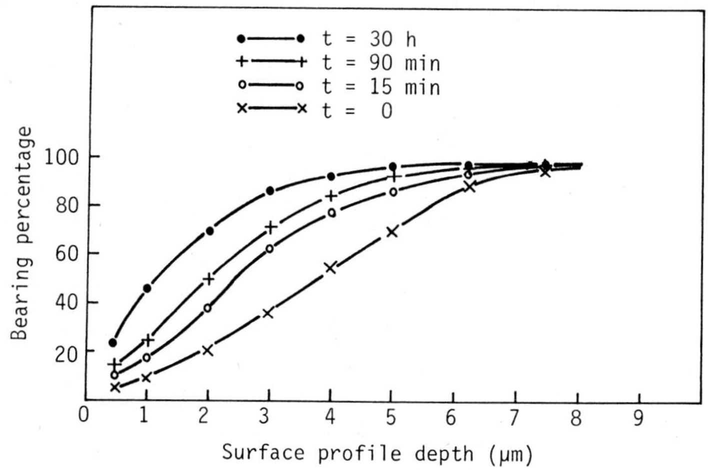

Figure 5.6 Bearing area curves of the cylinder liner of a Diesel engine at different stages of running-in (Ref. 5.10).

The surfaces are properly run-in if a certain bearing-area percentage at a certain profile depth is obtained and a stabilized (minimum) wear rate is reached. The marked influence of the operating conditions on running-in are illustrated in Figure 5.7.

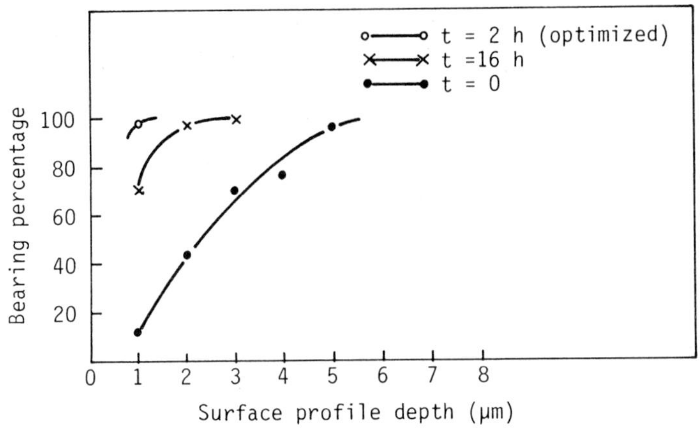

Figure 5.7 Bearing area curves of the cylinder liner of a Diesel engine after different running-in procedures (Ref. 5.10).

It can be seen that with an optimized running-in schedule the required surface quality was reached after a running-in duration of 2 hours, whereas in a conventional non-optimized running-in procedure for a comparable surface quality a running-in duration of more than 16 hours was needed.

5.2.2 SURFACE COMPOSITION CHANGES

This aspect of the tribo-induced changes of the structure of a mechanical system is extremely complicated both from the theoretical and experimental point of view. Theoretically, owing to the nature of surfaces (see Figure 4.5) and the complexity of the tribological materials-lubricant-atmosphere interactions as discussed in Chapter 4, it appears extremely difficult to develop models for the actual dynamic composition of tribologically stressed surfaces. Experimentally, owing to the inaccessible contact interface, it is not generally possible to study "in situ" the actual composition of surfaces in contact. Nevertheless, some progress has been made in recent years in applying the new physical surface-analytical tools, like Auger electron spectroscopy,or ESCA (see Section 7.6), in connection with sophisticated surface-preparation techniques. In the following, some of the basic aspects of tribo-induced changes of surface composition are discussed for systems of increasing complexity.

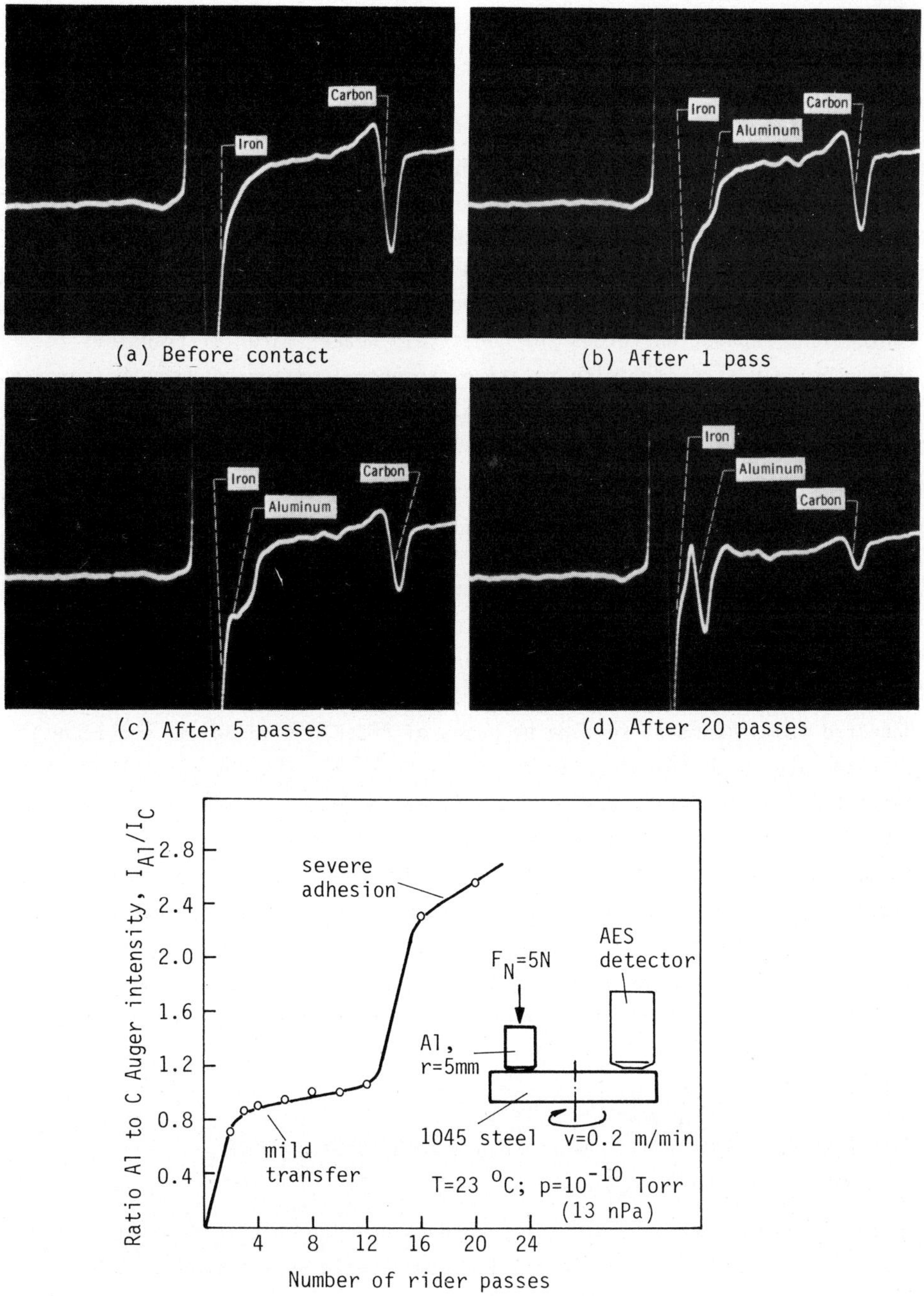

Figure 5.8 Aluminium adhesive transfer to steel surface during sliding (Ref. 5.11).

(a) SURFACE-COMPOSITION CHANGES OF A SLIDING SYSTEM UNDER ULTRA-HIGH-VACUUM CONDITIONS.

In this case, the tribological system consists of only two partners and the tribo-induced surface composition changes are mainly the material transfer from one partner to the other and vice versa. The physical causes for this material transfer, i.e., the adhesion mechanisms, have been explained in Sections 4.2.2 and 4.4.4. As mentioned there, in adhesion experiments of metal-metal pairs, it has been observed that under the action of pure normal contact and separation forces the cohesively weaker material transfers to the cohesively stronger. With relative tangential motion between the surfaces, as in sliding, rolling, or oscillating, interfacial materials transport becomes more complex. Under such conditions, plastic deformation, shear and high surface temperatures due to frictional heating are some factors which must be given attention.

In the last few years it has become possible to study the tribo-induced changes in the composition of the uppermost layers of surfaces by means of Auger electron spectroscopy (AES). As an example, the changes of the composition of a steel surface sliding against an aluminium counterface will be discussed (Ref. 5.11). Before sliding contact, iron and carbon peaks were detected from the steel surface by means of AES. In sliding, it was found that the aluminium transfers to the steel surfaces. The degree of the "contamination" of the steel surface with Al was detected by measuring the ratio of the aluminium to carbon AES peak intensities. This ratio, as a function of the number of passes of the Al rider over the steel surface, is shown in Figure 5.8. It can be seen that a mild transfer regime is followed by a severe transfer regime. This example shows the importance of surface-composition changes on the behaviour of tribo-mechanical systems: it is the severe adhesive wear, indicated in the right-hand side of the graph of Figure 5.8, which often causes the catastrophic failure of components of mechanical systems.

(b) SURFACE COMPOSITION CHANGES OF A SLIDING SYSTEM UNDER ATMOSPHERIC CONDITIONS.

In these cases, surface composition changes due not only to materials transfer but also to the interaction of the atmospheric environment with the sliding surfaces activated by the frictional energy, take place. Important examples are the tribochemical oxidation processes in dry sliding of steel surfaces. The corresponding changes in surface composition have often been investigated in studying the chemical composition of the surface debris.

Fink and Hoffmann observed in 1932 in a classic paper that under conditions of rolling with slip, the surface debris consisted of Fe, Fe_2O_3 and Fe_3O_4; FeO was not detectable (Ref. 5.12). In another fundamental paper, Dies in 1939 studied the connections between the composition of the debris and the wear of a steel specimen (0.04% C) sliding under atmospheric conditions at v = 1m/s against a hardened steel (0.86% C, 1.64% Cr) disc (Ref. 5.13). In Figure 5.9, the results obtained by Dies as a function of the contact pressure are shown.

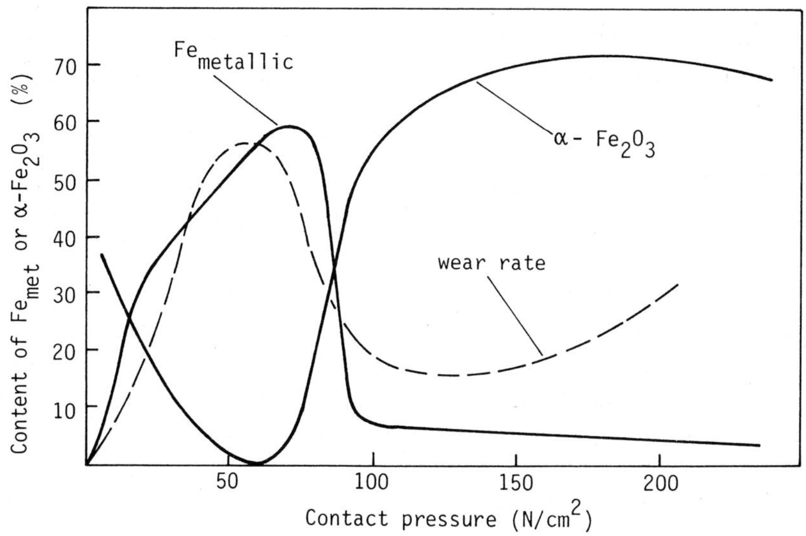

Figure 5.9 Composition of wear debris as function of contact pressure (Ref. 5.13).

The broken curves represent the measured wear rate in relative units. (The characteristic shape of the wear curve is discussed in the next section, in connection with the tribo-induced changes of surface strength properties.) The full curves of Figure 5.9 show the variation of the composition of the debris with increasing contact pressure. It can be seen that at the minimum of the wear rate, debris contained mainly α - Fe_2O_3 whereas at the maximum of wear, the debris consisted mainly of metallic iron.

Similar changes have been reported to occur also as a function of the sliding velocity (Ref. 5.14, 5.15). From this work it may be concluded that a tribo-chemical wear mechanism operates at low contact pressures and low sliding velocities whereas at higher values of this operating variables metal-metal adhesion becomes the dominant wear process. Quinn and Wooley (Ref. 5.16) were able to show that under mild conditions, the oxide Fe_2O_3 was generated but that a transition to Fe_3O_4 occurred as the load or velocity was increased above certain levels.

(c) SURFACE COMPOSITION CHANGES OF A SLIDING SYSTEM UNDER LUBRICATED CONDITIONS.

In these situations, the whole complex of solid-solid-lubricant-atmosphere interactions must be taken into consideration. Some of the main physico-chemical aspects of these processes have already been discussed in connection with the boundary lubrication mode (see Section 4.5.5). One of the advantages of the presence of a lubricant in a tribo-mechanical system is the possibility of influencing the actual composition of the sliding surface by lubricant additives. Through the interfacial tribo-chemical reactions, beneficial surface layers may be build up.

Obviously, as in dry sliding, oxygen and water play extremely important roles for the "in situ" composition of tribologically stressed surfaces and, consequently, for their functional behaviour. Increasing oxygen concentrations seem to have a favorable effect on the amount of wear, as well as on the load-carrying capacity of sliding surfaces. For systems lubricated with plain hydrocarbon lubricants, as the oxygen availability increases, the inorganic wear products from the sliding surfaces change from iron and iron carbide FeC to the iron oxides Fe_3O_4 and α - Fe_2O_3. If the lubricant contains "lubricity" agents, like fatty acids, carboxylate "soaps" are formed when oxygen or water is present. The inorganic films formed "in situ" from extreme pressure agents on metal surfaces, such as sulfides, chlorides, and phosphides, provide lubrication at very high temperatures (see also Section 4.5.5). These tribo-chemically formed surface films are more stable than any physically or chemically adsorbed film. Reacted films need not to be tribo-chemical, but the rate of reaction can be enhanced by the exposure of surface and the rubbing temperature.

The actual surface composition is not well known, and a study of the literature shows that only in a few cases have the actual compounds been identified on worn surfaces. This inability to define the chemical nature of the surface species has been due mainly to the facts that the layers are extremely thin (about 10^{-3} µm) and are frequently amorphous. Recently, in applying the newly developed sensitive X-ray photoelectron spectroscopy technique (ESCA or XPS, see Section 7.6) examinations of steel surfaces before and after wear testing have been performed (Ref. 5.17, 5.18). In these studies, the surface composition after the action of typical anti-wear (AW) additives, like zinc dialkyldithiophosphate (ZDDP) and typical extreme pressure (EP) additives, like dibenzyl disulphide (DBDS) have been studied. The results of these studies indicate that of the elements zinc, sulfur, phosphorus, carbon and oxygen in the AW additive, only sulfur underwent a detectable change in chemical nature during wear testing of the steel surface (Ref. 5.17). The sulfur surface species produced was identified as a metal sulphide. The results suggest that this surface sulphide is associated with boundary lubrication (see Section 4.5.5). Likewise, in the case of the EP additive DBDS, the rubbed surfaces had thick discontinuous layers of sulphide (Ref. 5.18). These findings are compatible with those of Coy and Quinn (Ref. 5.19) who also observed thick sulphide layers for the high EP region.

5.2.3 SURFACE STRENGTH PROPERTIES CHANGES

There are two possibilities of tribo-induced changes of the strength properties of interacting moving surfaces:

(i) an increase in strength as a consequence of effects like "strain hardening" due to contact stresses and the generation of internal stress fields,

(ii) a decrease in strength due to the "softening" effect of friction induced heat in the contact interface.

In this connection the question of the tribologically relevant strength properties arises. From a discussion of the various attempts to establish correlations between wear behaviour and mechanical properties - neglecting the "interaction effects" of adhesion and tribo-chemical reactions - it has been concluded that many of the relevant parameters can be expressed in terms of the ratio of the elastic modulus E to the hardness H (Ref. 5.20). It has already been explained above that the ratio E/H is part of the plasticity index ψ which is a satisfactory criterion by which to determine in which regime a given contact deformation process lies. It may therefore be sufficient here to restrict the discussion to the tribo-induced changes of E and H. The elastic modulus, E, of a polycrystalline metal is physically speaking a "structure-insensitive property", and is only little affected by both defects and impurities and temperature and strain rate (Ref. 5.21).

On the contrary, the indentation hardness, H, of materials is a markedly "structure-dependent property". It follows that the indentation hardness may be the most relevant strength property characterizing the tribo-induced changes of surface strength properties.

Burwell and Strang (Ref. 5.22) have shown that steel-steel sliding couples exhibit a transition from mild to severe wear when the apparent contact pressure reaches a value of about one-third of the hardness of the softer material. As this pressure corresponds quite closely to the uniaxial bulk flow stress of the metal, it was suggested that the wear transition was due to the interaction of plastic zones beneath contacting asperities and the onset of bulk plastic deformation. In studying the tribo-induced changes of the hardness of sliding surfaces, however, the whole stress field of normal and tangential stresses must be taken into consideration (Ref. 5.23). Since the tangential stresses depend on the frictional behaviour of the system, those properties which determine the friction (like surface contamination, atmospheric environment, etc) must also be included. A characteristic example of the changes of the hardness of sliding surfaces with relation to the various influencing factors is given in Figure 5.10. The figures shows cross-sections of pins of pure iron, which have been slid against a disc in dry synthetic air of different pressure, together with micro-hardness data (Ref. 5.24, 5.25). Under conditions of normal atmospheric pressure through tribo-chemical reaction of the iron with the oxygen an oxide layer is formed as can be seen in the upper part of the micrograph of Figure 5.10. The frictional shear stresses act mainly in the oxide layers with the result that the sub-surface layers are strain-hardened only to small degree ($H = 2000 \longrightarrow 3000$ N/mm^2). If, however, under the same operating conditions of contact pressure, sliding velocity, ambient temperature and sliding distance, the tests are performed in a vacuum of 10^{-3} Torr a different behaviour is found. Here, a lesser amount of oxygen for oxide layer formation is available and higher friction coefficient values are measured. The combined action of the normal forces and the increased tangential friction forces act mainly in the metallic parts and lead to a considerable increase in the sub-surface hardness ($H = 2000 \longrightarrow 6000$ N/mm^2).

The importance of the changes of surface hardness and surface composition on the wear behaviour of a tribo-mechanical system in dry sliding is illustrated in Figure 5.11. In this graph, the wear coefficient K' of 0.52 percent carbon steel as a function of the load F_N is plotted (Ref. 5.26, 5.27). The initial hardness of the specimen was 268 DPN. At low loads mild

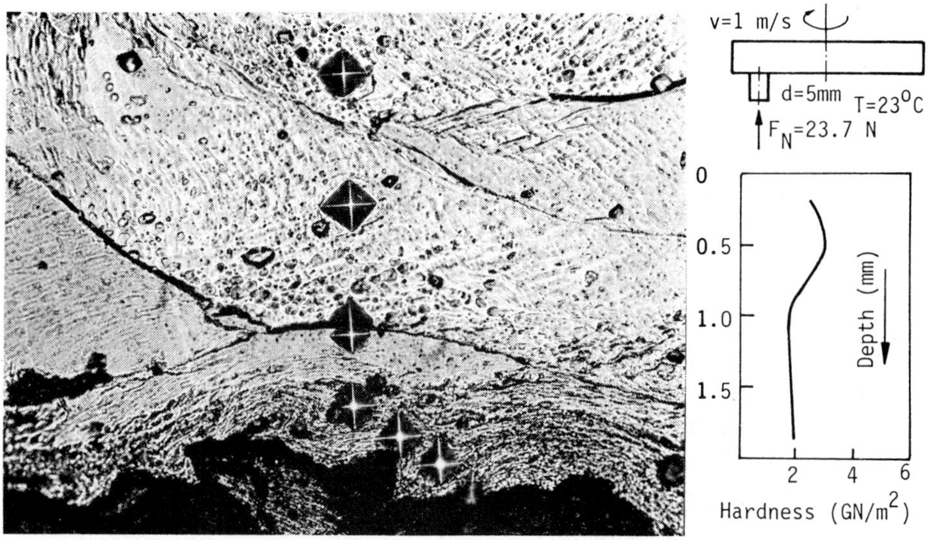

(a) atmospheric pressure : p = 760 Torr (0.1 MPa)
friction coefficient : f = 0.55 - 0.64
wear coefficient : K = 0.00002

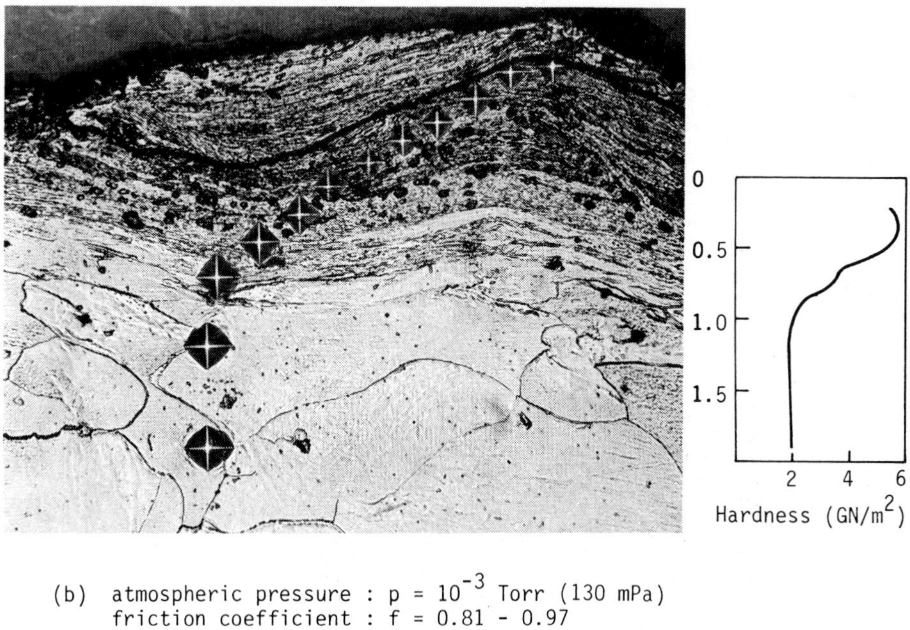

(b) atmospheric pressure : p = 10^{-3} Torr (130 mPa)
friction coefficient : f = 0.81 - 0.97
wear coefficient : K = 0.00058

Figure 5.10 Tribo-induced hardness changes.

wear occurs until at F_N = 1N the transition Tr 1 to severe wear occurs. The severe wear mode persists up to a load of about F_N = 25 N. At this point, a second transition Tr 2 occurs when the wear mode reverts. It was found, from metallography and micro-hardness tests, that above the Tr 2 transition the surface layers of the specimens were hardened by a phase change. The value of the Tr 2 transition load was determined for a number of sliding velocities and for steels of various carbon contents, and in each case the value of the flash temperature was calculated. Bearing in mind a number of uncertainties in these calculations, the overall pattern of results suggests that the surface hardening is caused by a martensic transformation, and that it is caused by the flash temperatures and their duration at the true area of contact. The overall effects illustrated in Figure 5.11 have been confirmed by Uetz and co-workers (Ref. 5.28). The results clearly indicate the importance of the combined effects of surface composition changes and surface strength changes on the tribological behaviour of sliding systems.

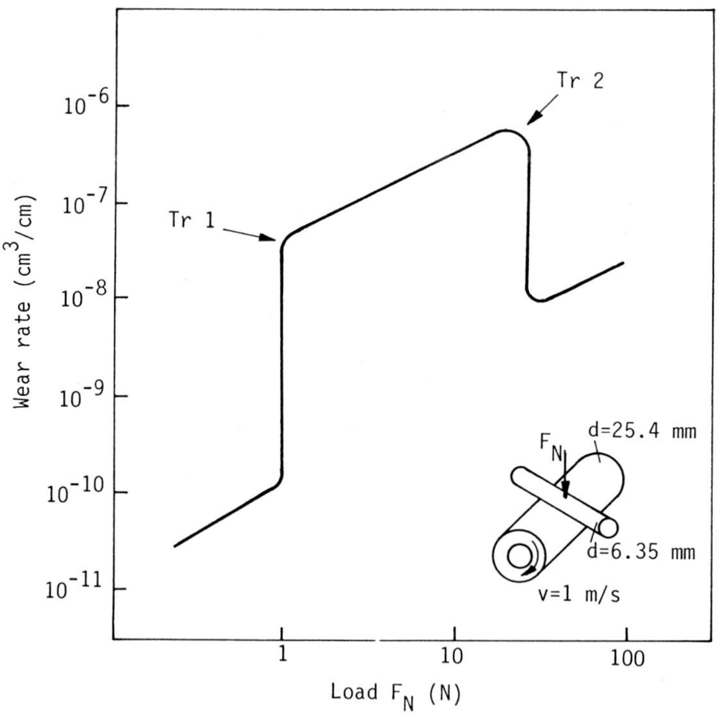

Figure 5.11 Wear modes as function of load (Ref. 5.26).

5.3 Material Losses

If, in a tribo-mechanical system it is not possible to eliminate wear processes (see Section 5.4) or to mitigate these processes to a "zero level" (see Section 6.6) the tribological processes between two moving surfaces generally lead to losses of material from the system. Phenomenologically, different regimes of material losses may be distinguished. Denoting the amount of the material loss by the output symbol Z^M, three simple time relations for Z^M may be expected which are indeed observed experimentally.

Firstly, the probability of the occurrence of elementary wear events may decrease with time. This case is given, for instance, if through changes of surface topography the interaction rate of surface asperity collisions decreases. In this "running-in" period the system self-accommodates. In the simplest case, the amount of loss-outputs will be given as a square-root function of time

$$\frac{dZ^M}{dt} = \frac{C_1}{Z^M} \Longrightarrow Z^M(t) = \left[2C_1 t\right]^{1/2}$$

Secondly, the structure of the system may acquire a relatively stable behaviour under the action of the tribological processes. In this case the probability of wear events remains constant. This "steady-state" situation is characterized by a constancy of the material-loss output per unit of time without change in the tribological processes. It follows that

$$\frac{dZ^M}{dt} = C_2 \Longrightarrow Z^M(t) = C_2 t$$

Thirdly, when the amount of material-loss output reaches a certain value, a qualitative transition in the state of the system may occur consisting of a significant change in the tribological picture of the phenomena. The changes in the state of the system are of a directed nature and the increments in the wear processes in this regime are mutually dependent. Here large increments in one interval of time cause even larger increments in the next time interval. It follows that

$$\frac{dZ^M}{dt} = C_3 Z^M \Longrightarrow Z^M(t) = C_o e^{C_3 t}$$

In this case the worn material-loss output increases rapidly with time. Depending on the values of C_0 and C_3 self-acceleration of the process occur which may lead to catastrophic damage of the structure of the system and, accordingly, to "failure" of the whole system (see also Section 6.4).

The three modes of material-loss outputs discussed may follow each other in time as shown in Figure 5.12. In Figure 5.12 Z^M_{lim} denotes a maximum level of admissible material loss-outputs. At this level, the structure of the system has changed as a consequence of the material loss-outputs, in such a way that the functional input-output-relations - which are performed via the structure of the system - are severely disturbed (see Section 6.4).

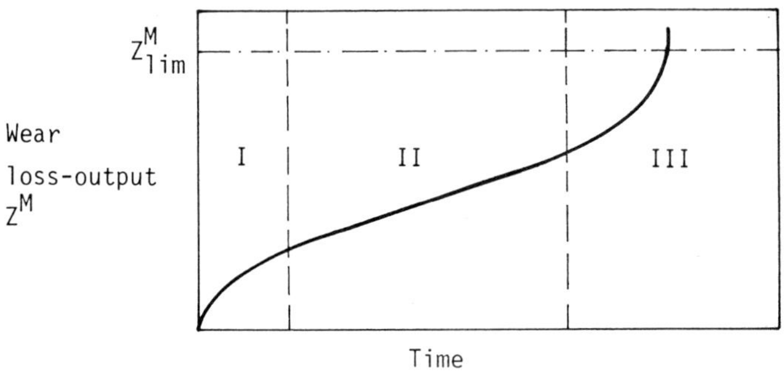

Figure 5.12 Schematic representation of the material loss-output curves of a tribo-mechanical system.

Relatively few workers in the field of tribology have attempted to derive correlations for wear-particle dimensions generated by the wear process. Rabinowicz proposed a correlation between mean adhesive wear, particle diameter and the "surface-energy criterion" which is the ratio of the work of adhesion between two metals to the hardness of the weaker material (Ref. 5.29). This theoretical approach has recently been modified (Ref. 5.20) and the following inequality has been suggested to describe the mean wear-particle diameter d

$$d \leqslant 15\ (E/H)^2 \quad (\text{nm})$$

E : elastic modulus

H : penetration hardness

Characteristics of wear regimes studied (schematic) :

Lubricated state
(Stribeck curve)

Dry state
(Transitions diagram)

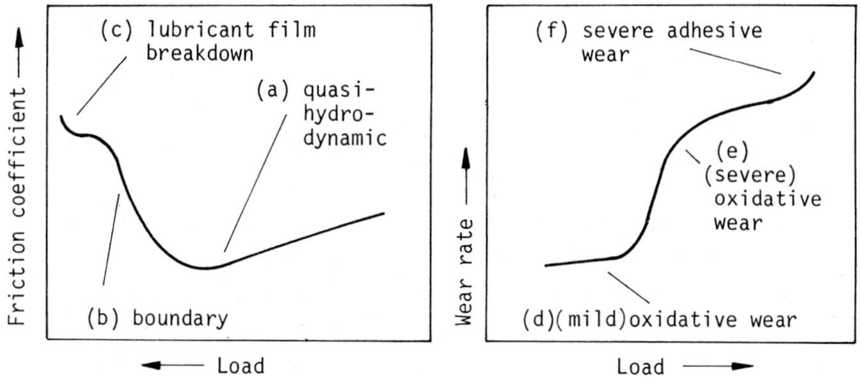

Regime	Particle description and major dimension, d	Surface description	Wear rate
(a)	Free metal particles; d < 5 μm	Varies between polished and very rough	→ Zero
(b)	Free metal particles; d < 15 μm	Stable, smooth layer with a few grooves	Low
(c)	Free metal particles; d < 150 μm	Ploughed with evidence of plastic flow and surface cracking	High
(d)	Red oxide particles, α - Fe_2O_3, d up to 150 μm	Ploughed with areas of oxides on the surface	High
(e)	Black oxide particles γ - Fe_2O_3, Fe_3O_4, FeO, d up to 150 μm	Ploughed with areas of oxides on the surface	High
(f)	Free metal particles, d up to 1 mm	Severely ploughed, gross plastic flow and smearing	Catastrophic

Figure 5.13 Characteristics of wear particles and tribo-induced surface features (Ref. 5.30).

The dependence on $(E/H)^2$ is assumed to be an upper bound, obeyed under conditions of clean surfaces. When normal laboratory conditions or more contaminated environments pertain, the dependence for particle size reduces empirically to (E/H).

Experimentally, the following findings on the chemical composition and the size of the material loss debris appear to have been observed quite generally for sliding systems operating under normal atmospheric conditions:

(i) In the so-called mild wear regime, the worn debris consists of small particles with a diameter of less than about 5 μm. These particles are formed mainly in tribo-oxidative wear processes and for sliding metal couples often consist of oxides. For example, the wear debris of iron, cobalt, magnesium sliding against themselves under dry conditions consists of α - Fe_2O_3, CoO and MgO. The formation of these oxides can be explained in accordance with the thermodynamics of the oxidation processes if the friction induced temperatures and the corresponding data of free energy of the materials are used.

(ii) In the so-called severe wear regime metallic particles ranging in size up to approximately 1 mm are found. These particles are mainly observed under severe metallic adhesion processes. There is some experimental evidence that for like metallic sliding couples the size of loose wear particles is larger than for unlike metal-metal pairs.

The details of the characteristics of the wear particles generated at the interface of lubricated and dry sliding concentrated AISI 52100 steel contacts have been studied (Ref. 5.30) in laboratory tests using the Ferrograph technique described in Section 7.5. In Figure 5.13 the characteristics of the worn material loss output and the wear-induced surface changes are compiled. As can be seen, each regime produced wear particles of characteristic morphology and composition. Since all the types of particles have been found likewise in the lubricating oil of field operating machines, the compilation of Figure 5.13 appears to represent a good survey of the nature of the wear-induced materials losses and the corresponding changes of the operating surfaces of mechanical systems. The data of Figure 5.13, which are valid for condition of pure sliding, show that from the study of the size distribution and the shape and composition of wear particles important information on the operating wear modes and the actual condition of the structure of a tribo-mechanical system can be obtained (Ref. 5.31). Based on this information, a prognosis of the imminent behaviour of machinery can then be undertaken.

5.4 Solutions for No-Wear Conditions

In the previous sections, we studied some of the basic aspects of the changes of the structure of mechanical systems occurring as a consequence of the tribological processes between the moving components of the system. Now, turning from diagnosis to therapy, the question arises how to avoid detrimental effects and how to realize tribo-mechanical systems with unchanging structures thus providing the prerequisite for a proper functional behaviour of the whole system. The results of the previous sections have shown that in almost every case the detrimental changes of the system's structure are due to the interfacial wear processes. It follows that the wear-induced changes of the system's structure are eliminated if a suitable interfacial element entirely prevents the solid-solid contact between the moving components of a mechanical system. There are four basic engineering solutions to this problem, as illustrated schematically in Figure 5.14.

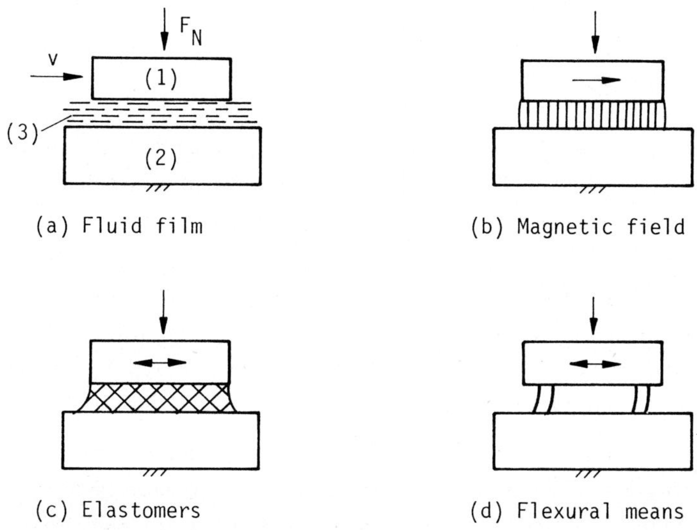

Figure 5.14 Schematic representation of solutions for no-wear conditions.

The most common method for the elimination of wear processes from a tribo-mechanical system (with the exception of surface-fatigue wear or, perhaps, fluid erosion or cavitation) is by the use of liquid or fluid lubricants, Figure 5.14 (a). In the cases where a converging gap exists, the relative motion of the elements (1) and (2) drags the fluid into this gap thus generating a load-carrying pressure within the fluid film. Alternatively, the pressure may be provided by an external pumping system as in a hydrostatic or aerostatic bearing. Another possibility, indicated in Figure 5.14 (b), is the use of magnetic or other repulsive force fields carrying the normal load and separating the moving surfaces (1) and (2). If the relative motion between the elements (1) and (2) is of fairly small amplitude, the solutions shown schematically in Figures 5.14 (c) and (d) may be applicable. Here, the moving surfaces are separated by an elastically deforming interfacial element (3), and the resistance to a relative motion between (1) and (2) is given by the internal deformation (internal friction) of element (3).

The principles shown in Figure 5.14 have the advantage that they eliminate wear processes altogether from a tribo-mechanical system thus providing an "invariant structure" of the system. Due to their great importance, the main features of the principles illustrated in Figure 5.14 will be discussed briefly.

5.4.1 Fluid films

The advantages of hydrodynamic lubrication as described in Section 4.5.2 are utilized not only in bearing applications but also in other tribo-mechanical systems like hydrodynamic seals or in manufacturing processes like hydrodynamic lubricated wiredrawing systems. Besides the self-acting hydrodynamic lubrication mode, hydrostatic or aerostatic lubrication is applied in externally pressurized bearings. Although these hydrostatic bearings may require relatively expensive and complex fluid supply equipment, they offer the following advantages:

(i) No solid-solid contact, i.e. no wear at any operating velocity (including $v = o$) or load for all types of relative motion.
(ii) Low friction (determined by the internal friction of the fluid).
(iii) Predictable functional performance (see Section 6.1).

In Figure 5.15 a typical hydrostatic thrust bearing is shown schematically.

The bearing pad, in this instance the lower member, contains a recess or pocket with a depth considerably greater than the lands or sills. (Sometimes the pad is above, e.g., on skids for moving heavy machines on plates let into the floor.) As fluid is supplied to the recess at a high pressure a pressure distribution builds up, whose integral must equal the load which is applied to the bearing. Figure 5.15 indicates that hydrostatic bearings can operate with a uniform film thickness h = const (in contrast to hydrodynamic bearings) and also with no relative tangential motion. Thus, in the Reynolds equation, treated in Section 4.5.2, the right-hand side equals zero (since v_o, v_1, $\frac{\partial h}{\partial t}$ may be put equal to zero). The Reynolds equation then reduces to

$$\frac{\partial^2 p}{\partial x^2} + \frac{\partial^2 p}{\partial z^2} = 0$$

This is Laplace's equation in two dimensions which must be solved for the particular geometry of the bearing under consideration.

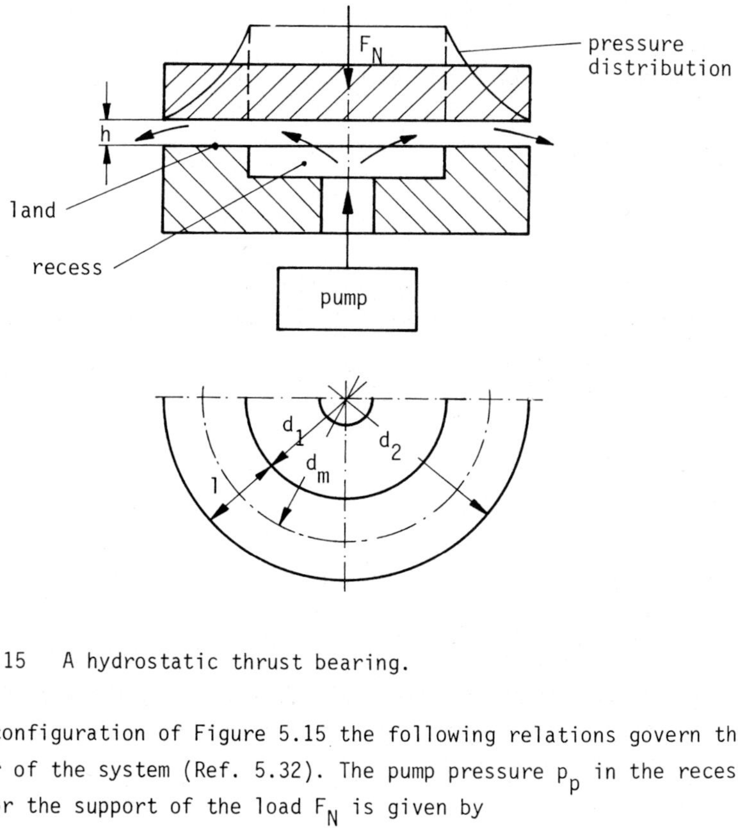

Figure 5.15 A hydrostatic thrust bearing.

For the configuration of Figure 5.15 the following relations govern the behaviour of the system (Ref. 5.32). The pump pressure p_p in the recess needed for the support of the load F_N is given by

$$p_p = \frac{2F_N}{\pi d_m l} \ln \frac{d_2}{d_1}$$

Flow rate q_1 and film thickness are related via the equation

$$q_1 = \frac{F_N h^3}{3 \eta d_m l} \qquad \eta : \text{ lubricant viscosity}$$

The term $q_1 p_p$ determines the necessary power of the pump. If the upper part of the hydrostatic bearing rotates with an angular velocity ω, a frictional torque, M_F, has to be overcome which is given by the internal friction of the lubricant film of equal thickness h to be calculated from the shear stress. It follows that

$$M_F = \frac{\pi \eta \omega d_m l}{4h} (d_m^2 + l^2)$$

This corresponds to a frictional power $M_F\omega$ and a friction coefficient

$$f = \frac{M_F \omega}{v F_N} = \frac{2M_F}{d_m F_N} = \frac{\pi \eta \omega l}{2F_N h} (d_m^2 + l^2)$$

Since wear processes are eliminated from hydrostatic bearings the tribological behaviour can be described entirely in terms of this data and equations. For other bearing geometries the relevant equations may be of more or less different form (Ref. 5.33, 5.34).

5.4.2 Magnetic fields

Tribological systems employing the principles of magnetism for the separation of the surfaces in relative motion can be divided broadly into two classes (Ref. 5.35):

(i) Systems which utilize the effect of a load-carrying force generated by the flow of conducting fluids within a magnetic field (Magnetohydrodynamic, MHD, bearings).

(ii) Systems with no lubricant, deriving their load-carrying ability from the attraction and repulsion associated with magnetic fields (Magnetic supension bearings).

Magnetohydrodynamic bearings (i) operate through mechanical forces produced by magnetic fields on a conducting fluid. From the laws of electromagnetics, it is well known that an electric charge q moving with a velocity v within a magnetic field B experiences a force F, the Lorentz force, acting perpendicular to the direction of motion and the direction of the magnetic field:

$$\vec{F} = q_e (\vec{v} \times \vec{B})$$

For MHD lubrication, it follows that when an electrically conducting fluid flows across a magnetic field the pressure developed in the fluid can exceed the ordinary hydrodynamically generated pressure. The theoretical treatment of MHD bearings starts from a combination of the fluid-mechanics equations and the electromagnetic equations leading to a magnetohydrodynamic form of Reynolds equation (Ref. 5.36). A considerable amount of theoretical work has been undertaken indicating the possibilities of MHD bearings. However, the impracticabilities of the magnet and associated equipment size have as yet restricted applications.

The other class (ii) of magnetic bearings needs no lubricant at all and derives its load-carrying ability from the attraction and repulsion associated with magnetic fields. The mechanical forces produced by either permanent magnets or by electromagnets can be used to freely support one partner of a bearing. The principle is illustrated in Figure 5.16.

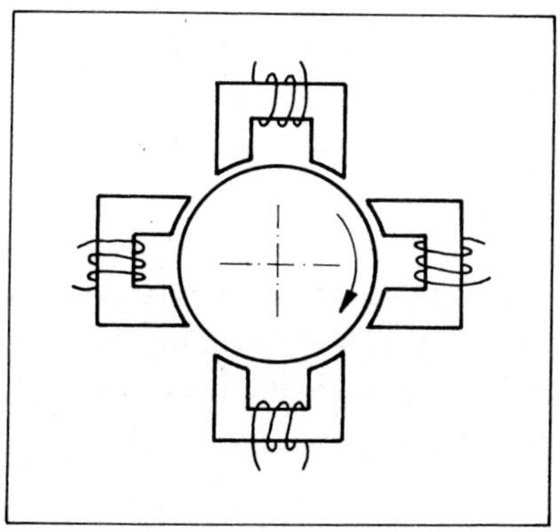

Figure 5.16 Principle of the support of a bearing shaft by four electromagnets.

One problem with bearings having a member freely supported in a magnetic field is stability. This problem can be solved with a servo-control system. If an external disturbing force acts on the "free" body, the corresponding displacement of the body detected by non-contacting sensors thus producing an electrical current which in turn is fed to the electromagnets generating an additional force compensating the disturbing displacement.

Owing to the advantageous tribological structure, (no solid-solid contact, no lubricant, low friction), magnetic bearings are able to operate under a broad spectrum of operating conditions. Charateristic data of commercially available bearings are (Ref. 5.37):

load capacity	:	0.2 N - 15 kN
velocity	:	up to 800,000 r.p.m.
temperature	:	up to 800 K

The lifetime of the bearing depends on the bulk materials properties; operating lifetimes up to 10^5h have been obtained independently of rotation velocity. The advantages of the tribological structure of these bearings are utilized in technical applications like the transmission of motion into closed chambers (for instance, vacuum chambers) or high-speed tracked-vehicle support.

5.4.3 INTERFACIAL ELASTOMERS

In this solution, wear processes in a tribo-mechanical system are eliminated by replacing sliding surfaces with the internal mobility of elastomers (Ref. 5.38). The elastomer act as a kind of lubricant and the resistance to motion is due to the internal friction of the elastomer. The most common elastomeric "lubricant" is natural rubber, because of its excellent strength and fatigue properties, but many synthetic elastomers can be used as well.

The principle of a typical elastomeric bearing is illustrated in Figure 5.17. Basically, the elastomeric bearing is comprised of alternating layers of rubber and metal laminates. The elastomer is vulcanized and bonded to the metal laminates as well as to the attachment metal components - inner and outer "races". The design of an elastomers bearing depend on types of load and motion (radial, axial, spherical, etc.). For instance, the "sandwich-type" bearing shown in Figure 5.17 is designed to support high axial loads F_N while accomodating torsional ($\Delta\varphi$) or lateral motions (Δs) through the compression and shear characteristics of the elastomer, respectively.

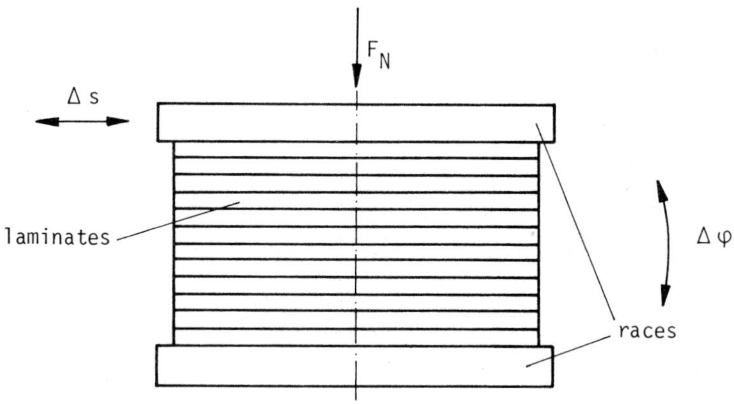

Figure 5.17 Typical elastomeric axial bearing.

This type of bearing is very stiff in the axial modes and soft in shear. In general, the design of elastomeric bearings must take into consideration factors such as the type and frequency of loads and motions as well as space-envelope limitations and life requirements. The advantage of the elastomeric bearing is that the various characteristics of the guidance of motion and the transmission and dissipation of mechanical work can be determined, at least in principle, from the bulk material properties of the interfacial material.

5.4.4 Flexural means

For tribo-mechanical applications in which only limited translatory or angular motions are required, flexural bearings, i.e., bearings which flex or twist in an elastic member to support the load, possess a wear-free invariable tribological structure. These bearings do not require a lubricant and the frictional resistance to motion is simply the elastic stress in bending or torsion integrated to give the force. Depending on the modes of load and motion, different types of flexural bearings have been designed (Ref. 5.39). As an example, a parallel spring support for translatory motion is shown in Figure 5.18 (Ref. 5.40).

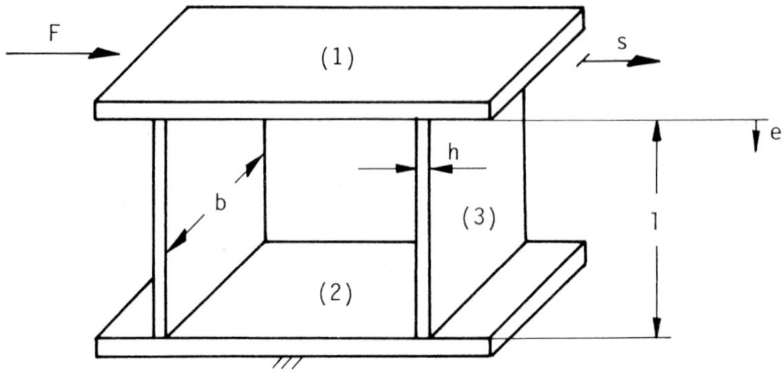

Figure 5.18 A flexural bearing.

The moving parts of the bearing, (1) and (2), are separated by two parallel flexure strips (3) which are made of heat-treated steel or phosphor bronze. For a translatory motion, s, of the upper part (1) of the bearing, a force, F_F, is necessary which is determined by the internal friction of the flexure strips:

$$F_F = \frac{2\ E\ b\ h^3}{l^3}\ s$$

E : elastic moduls
b, h, l : dimensions of the strips
s : displacement

The contraction perpendicular to the motion is given by

$$e = \frac{3\ s^2}{5l}$$

Summing up, the above examples show that there are different ways of eliminating solid-solid wear processes from mechanical systems under certain circumstances, thus providing the prerequisite for a constant, invariant structure of the tribo-mechanical system. The advantages of these solutions are:

(i) The tribological processes in such systems are due to internal processes (internal friction and fatigue) of the interfacial components (fluids, liquids, elastic materials).

(ii) The frictional losses and the lifetime properties can be estimated from the bulk properties of the interfacial components.
(iii) The functional behaviour of these systems can be described by applying analogue models of the well established theories of engineering systems analysis (see Section 6.1).

5.4.5 Wear resistant coatings

With the solutions described in the foregoing sections, it is possible to separate completely the moving surfaces of a tribo-engineering system eliminating any wear process from the system.

If the solutions described above are not applicable, an advantageous structure of a tribo-mechanical system may be established by applying a suitable surface treatment of the surfaces in order to mitigate tribo-induced changes of surface properties. There are several different surface treatments available which are suited for tribo-engineering applications (Ref. 5.41, 5.42). A compilation of these methods is given in Table 5.1. In the choice of a suitable coating for a given problem two basic aspects should be considered, in addition to a cost-benefit analysis:

(a) The technical function of the system under consideration and the values of the operating conditions of load, velocity and temperature.
(b) The type of wear mechanism or mechanisms expected to act in the interface of the moving components.

Depending on the type of the dominant interfacial wear mechanisms, different suitable surface compositions have been suggested. In considering this aspect from the view of the "delamination theory of wear" for low-speed dry-sliding wear conditions, Suh and co-workers (Ref. 5.43) suggested that a soft metal surface layer should be thinner than a critical thickness to prevent the accumulation of dislocations in the plated layer and the formation of the delaminated layer. This suggestion was investigated by plating annealed AISI 1018 steel with 0.1 μm cadmium layers on both of the contacting surfaces. In a special case this has led to a reduction of the wear rate of about three orders of magnitude.

Some general advice for the design of surface coatings in relation to the interfacial wear mechanisms is compiled in Figure 5.19 (Ref. 5.44): If in a sliding system the adhesive wear mechanism is expected to dominate, surface layers of low tendency for "cold welding" and with low shear strength

TREATMENT	CONSTITUENTS
DEPOSITED COATINGS	
Electroplating	Cr, Rh, Ni, Sn-Ni, Ag
Electrochemical Co-deposition	Co/Cr_2C_3, Pb/PTFE
Chemical Vapour Deposition (CVD)	TiC, TiN, WC
Arc Deposition Powder/Wire Spraying	Co-Cr-Ni Alloys
Plasma Spraying Detonation Gun	Ceramics, Cermets
Spraying/Brushing	MoS_2 - Resin Bonded
Sputtering	MoS_2 Films
Porous Sintered Layers	Bronze/PTFE, Co/PTFE
Porous Anodic Films	Ti/PTFE, Al/PTFE
CHEMICAL CONVERSION COATINGS	
Phosphate	Ferrous Alloys
Anodize	Titanium & Al-Alloys
Oxalate	Copper Alloys
Dichromate	Magnesium Alloys
DIFFUSION COATINGS	
Carburizing	C (in Ferrous Alloys)
Nitriding	N,C (in Ferrous Alloys)
Sulfonitriding	N,S (in Ferrous Alloys)
Siliconizing	Si (in Ferrous Alloys)
Boriding	B, Fe_2B (in Ferrous Alloys)
Chromizing	Cr (in Ferrous Alloys)
Beryllliding	Be in Titanium

Table 5.1 Surface treatments for tribo-engineering applications.

may be applied. Depending on the material of the counterpartner, thin surface layers of soft materials with a hexagonal crystallographic structure may for example be suitable. (Thin soft coatings are also useful from the point of the delamination theory; they may also - in a broader sense - positively influence surface fatigue wear processes.) If the abrasive wear component dominates, very hard surface coatings, like titanium carbides, may lead to low wear rates. Hard coatings, e.g., TiN, will also reduce adhesion of steel as well as the wear rate. If superposition of various wear processes is to be expected, multi-layered surface coatings may be beneficial (Ref. 5.45, 5.46). For certain cases of application, the following rules for the design of a tribologically stressed surface as illustrated in the lower part of Figure 5.19, may be favourable:

(i) A thin outmost surface layer should prevent adhesional cold welding effects with the counterpartner.

(ii) Below the thin layer a hard layer of sufficient ductility should bear the maximum of Hertzian contact stresses. This layer should exhibit a gradient of mechanical strength properties increasing from the outside to the inside of the specimen.

(iii) Below the hard intermediate layer the strength decreases progressively until the lower values of the base material are reached.

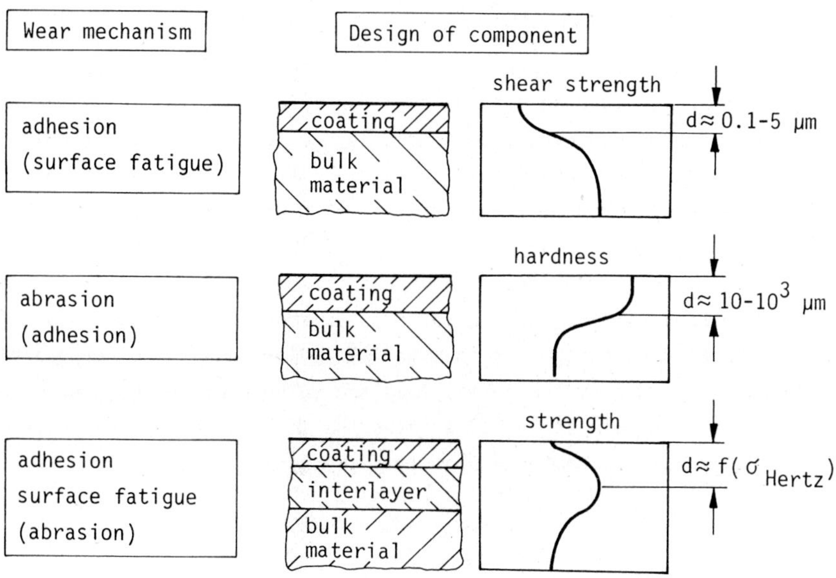

Figure 5.19 Suitable composition of surface coatings.

These rather qualitative points should be supplemented by an example of the design of an actual tribo-engineering component. Consider as a characteristic example the design of a bearing bushing (Ref. 5.47), as shown schematically in Figure 5.20.

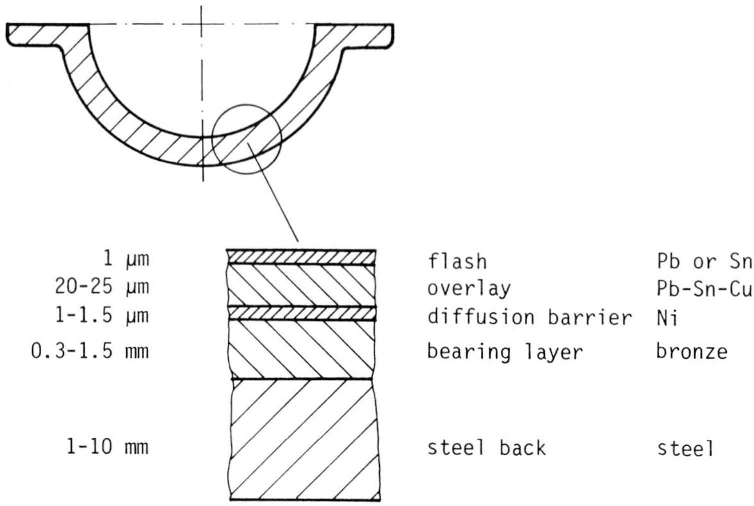

Figure 5.20 Design of a bearing bushing.

It can be seen that a contemporary bearing may consist of five layers to fulfill its technical purpose. The protection against (corrosive) environmental attack is provided by the outermost layer, the flash. The tribological behaviour is determined mainly by the overlay and the bearing layer. (Both layers are separated by a thin nickel layer which prevents the diffusion of tin from the overlay to the copper of the bearing layer which may occur at elevated temperatures.) Whereas the overlay forms the sliding surface, the purpose of the bearing layer is to embed hard particles and to replace the sliding surface if the overlay is worn away. The purpose of the steel back is to carry the normal load and to connect the bushing with its housing. This example again illustrates the importance of a suitable structural design of the components of tribo-mechanical systems.

6 Influence of tribological processes on the function of mechanical systems

6.1 General Considerations

Whereas in the preceding chapter the influence of tribological processes on the internal structure of mechanical systems has been discussed, in this chapter the influence of tribological processes on the external function of mechanical systems will be studied. It has been shown in Chapter 3 that the technical function of the various tribo-mechanical systems can be described formally by transformations of certain system inputs, such as motion and work, into the outputs which are used technically. Detrimental influences of the friction and wear processes described in Chapter 4 may then lead to disturbances of the functional input-output relations, for example friction-induced stick-slip effects, unfavourable mechanical efficiencies or even gross functional failures. Although the detrimental influences of the various tribological processes on the manifold mechanical systems (as compiled in the Appendix A) may be apparently rather different, from a systems point of view there are some features which appear to be of general importance. These basic aspects will be studied in this chapter and conclusions for a proper functional behaviour of tribo-mechanical systems will be drawn.

To obtain a starting point for a systems approach to the study of the main influences of tribological processes on the function of mechanical systems, first the functional behaviour of mechanical systems with negligible detrimental tribological influences will be discussed in this section.

If detrimental influences of tribological processes are negligible and if a mechanical system possesses a constant, invariant structure (see Section 5.4), it is possible to apply the network approach of engineering systems analysis to the description of the functional behaviour of the system. As outlined in Section 2.3, the characterization of the functional behaviour of a mechanical system by means of the network approach of engineering systems analysis starts with the identification and modelling of the elements of the system. Next, the input and output variables of the system are identified and classified into "across" and "through" variables or "effort" and "flow" variables. Then network graphs and signal flow graphs are drawn and the state equations are formulated. For this purpose, the Kirchhoff's mesh and node laws are often utilized. The resulting equations are then solved applying suitable methods, like the Laplace transform, or by utilizing digital or analogue computers in order to characterize the functional behaviour of the system.

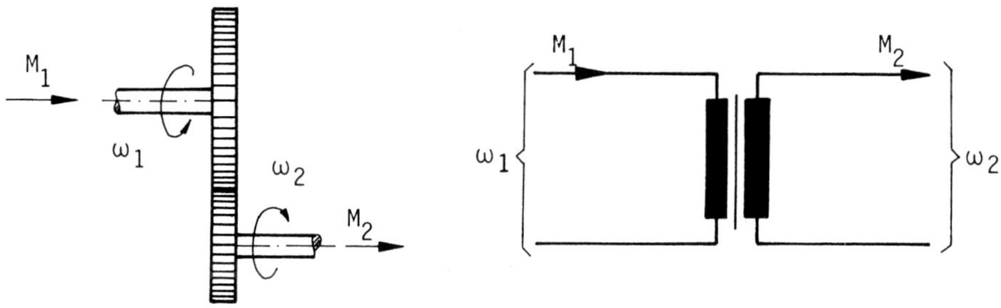

Figure 6.1 Simple gear train and its schematic representation.

A mechanical system which is frequently used to illustrate the application of the network technique is the "ideal" gear train (no friction losses, no wear effects). A simple gear train is shown in Figure 6.1. The gear train consists of two gears with radii r_1 and r_2, and the ratio of its radii is called the gear ratio $N = r_2/r_1$. The gear train is transmitting both torque $M_1 \rightarrow M_2$ and angular velocity $\omega_1 \rightarrow \omega_2$ and it is sometimes called a mechanical-mechanical energy converter. The functional input-output relations of the gear-train system can be found by an electrical analogue from the generalized Kirchhoff's laws (see Section 2.3):

(i) From force equilibrium, the force exerted mutually by the interacting gear teeth must be equal, i.e. $F_2 = F_1$. Introducing the gear ratio, N, it follows that the output torque M_2 is equal to the input torque M_1

times the gear ratio N:

$$M_2 = N\,M_1$$

(ii) From geometrical compatibility, the surface velocities must be the same, i.e. $v_2 = v_1$. This leads to a relation between the input and output angular velocities ω_1, and ω_2:

$$\omega_2 = \frac{1}{N}\,\omega_1$$

The relations between input and output torque and angular velocity can be combined using a matrix representation. Input torque M_1 and input angular velocity ω_1 are described together as an input vector

$$X = \begin{bmatrix} M_1 \\ \omega_1 \end{bmatrix}$$

This input vector is transformed into the output vector

$$Y = \begin{bmatrix} M_2 \\ \omega_2 \end{bmatrix}$$

by the transformation matrix

$$T = \begin{bmatrix} N & 0 \\ 0 & 1/N \end{bmatrix}$$

The functional behaviour of an ideal gear train can be then symbolized by the representation shown in Figure 6.2.

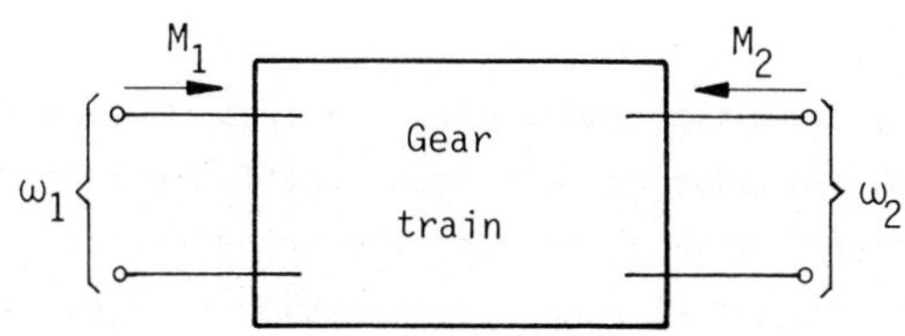

$$\begin{bmatrix} M_2 \\ \omega_2 \end{bmatrix} = \begin{bmatrix} N & 0 \\ 0 & 1/N \end{bmatrix} \begin{bmatrix} M_1 \\ \omega_1 \end{bmatrix}$$

Figure 6.2 Network representation of gear train system.

This is the typical network representation of a (linear) "two-port" system. If more than two input and output variables are connected with the system, the generalization of the modelling technique leads to the general characterization of a n-port system as shown in Figure 6.3.

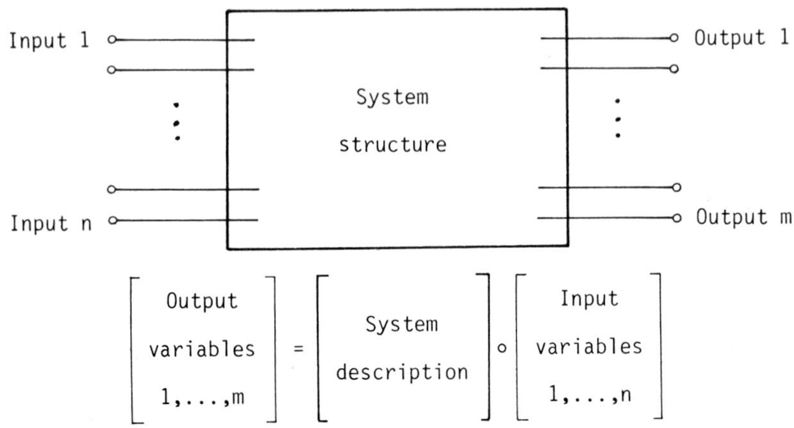

Figure 6.3 Network representation of n-port system.

The technique illustrated in Figures 6.2 and 6.3 only works if the elements of the actual system can be modelled by simple physical elements with linear cause-effect interrelationships between the input and output variables. Moreover, the network technique assumes a loss-free steady-state function of the system and a constant system structure. Due to these restrictions the conventional network technique, as illustrated by the simple gear train example, has to be modified for the functional description of tribo-mechanical systems. An extension of the network method for the steady-state case of power transmission systems which takes into account frictional energy losses has been proposed by Schlösser (Ref. 6.1). In this model a Coulomb-type of friction is assumed, introducing a constant coefficient of friction for a given set of conditions. With this model the transfer and loss-characteristics of the various mechanical power-transmission systems can be described. In order to study the dynamics of tribo-mechanical systems, variable friction conditions are to be taken into consideration as discussed in the next section.

6.2 TRANSMISSION OF MOTION AND STICK-SLIP EFFECTS

6.2.1 DYNAMICS OF TRIBO-MECHANICAL SYSTEMS

The functional behaviour of any tribo-mechanical system is connected, by definition, with relative motion of one or more of the components of the system. As described in Chapter 3, this motion may constitute a transfer of work, information or material through the tribo-mechanical system. In any case, the motion and the dynamics of the whole system are influenced by the interfacial friction processes between the moving components. Clearly, the details of the influences of tribological processes on the dynamics of motion may be quite different for the various tribo-mechanical systems, for example, ball bearings (Ref. 6.2), slideways (Ref. 6.3) or metal machining systems (Ref. 6.4). Considering the functional transmission of motion through the various types of tribo-mechanical systems, the influences of the tribological processes described in Section 4.3 may lead to unwanted vibrations of the moving parts, and to "stick-slip" motion. These disturbing influences on the functional behaviour can be observed in various tribo-mechanical systems: from the squealing of brakes to the chattering of machine tools during cutting processes. Since the occurrence of stick-slip effects may influence the functional behaviour of any tribo-mechanical system, this effect will be studied in some detail.

Many tribo-mechanical systems whose functional purpose is connected with the transmission of motion can be modelled in a simplified manner by the configuration shown in Figure 6.4. The model system consists of a body (1) of mass m_1, moving relatively to its counterpart (2) of mass m_2 fixed to the ground via a spring with a spring constant C_{s2} and a damper with a damper constant C_d. The body (1) is driven via the spring C_{s1} at constant velocity $v_o = s/t$.

The motion of body (1) of velocity v_1 and distance x relatively to body (2) of velocity v_2 and distance z is influenced by the friction force F_F acting in the interface (3) between body (1) and body (2). From the following simple qualitative consideration (Ref. 6.5), it follows that the type of motion is determined by the value of the friction force at $v_{rel} = 0$ and the dependence of the friction force on the velocity $F_F = f(v)$.
Let the initial state of the system shown in Figure 6.4 be such that the springs C_{s1} and C_{s2} are uncompressed and m_1 and m_2 are at rest. When the motion of velocity v_o is introduced there will be no movement of m_1 relative

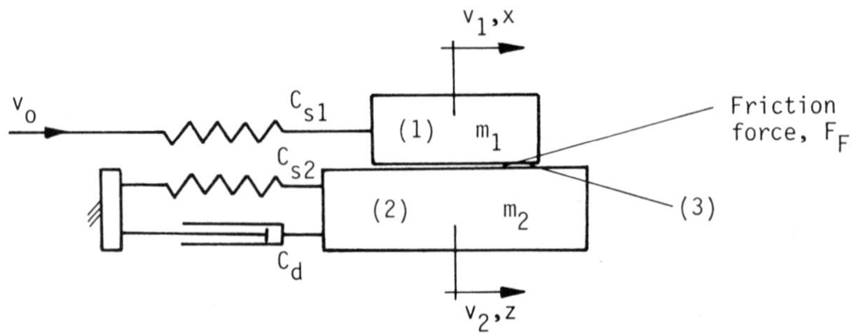

Figure 6.4 Model of a tribo-mechanical system.

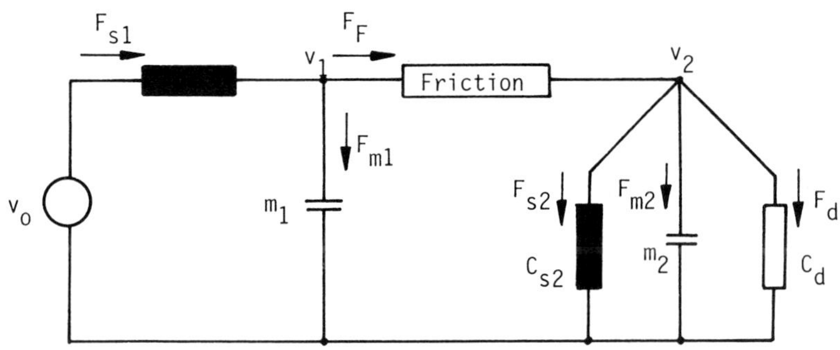

Figure 6.5 Network representation.

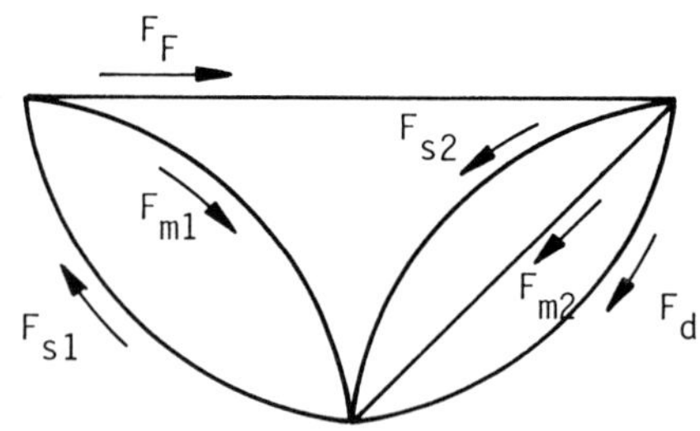

Figure 6.6 Signal flow graph.

to m_2 ("stick" phase) until the driving force on m_1 is high enough to overcome the (static) friction force between m_1 and m_2. If then the motion of m_1 relative to m_2 starts ("slip" phase) the springs decompress. Thus the driving force is lowered by a certain amount. If now the driving force on m_1 falls below the (kinetic) friction force, a second "stick" phase may evolve.This in turn leads to an increase of the driving force until the motion of the second slip phase starts, and so on.

The basic equations which govern the functional behaviour of the mechanical transmission system shown in Figure 6.4 can be formulated by utilizing the network approach of systems analysis (see Section 2.3).

Mechanical	Electrical
mass	capacitor
spring	inductor
damper	resistor

Table 6.1 Analogue mechanical and electrical system elements.

In Figure 6.5, the network graph of the system of Figure 6.4 is drawn. The network graph is based on the electro-mechanical analogue compiled in Table 6.1. In applying the velocity-effort, force-flow analogy described in Section 2.3, from a loop point of view the velocity can be seen in analogy to a "voltage driver" of the circuit. The forces F are then the "flow" variables through the elements of the circuit, i.e. the masses m_1, m_2, the springs s_1, s_2 and the damper d. The corresponding signal flow graph of the network graph of Figure 6.5 is shown in Figure 6.6. From the network and signal-flow graphs, the equations describing the functional behaviour of the tribo-mechanical system can be easily derived. Utilizing Kirchhoff's node law the following equations result:

(I) $F_{s1} = F_F + F_{m1}$ (node 1)

(II) $F_F = F_{s2} + F_{m2} + F_d$ (node 2)

These equations can be written as:

$$\text{(I)} \qquad C_{s1}\,(v_o t - x) = F_F + m_1\ddot{x}$$

$$\text{(II)} \qquad -F_F = C_{s2}z + m_2\ddot{z} + C_d\dot{z}$$

or

$$\text{(I)} \qquad \ddot{x} = -\frac{C_{s1}}{m_1}x + \frac{C_{s1}}{m_1}v_o t - \frac{F_F\,(|\dot{x} - \dot{z}|)}{m_1}$$

$$\text{(II)} \qquad \ddot{z} = -\frac{C_d}{m_2}\dot{z} - \frac{C_{s2}}{m_2}z - \frac{F_F\,(|\dot{x} - \dot{z}|)}{m_2}$$

For the solution of these differential equations, different methods can be used. (For a review of analytical methods see Ref. 6.5.) In the following, these equations and the stick-slip behaviour of the tribo-mechanical system shown in Figure 6.4 will be studied by means of an analogue computer.

6.2.2 Simulation of stick-slip behaviour

If a system can be represented by differential equations, the behaviour of the system can be studied with an analogue computer. The systems elements can be represented with reasonable accuracy by standard computer elements. Whereas for the tribo-mechanical system modelled in Figure 6.4 the values of the springs and the damper C_{s1}, C_{s2} and C_d, respectively, can be easily adjusted by potentiometer settings, the modelling of the friction characteristics requires an appropriate function generator. The above qualitative discussion of the stick-slip effect has indicated that the stick-slip motion may be determined by the type of the dependence of the friction force F_F (or the friction coefficient f) on the velocity. In order to study the stick-slip motion with sufficient generality, in this study, for the friction-velocity characteristic the shape of the Stribeck curve has been chosen and the behaviour of the tribo-mechanical system at the different parts of the Stribeck curve has been studied (Ref. 6.6). The analogue computer flow diagram for the system is shown schematically in Figure 6.7.

It can be seen that the upper and the lower parts of the computer flow diagram correspond to the above differential equations (I) and (II) and that the coupling of these differential equations via the friction force (or the friction coefficient) is simulated by the function generator in the middle of Figure 6.7. The appropriate scaling factors are determined from the design parameters of a pin-on-disc tribometer modelled by Figure 6.4.

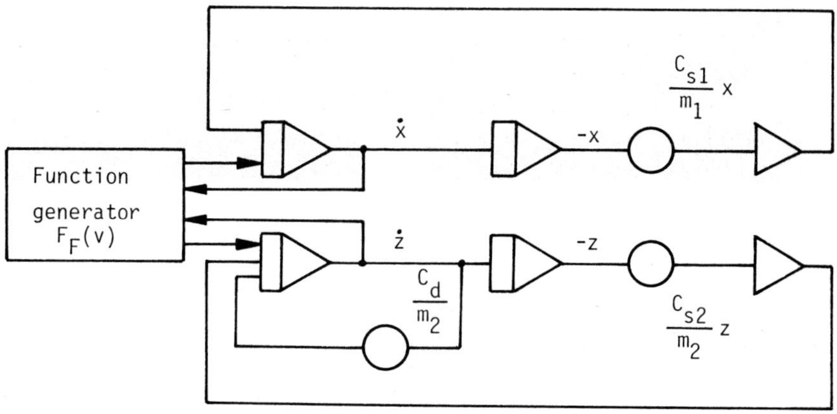

Figure 6.7 Analogue computer flow diagram.

For given constant data of m_1, m_2, C_{s1}, C_d, depending on the location of the friction coefficient within the Stribeck curve, a different dynamic behaviour of the tribo-mechanical system can be observed. Without going into numerical details, the three following different general patterns of the motion behaviour of the modelled tribo-mechanical system of Figure 6.4 can be distinguished:

(i) For the condition of friction around the minimum of the Stribeck curve, the system is unstable and the motion following a disturbance is divergent, i.e., the system excites itself to vibrations, as shown in Figure 6.8.

(ii) For the conditions of friction on the left part of the Stribeck curve the typical stick-slip motion diagram results as in Figure 6.9.

(iii) For the conditions of friction on the right part of Stribeck's curve the system is stable, i.e. vibrations introduced to the system are damped automatically. This behaviour can be seen in Figure 6.10 for five different slopes of the right part of the Stribeck curve and five operating values of the friction coefficient f.

The results confirm the experimental observation that stick-slip effects are likely to occur if the slope of the friction-velocity curve is negative

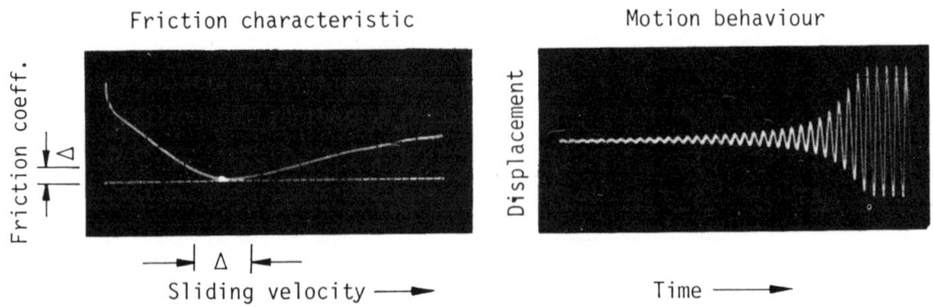

Figure 6.8 Instable motion behaviour for conditions at the minimum of the Stribeck curve (Δ : range of variation of f, v).

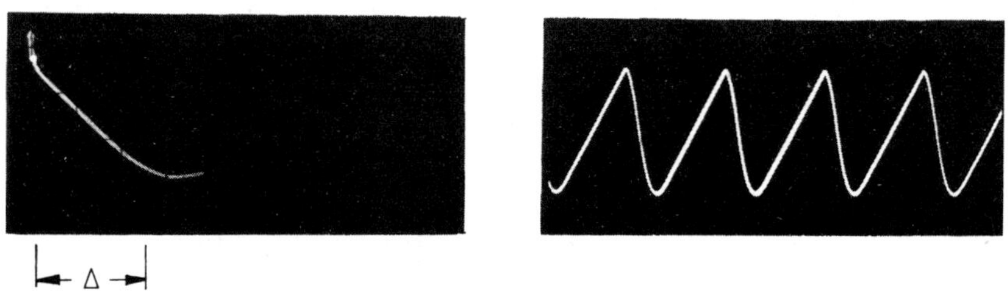

Figure 6.9 Stick-slip oscillations for conditions on the left part of the Stribeck curve (f_0=0.6).

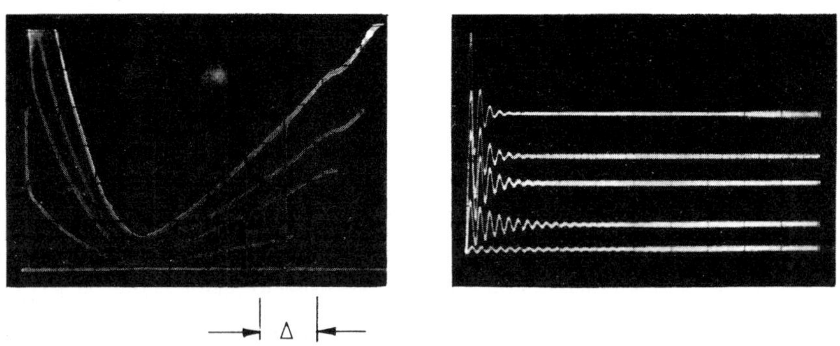

Figure 6.10 Stable motion behaviour for conditions on the right part of the Stribeck curve (f=1.0; 0.7; 0.5; 0.3; 0.02).

or equal to zero, $\frac{df}{dv} \leq 0$, as in the left part of the Stribeck curve. Thus, stick-slip effects may occur only under conditions of solid friction or boundary or mixed lubrication but are unlikely to occur under conditions of hydrodynamic lubrication.

These results show that the stick-slip behaviour of tribo-mechanical systems can be conveniently studied by analogue computer simulation. If an actual tribo-mechanical system is modelled in this way, it is also possible to use this model as a basis for attempts to eliminate stick-slip effects and to optimize the dynamic behaviour of the system. The optimization data can be obtained by varying the values of the analogue computer settings, thus simulating a change in the design of the elements of the system (masses, springs, dampers) or a change of the friction conditions. Further, critical operating conditions of the modelled system may be studied. It follows that the application of a combination of system modelling techniques and analogue computer studies may be of great help for obtaining a proper functional behaviour of a given tribo-mechanical system.

6.3 Mechanical Efficiency

In Chapter 3 the functional transfer of mechanical work through a tribo-mechanical system has been described as a process of translation and transformation of mechanical energy. This can be characterized in a simplified manner by the diagram shown in Figure 6.11.

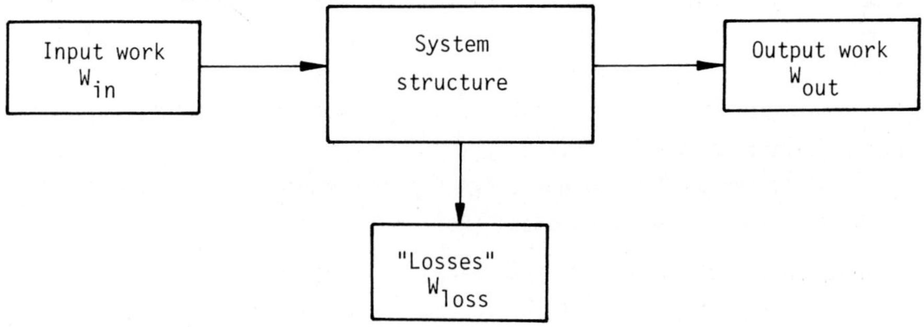

Figure 6.11 Work transactions of a tribo-mechanical system.

The ability of a mechanical system to transfer work, mechanical energy or power into a technically useful output is usually expressed by the efficiency η of the system defined as

$$\text{efficiency} = \frac{\text{useful output work}}{\text{input work}}$$

$$\eta = \frac{W_{out}}{W_{in}}$$

Because of its economic importance, the efficiency of a mechanical system is one of its basic engineering characteristics and most standard books of mechanical engineering contain some data on the efficiency of machine elements. (See for instance Ref. 6.7.) However, due to the great variety of tribo-mechanical systems and the complexity of tribological processes, there exists no general theory correlating the efficiency of mechanical systems with the tribological processes occurring within the system as discussed in Section 4.3.4. In the following, the influence of friction on the efficiency of a tribo-mechanical system will be studied phenomenologically in discussing a simple technical example. From the results of this example some general conclusions will be drawn.

Consider a simple "wedge drive" as illustrated in the upper part of Figure 6.12. The system consists of a horizontally moving wedge (1) and a vertically moving wedge (2). The technical purpose of the system is to transform the input motion, i.e. the input translation x into an output motion y at a right angle to it, and to lift a load F_y at the output by introducing an input force F_x. The basic principle of the transformation of inputs of force and motion (F_x, x) into technically useful outputs of force and motion (F_y, y) by means of a "wedge effect" is utilized in various tribo-mechanical machine elements, including translation screws, worm-wheel drives, cam-and-follower sets, etc.

The efficiency of the wedge drive system shown in Figure 6.12 can be calculated with the following steps (Ref. 6.8):

(a) draw free-body diagrams of the elements of the system,

(b) draw vector diagrams of the forces acting on the system's elements,

(c) determine the conditions of force equilibrium,

(d) calculate the efficiency η defined as $\eta = F_y \cdot y/F_x \cdot x$.

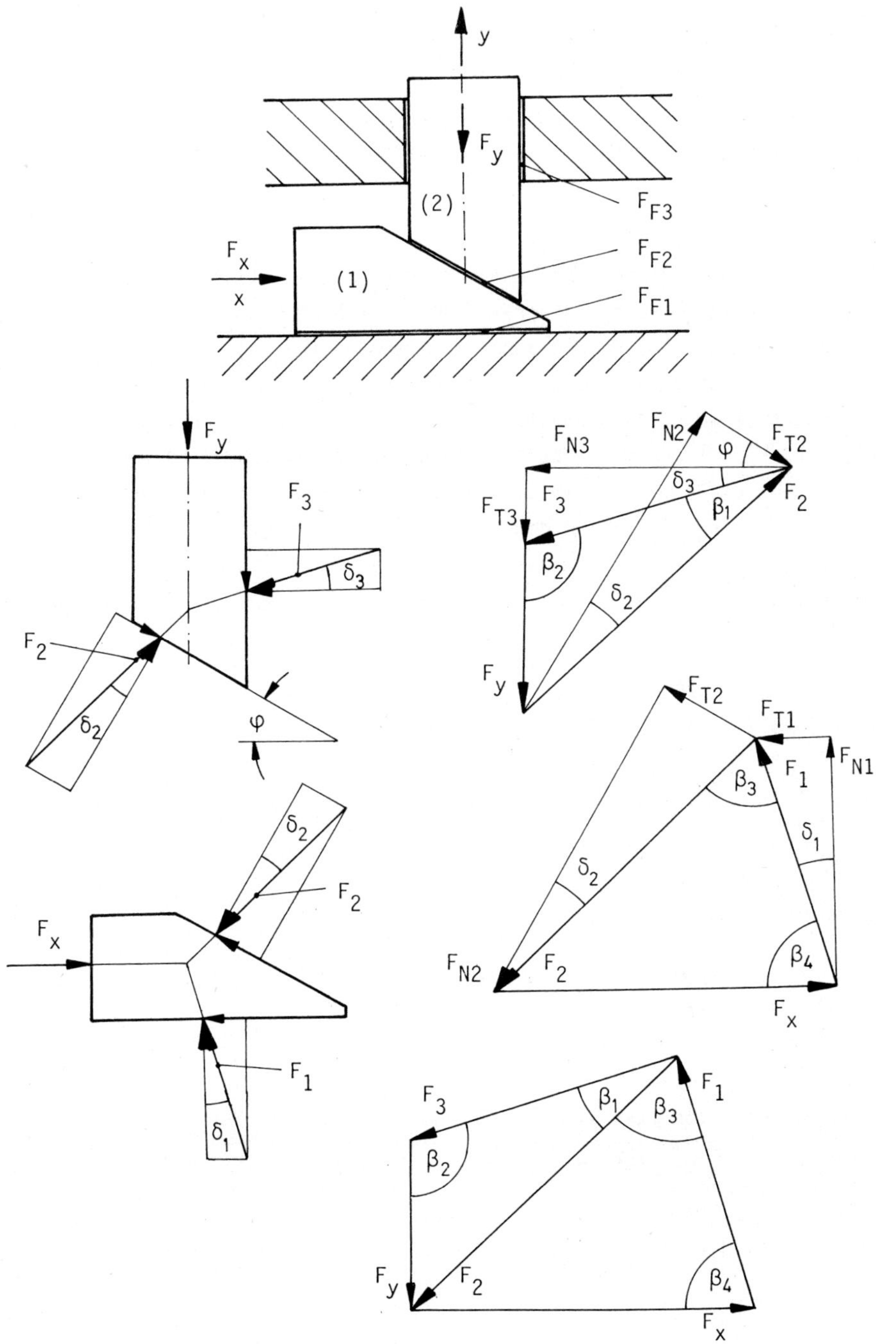

Figure 6.12 Characteristic of "wedge drive" system.

As indicated in Figure 6.12, the transmission of the input force F_x into the output force F_y is influenced by three sources of friction:

(i) F_{F1} is the friction force acting at the support of wedge (1),

(ii) F_{F2} is the friction force acting at the interface between wedge (1) and wedge (2),

(iii) F_{F3} is the friction force acting in the guides of wedge (2).

Splitting up these forces into components normal and tangential to the surfaces, the force vector diagrams shown in the right-hand part of Figure 6.12 can be drawn. In these diagrams, δ denotes the angle between the tangential (friction) force component F_T and the normal force component F_N, the tangent of which is equivalent to the coefficient of friction, i.e., $\tan\delta = F_T/F_N = f$. From the partial force vector diagrams the resulting force vector diagram, describing the equilibrium condition between the forces F_x, F_1, F_3, F_y can be determined, see Figure 6.12 below. From the trigonometric relations for the two triangles within this diagram, it follows that

$$\frac{F_y}{\sin\beta_1} = \frac{F_2}{\sin\beta_2}$$

$$\frac{F_y}{F_2} = \frac{\sin\left[90^0 - (\varphi + \delta_2 + \delta_3)\right]}{\sin(90^0 + \delta_3)} = \frac{\cos(\varphi + \delta_2 + \delta_3)}{\cos\delta_3}$$

and

$$\frac{F_x}{\sin\beta_3} = \frac{F_2}{\sin\beta_4}$$

$$\frac{F_x}{F_2} = \frac{\sin(\varphi + \delta_1 + \delta_2)}{\sin(90^0 - \delta_1)} = \frac{\sin(\varphi + \delta_1 + \delta_2)}{\cos\delta_1}$$

Thus

$$\frac{F_y}{F_x} = \frac{\cos(\varphi + \delta_2 + \delta_3)\cos\delta_1}{\sin(\varphi + \delta_1 + \delta_2)\cos\delta_3}$$

If it is assumed that the conditions of friction (i), (ii), (iii) are equal, i.e. that

$$f_1 = f_2 = f_3 = f,$$

it follows that

$$\frac{F_y}{F_x} = \frac{1}{\tan(\varphi + 2\delta)}$$

Since the relation between the distances x and y is given by

$$y = x \tan\varphi ,$$

then for the efficiency

$$\eta = \frac{F_y \cdot y}{F_x \cdot x}$$

it follows that

$$\eta = \frac{\tan\varphi}{\tan(\varphi + 2\delta)}$$

This relation indicates that for a given value of the wedge angle φ, the efficiency is determined by the friction coefficient $f = \tan\delta$ at the three frictional interfaces shown in Figure 6.12. Similar expressions are also obtained for other tribo-mechanical systems. For instance, the efficiency of a translation screw (Ref. 6.9) is given by the relation

$$\eta = \frac{\tan\varphi}{\tan(\varphi + \delta)}$$

which is quite similar to that of the wedge-drive system. A graph of the efficiency of the wedge-drive system for a wedge angle $\varphi = 30^{\circ}$ as a function of the coefficient of friction is plotted in Figure 6.13.

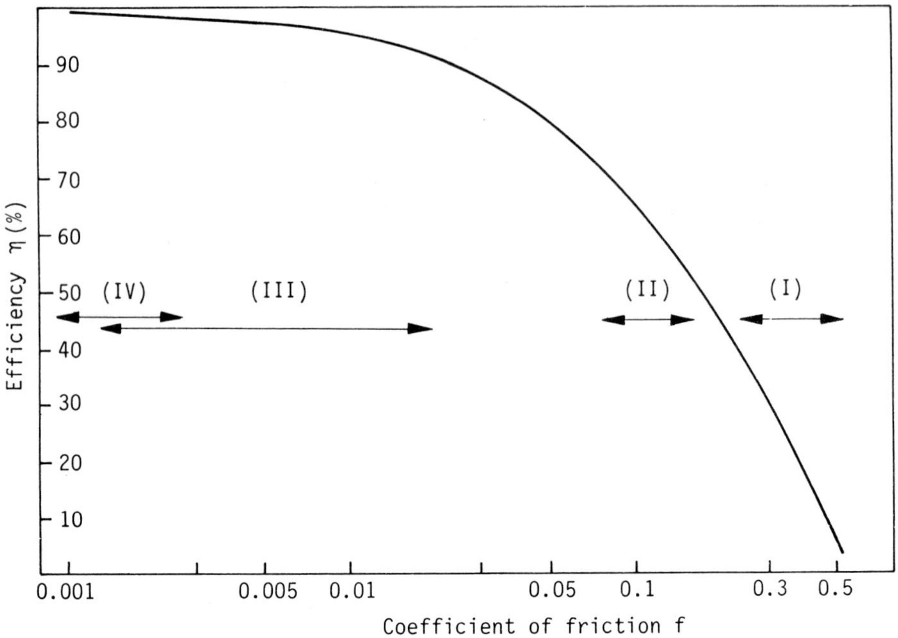

Figure 6.13 Efficiency of a wedge drive system as function of friction.

It can be seen that the efficiency varies over a broad range as a function of the coefficient of friction. For typical friction and lubrication regimes the following ranges are found:

(I) dry sliding friction : $\eta \approx$ 5 to 40%
(II) boundary lubrication : $\eta \approx$ 60 to 70%
(III) hydrodynamic lubrication : $\eta \approx$ 90 to 98%
(IV) rolling friction : $\eta \approx$ 97 to 99%

These data indicate that for sliding situations, efficiencies greater than 90 percent can only be obtained with full fluid lubrication, and that highest efficiencies can be realized by substituting sliding friction by rolling elements. Although these figures are derived for the specific wedge-drive system, they characterize broadly the typical efficiency ranges of tribo-mechanical systems operating under different tribological conditions.

If exact efficiency data of a given tribo-mechanical system are needed, experimental measurements should be used in addition to an analytical estimation. For example, for the accurate determination of the efficiency of gear systems, both experimental measurements and mathematical model simulations were performed in order to determine the influence of such factors as diametral pitch, contact ratio, magnitude of transmitted load, and speed upon the efficiency (Ref. 6.10). If different tribo-mechanical systems act together, the efficiencies must be multiplied. For instance, for a spur gear system of efficiency η_g, which is supported by two bearings of efficiency η_b the total efficiency is given by

$$\eta_{total} = \eta_g \cdot \eta_b^2$$

In generalizing these results, it follows that a multi-component mechanical system of high total efficiency can only be realized if all individual subsystems are designed with efficiencies as high as possible. For the determination of the total efficiency of a multi-component system the procedure of "systems tearing" as explained in Chapter 2 may be used.

6.4 Functional Failures

6.4.1 Causes of failure

The discussions in the preceding sections have shown that tribological processes determine the efficiency of mechanical systems and may also disturb the functional input-output relations. If in the performance of a given system, the functional transformation of the system's inputs into the technically used outputs is disturbed in such a way that the operating conditions exceed certain tolerance limits it is said that a "functional failure" of the system has occurred. Since in any system, the inputs are transformed into the outputs via the structure of the system, in general the effect of a functional failure is associated with, or caused by, significant changes of the structure of the system (see Chapter 5). These structural changes in turn are influenced or caused by the tribological processes occurring within the system during the functional transformation of the inputs into the outputs. Before studying the role of tribological processes in the failure of mechanical systems, some general theoretical considerations of the aspects of safety and failure of engineering systems will first be made.

Why do failures and accidents happen and what should be done to prevent them? According to Stanton (Ref. 6.11) there are three important theories in this field.

The first is Heinrich's domino sequence theory which postulates that five factors, occurring in sequence, lead to an accident (Ref. 6.12). These five factors are: (a) personal traits, either inherited or acquired from the environment; (b) a personal deficiency, i.e., these traits predispose a person to commit an unsafe act (or omission) or to allow a physical or mechanical hazard to exist; (c) an unsafe act (or omission) or a physical or mechanical hazard; (d) the accident; and at the end of the sequence, (e) the injury. Each factor is like a domino standing on end. As one falls, it knocks the others down. To prevent accidents, Heinrich suggests removal of one of the dominoes, preferably the domino labelled "unsafe act or mechanical or physical hazard".

The second important theory is Haddon's abnormal energy exchange theory (Ref. 6.13). According to this theory, damage occurs when a system exchanges energy with its environment beyond some acceptable or normal level. "Level" can mean quantity, quality, or rate of transfer. "Energy" can be of its different forms, e.g. electrical, mechanical, acoustical, radiation, thermal, etc. Applying Haddon's theory to the prevention of failure requires that we know the levels of energy input from or the output to its environment which will initiate damage to the object under study.

The third important theory in the field of safety is Grimaldi's risk acceptability theory (Ref. 6.14). He recognizes that the ideal of eliminating or neutralizing all hazards can never be achieved in real life. The "acceptability" of a hazard is determined by the tolerability of the risk associated with that hazard. The magnitude of the risk's tolerability is conceived as a three-dimensional space. The dimensions of this space are (i) sensitivity to exposure, (ii) probability of occurrence, and, (iii) seriousness of the effects. Sensitivity of exposure involves such matters as legality, confidence, and ethical responsibility, or values. To control accidents by applying Grimaldi's theory, we reduce the probability of occurrence by using fail-safe devices, reduce the seriousness by exposing fewer people for shorter periods of time, and reduce the sensitivity by making our designs more acceptable to the changing values of our society (Ref. 6.11).

Causes of failure	Occurrence (%)	
	rolling bearings	sliding bearings
manufacturing faults	14.4	10.7
design and calculation faults	13.8	9.1
materials faults of components	1.9	3.6
service faults, maintenance faults, failure of monitoring equipment	37.4	39.1
wear	28.5	30.5
failure through external causes	4.0	7.0

Table 6.2 Causes of failure of rolling and sliding bearings (Ref. 6.16).

Damage types	Occurrence (%)
breakages due to overstressing	60
scuffing, seizure	18
mechanical and corrosive surface damage	15
cracks	5
deflections, deformations	2

Table 6.3 Types of damage of mechanical clutches (Ref. 6.16).

Turning our attention now to the field of mechanical faults, in addition to the general theories of safety and failure, it is necessary to consider the causes of failure observed experimentally.

In reviewing the causes of mechanical failure, Collacott in his recent book on "Mechanical fault diagnosis and condition monitoring" (Ref. 6.15) distinguishes between the following main aspects:

(a) Service failures

(b) Fatigue

(c) Excessive deformation

(d) Wear
(e) Corrosion
(f) Blockages
(g) Design, manufacturing and assembly causes of failure

This compilation of the main classes of failure already indicates that there are several non-tribological causes which may lead to failure of mechanical equipment. This can also be seen from the data of Tables 6.2 and 6.3 in which failure characteristics of typical tribo-mechanical systems, namely rolling and sliding bearings and mechanical clutches, are listed. The data have been compiled by the insurance company ALLIANZ (Ref. 6.16). Table 6.2 contains the results of investigations of the causes of 1400 rolling-bearing failures and 530 sliding-bearing failures. It can be seen that for these tribo-mechanical systems, about 30 percent of the functional failures are due to wear processes. Table 6.3 contains the percentage of the main types of damage of mechanical clutches. Also in this case the damage occurring at the load-transmitting surfaces has been found to constitute about 30 percent of the total failure causes.

These examples show that besides the tribo-induced causes of failure, various non-tribological causes may lead to failure of mechanical equipment. Turning now to tribological causes of failure, in the next section the systems approach to the study of the failure of mechanical systems will be outlined by discussing an example.

6.4.2 A CASE STUDY: FAILURE MODES OF GEARS

In Section 6.1, an ideal gear train was studied by the systems analysis network method to illustrate "ideal" functional behaviour. In this section the main aspects of the failure modes of gears will be considered, as an example of a systematic study of the failure of a tribo-mechanical system.

In a recent review, P.M. Ku outlined the many factors influencing gear failure, emphasizing the importance of both gear mechanics and gear lubrication on gear failure modes (Ref. 6.17). He pointed out that any rational approach to gear design must consider, as a minimum, the impact of the various modes of gear-tooth failure on gear performance, operating life, reliability, size, weight, and cost.

It has been emphasized above that failure of a tribo-mechanical system occurs through the action of the operating variables on the structure of the system characterized by a severe disturbance of the functional input-output relations. Accordingly, a study of the failure of a gear system must, by necessity, include a consideration of the total effect of the following participating factors:

I OPERATING VARIABLES

- (a) type of motion, i.e. the kinematics,
- (b) forces which may be studied under steady-state conditions ("pseudo-statics") and under dynamic conditions,
- (c) velocities,
- (d) ambient temperature,
- (e) operating duration.

II STRUCTURE OF THE SYSTEM

Restricting the discussion to the case of a simple two-gears system, the following aspects of the system structure must be taken into consideration:

(a) System components

- (1) the first gear
- (2) the second gear
- (3) the lubricant
- (4) the atmospheric environment

(b) Properties of the system elements

- (i) material and surface characteristics of gear (1) and gear (2),
- (ii) lubricant (3) characteristics, both physical and chemical,
- (iii) characteristics of the surrounding atmosphere (4), both physical and chemical.

(c) Interactions of the system elements

The tribological interactions embrace the various processes discussed in detail in Chapter 4, namely:

(i) Contact processes

contact mechanics (contact deformation, tooth deflections, tooth misalignment),

(ii) Lubrication modes

elastohydrodynamic lubrication with some special conditions, like unsteady EHD conditions (dynamic tooth load and squeeze-film effects), lubricant starvation, non-isothermal flow conditions, elliptic gear-mesh conjunctions, surface-roughness and surface-texture effects, non-Newtonian flow behaviour, boundary lubrication including the various physical and chemical aspects of the material-lubricant-atmosphere interactions.

(iii) Friction processes
sliding/rolling friction of the counterformal gear-tooth contact under dynamic loads and velocities.

(iv) Wear processes
gear-tooth wear occurring through the action of one or more of the basic wear mechanisms, namely surface fatigue, abrasion, adhesion, tribo-chemical reactions.

The nomenclature of the American Society of Gear Manufacturers (AGMA) cites 21 modes of gear-tooth failure. According to Ku (Ref. 6.17), it is thought to be more logical to classify the gear-tooth failure modes in two basic categories, namely, strength-related modes and lubricant-related (or more generally tribo-related) modes. Major modes of strength-related failure are plastic flow and breakage. Major modes of tribo-related failure are rubbing wear, scuffing (or scoring), and pitting. From a systems point of view these processes are primarily connected with changes in the structure of the system under consideration (see Section 5.2). These tribo-induced changes of the system structure then determine the functional transmission of power through the system: in the absence of strength-related failures, the maximum power that can be transmitted through a set of homologous gears is primarily limited at low velocities by rubbing wear, at intermediate velocities by pitting or scuffing and at high velocities by scuffing. Some of the basic aspects of these failure-inducing tribological processes have already been studied in Chapter 4. In the following, some aspects of the most pronounced gear-failure mechanisms, namely pitting and scuffing will be discussed.

Being a surface fatigue phenomenon, pitting is the consequence of repeated stress cycling of the surfaces beyond the material's endurance limit. It takes some time to develop and leads to surface or subsurface cracks and eventually to the detachment of fragments from and the formation of pits on one or both surfaces. Like all surface fatigue effects pitting occurs even if the surfaces are completely separated by an intact oil film. This is because the presence of an oil film merely modulates the intensity of the repeated surface stressing, but does not eliminate it altogether. The lubricants does however have two specific influences:

(i) it has been observed that the pitting life is largely a function of the ratio of the oil film thickness to the composite surface roughness

(ii) there is some evidence that with oils of different chemical compositions, the chemical effect on pitting life cannot be ignored.

From the material point of view, the steel composition and the metallurgical structure exert an influence on pitting life. This is influenced also by load-sharing and tooth-profile modification. The effect of pitting occurs with gears of low contact ratios usually near the pitchline where the motion is nearly pure rolling but the Hertzian stress is a maximum. For gears with higher contact ratios, the load sharing pattern is quite different and the maximum Hertzian stress may very well occur in the mesh cycle where the sliding velocity may be high.

Scuffing is the other major failure mode of gears, and is a severe form of adhesive wear which cannot occur if an oil film of sufficient thickness separates the surfaces. In contrast to the time-dependent failure mode of pitting, scuffing failures may occur quite precipitously. It is believed that, in order to ensure full EHD lubrication the operating conditions must be such that for a given system the ratio of oil film thickness to composite surface roughness, λ , is in the order of three. As has been explained in Section 4.5.6, failure of a full EHD lubricated systems occurs only if two conditions are met:

(i) the necessary condition for the breakdown of the EHD film,
(ii) an additional condition for the breakdown of the surface films.

Whereas it appears possible to tackle condition (i) from basic EHD principles taking into account the dynamic influences of temperate and pressure-induced viscosity changes and surface roughness effects, there is still some discussion on the physical mechanism of the condition (ii). It appears that there might be a critical interfacial "surface temperature" or a critical "power input" or a critical "power intensity input" that a given boundary lubrication film may withstand. Clearly, the critical failure condition depends on the materials-lubricant-atmosphere interaction under the action of the set of the operating load-velocity-temperature-time variables.

This survey on gear failure is based on the review article by Ku (Ref. 6.17), and is presented here as an example of the failure modes of a tribo-mechanical system because it clearly demonstrates the multiplicity of failure-inducing parameters and failure-inducing effects. It also indicates the close interrelations between tribo-induced changes of the system's structure and the functional behaviour of the system. Although the various aspects of failure have been discussed elsewhere for other tribo-mechanical systems, for example plain bearings (Ref. 6.18), rolling contact bearings (Ref. 6.19) or piston rings (Ref. 6.20), it is somewhat difficult to generalize these findings. An approach to dealing with the various aspects of failure in a generalized way is provided by the probabilistic reliability estimation for components and systems as discussed in the next section.

6.5 Mechanical Equipment Reliability

Reliability is defined as "the probability of a device performing its purpose adequately for the period of time intended under the operating conditions encountered". This is the classic definition of reliability given by the Radio Electronics and Television Manufacturers Association in 1955. Since that time, the reliability of electronic equipment has received increasing attention, as apparatus has grown more complicated and has been applied to a great variety of important tasks. Now, the theory and practice of electronic equipment reliability is of a fairly high standard and methods for reliability prediction and designing for reliability have been developed including the aspects of redundancy (Ref. 6.21, 6.22, 6.23). Attempts of well-founded considerations of mechanical equipment reliability are of only recent origin (Ref. 6.24, 6.25). The differences in the state of the art of reliability considerations of electrical and mechanical systems may be due to the differences in their "internal structures" as discussed in Section 3.1 (see Figure 3.1). Consider, as a starting point for the reliability considerations of tribo-mechanical systems, the wear behaviour as a function of time. In Section 5.3 it was explained that for the loss-output wear rates of a tribo-mechanical system three main different characteristics may be distinguished (Ref. 6.26, 6.27):

(I) self accommodation ("running in")
(II) steady state
(III) self acceleration ("catastrophic damage")

These three modes of change in the system behaviour may follow each other in time, as indicated in Figure 6.14. In Figure 6.14, Z^M_{lim} denotes a maximum admissible level of the wear losses. At this level the system structure has changed in such a way that the functional input-output relations of the system are disturbed severely. Repeated measurements show random variations in the data, as indicated by the dashed lines in Figure

6.14. From sample functions of the wear process a distribution of the "lifetime" of the system, i.e., a failure distribution can be derived.

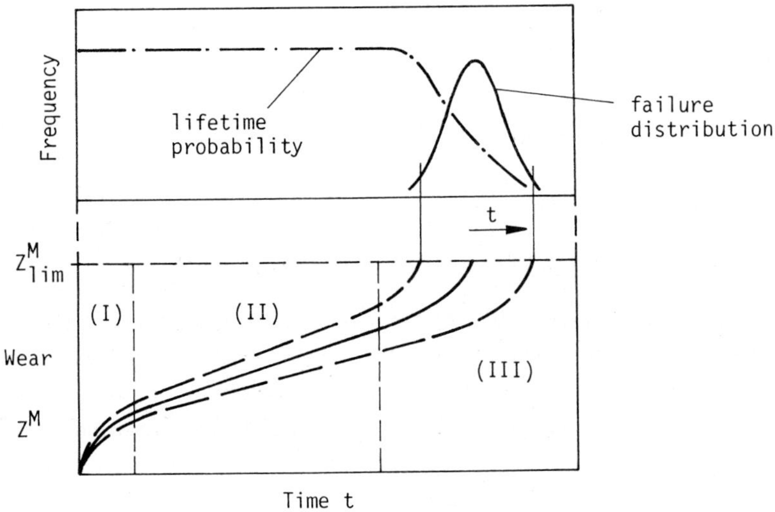

Figure 6.14 Wear curves and failure distribution.

In a quantitative way the reliability of a mechanical system may be characterized mathematically as follows:

Generally, the reliability of a mechanical system is expressed by a probabilistic function R(t) based on the following definitions:

$F(t)$	:	probability distribution function of the time to failure
$f(t) = \frac{dF(t)}{dt}$	:	density function
$\lambda(t) = \frac{f(t)}{1-F(t)}$	:	failure rate ($\lambda(t)dt$ is a conditional probability that the system will fail during the time t + dt under the condition that the system is safe until the time t)
$R(t) = 1-F(t)$	:	reliability function
MTBF	:	mean time to failure (measure of reliability for repairable equipment)

In some cases, the failure rate $\lambda(t)$ of a component in a system can be estimated from the point of view of the physical behaviour of the material used. Empirically, and sometimes theoretically, the following probabilities are proposed (Ref. 6.28):

(a) EXPONENTIAL DISTRIBUTION

$$\lambda(t) = \text{constant} = C$$

$$f(t) = C \cdot \exp(-Ct)$$

$$R(t) = \exp(-Ct)$$

In this case the failure rate is constant. It means physically that any failure occurs accidentally without any accumulation of fatigue-like effects during its service time under certain stresses. Many kinds of electronic components follow this type of failure. Components in a machine break down in this mode when the failure is brittle fracture. As an example, in Figure 6.15 the density function of the failure of a Diesel engine control unit is plotted showing an exponential distribution (Ref. 6.29).

(b) RAYLEIGH DISTRIBUTION

$$\lambda(t) = Ct$$

$$f(t) = Ct \cdot \exp\left(\frac{-Ct^2}{2}\right)$$

In this case the failure rate increases with time. The constant, C, indicates the rate of deterioration of the component which depends upon the stress level applied to it.

(c) NORMAL DISTRIBUTION (TRUNCATED)

$$f(t) = (1/s\sqrt{2\pi})\exp\left\{-\frac{1}{2}\left(\frac{t-\mu}{s}\right)^2\right\}$$

Many components of machines obey this distribution, especially if the failure occurs due to wear processes. The failure rate of this distribution cannot be expressed in a simple form.

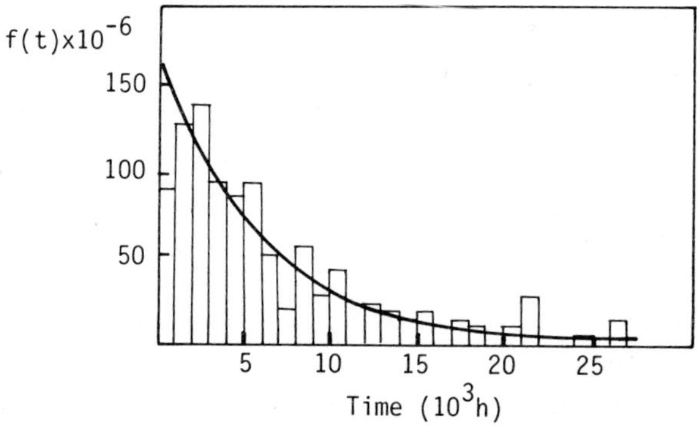

Figure 6.15 Density function f(t) of the failure of Diesel engine control units (Exp. distribution) (Ref. 6.29).

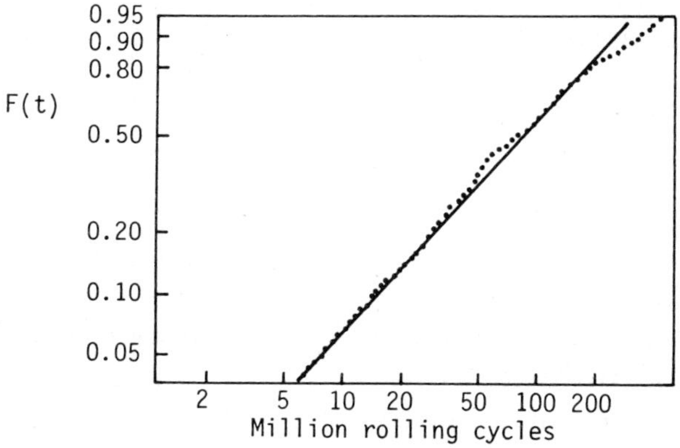

Figure 6.16 Failure distribution function F(t) of ball bearings (Weibull distribution) (Ref. 6.30).

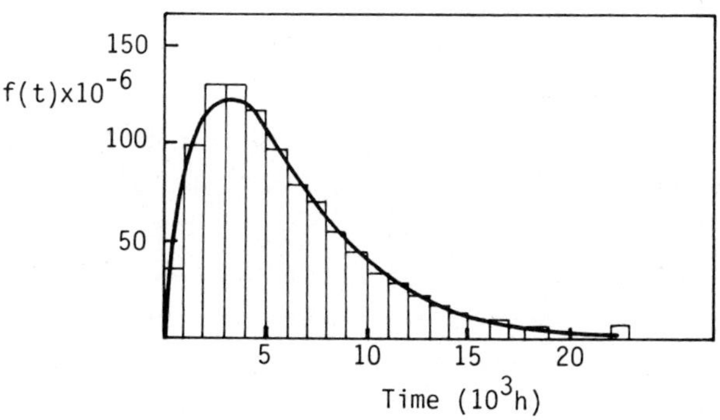

Figure 6.17 Density function f(t) of the failure of Diesel engine pistons (Gamma distribution) (Ref. 6.29).

(d) WEIBULL DISTRIBUTION

$$\lambda(t) = \frac{C}{t_o} t^{C-1}$$

$$f(t) = \frac{C}{t_o} t^{C-1} \exp(-t^C/t_o)$$

This is a distribution with two parameters, t_o, the nominal life and the constant C. The distribution is found to represent failure of many kinds of mechanical systems, such as fatigue in ball bearings. As an example, in Figure 6.16 the probability distribution function of the time to failure, F(t), as determined by testing 500 grease lubricated ball bearings at 1000 r.p.m., is shown (Ref. 6.30).

(e) GAMMA DISTRIBUTION

$$f(t) = C \frac{(Ct)^{x-1}}{\Gamma(x)} \exp(-Ct)$$

where $\Gamma(x)$ is a Gamma function. This is also a distribution with two parameters. Theoretically, the importance of this distribution is attributed to the fact that the equation is an x-fold convolution of the exponential function. It means physically that a component fails at x-th shock which occurs as a Poisson statistical process. As an example, in Figure 6.17 the density function of the failure of the pistons of Diesel engines is plotted showing a Gamma distribution with x = 2 (Ref. 6.29).

These distributions are representative ones which appear in the failure process of various components and systems. As a general overview, in Table 6.4 a compilation of the phenomena of deterioration and the mode of failure in connection with the underlying physical processes is given. The table is due to Yoshikawa (Ref. 6.28).

It can be seen that different failure modes and different elementary failure processes are associated with different types of failure distribution functions. It follows, that in turn from the experimental determination of failure distribution curves conclusions on the type of failure mechanism may be drawn. For most tribo-mechanical systems failing as a consequence of wear processes, the failure behaviour is characterized by the normal distribution or the Weibull distribution. If for a given type of tribo-mechanical system the failure mode and the type of failure distribution are known, this knowledge can be used to improve the reliability of the system.

Physical process	Deterioration (Mode of failure)						Probability distribution
	Topological change		Geometrical		Physical property		
	Fixing	Separation	Micro	Macro	Bulk	Sur-face	
Fracture		●					Exponential
Yielding				●			Exponential
Fatigue		●	●		●		Weibull
Creep		●		●	●		Normal
Diffusion					●		Normal
Corrosion						●	Rayleigh
Erosion						●	Rayleigh
Rusting	●					●	Rayleigh
Wear			●			●	Normal
Adhesion	●						Gamma
Staining			●			●	Exponential

Table 6.4 Phenomena of deterioration and mode of failure (Ref. 6.28).

For instance, this approach can be used to select the type of a ball or roller bearing system to operate under a given set of operating conditions with high operational safety (Ref. 6.31, 6.32). In this connection, the importance of lubrication technology on the reliability of tribo-mechanical systems has been emphasized (Ref. 6.33, 6.34).

To conclude the discussion on the failure and reliability of tribo-mechanical systems, the dependence of the failure rate on the operating duration of a system should be considered. If the failure rate is plotted as function of time, a curve, known as the "bath-tub-curve", is found, as shown in Figure 6.18.

In this curve three regimes can be distinguished: (a) early failures, (b) random failures, (c) wear-out failures. None of the distribution curves discussed above have this bath-tub-shaped failure curve but an approximation may be obtained by selecting an appropriate probability density function for each of the three regimes (Ref. 6.35). Regime (a) describes the region of the "infant death" of the system. This regime is characterized by a decrease of the failure rate with time for example in effective running-in. The regime (b) of constant failure rate is the region of normal running. Here, failure occurs as a consequence of statistically independent factors. Regime (c) is characterized by an increase of the failure rate with time. Here failure may be due to ageing effects. As described above, for a great deal of tribo-induced failures the failure rate increases with time. Thus region (c) of the "bath-tub curve" of Figure 6.18 appears to be relevant for the normal mode of wear induced failure of mechanical systems.

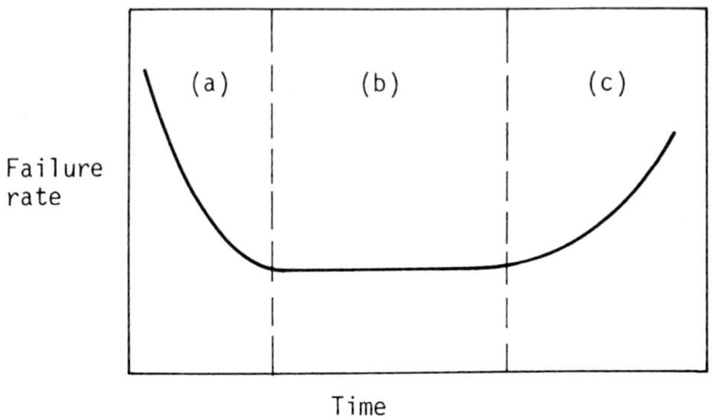

Figure 6.18 "Bath-tub" failure rate curve.

6.6 Requirements for Proper Functional Behaviour

Having studied in some detail the influences of tribological processes on the function of mechanical systems, finally the requirements for a proper functional behaviour of tribo-mechanical systems should be considered. The four main aspects that may be important in this connection will be discussed in the following sections (see also Chapter 8).

6.6.1 Study of alternative solutions

Before designing a mechanical system, the possibility of eliminating tribo-induced difficulties by using an alternative technical solution should first be considered. This requires careful consideration of the technical purpose of the system under question and the compilation of the technically required inputs and use-outputs (see Table 3.1 and 3.2). If this analysis shows that the system's inputs and outputs needed to fulfil the required technical purpose are essentially mechanical in nature, i.e. comprise the transmission of motion, mechanical work and/or materials, the technical purpose can, of course, be fulfilled only by a mechanical system. There are, however, some technical systems that are not intrinsically mechanical in nature but are realized by mechanical means only for historical reasons, since mechanical methods have often, in the past, been the first means available for the solution of technical problems. The function of such systems is connected mainly with the generation, transmission or reproduction of "information". Examples of this class of (pseudo-mechanical) systems are the portable mechanical clock (the prototype was invented and built by P. Henlein around 1510), the mechanical calculating machine (the prototype was invented by B. Pascal and G.W. Leibniz between 1640 and 1670), or mechanical devices for the control of machinery or production processes. The technical function of these engineering systems is essentially the generation or reproduction of information. Whereas the traditional mechanical realizations of these technical systems, like the clock, are based on the motion of macroscopic bodies, such as gears or levers, the contemporary electronic realizations of these systems are based on the motion or impulses of sub-microscopic particles, namely electrons. Although in both types of clock the external technical function is identical, the internal system structures are entirely different. Clearly, the replacement of the motion of macroscopic bodies by the motion of sub-microscopic particles not only eliminates tribological processes but also leads to a micro-dimensional design and to a fast and reliable function of the system. In the future some other tribo-mechanical systems traditionally used for the generation, transmission or reproduction of information may be replaced by technically more suitable electronic solutions.

The replacement of conventional mechanical systems by electronic systems, consisting of mini-chips, large scale integrated circuits, microprocessors, etc., and the corresponding industrial changes have some far-reaching scientific, technical, economical and sociological consequences. (In the journalistic language this has been already termed the "third industrial revolution", Ref. 6.36.) Cramer has recently pointed to some of these consequences in discussing the electronic teletype as an example for the "structural changes" in precision engineering (Ref. 6.37).

In the study of alternative engineering solutions for tribo-mechanical systems, however, in addition to the functional input-output relations the influences of disturbing inputs from the environment should be taken into account. Special care is then required for example, in the design of electronic components suited for applications in a vibrating environment, as in a motor-car or an aircraft. From this point of view, a robust mechanical solution may be more appropriate than a fragile electronic solution.

6.6.2 Proper design of system structure

The main aspects of the appropriate design of the structure of tribo-mechanical systems have already been discussed in Section 5.4. Some further suggestions for the design of bearing systems can be found in Ref. 6.38, 6.39, 6.40.

Discussing the requirements of a correct design of tribo-mechanical systems, Peeken pointed out that the conventional production-oriented and kinematics-oriented design of mechanical systems is often not in accordance with tribological requirements (Ref. 6.41). A designer who tries to design a mechanical system in a tribologically correct manner should attempt to solve the basic engineering task of tribology: to design interacting surfaces in relative motion in such a way that forces can be transmitted through the interface with a minimum of wear. Peeken mentioned three points which may be taken into consideration in order to improve the design of tribo-mechanical systems:

(i) Influence interfacial bearing geometry through elastic and thermo-elastic processes by changing the shape of the moving surfaces.

(ii) Lower the forces to be transmitted. Equalize the force and pressure distributions and influence the velocities by changing the geometrical design.

(iii) Change the kinematics, for example by changing the type of interfacial friction (e.g., sliding friction or rolling friction) taking into consideration the consequences on production.

In designing tribo-mechanical systems where reliability is at a premium, certain basic principles of design should be carefully taken into consideration. These are the use of diversity, redundancy and designing with a preferred mode of failure in mind such as always endeavouring to produce a "fail-safe" condition. The principle of equipment redundancy implies that the system will generally survive a single equipment fault and continue with its intended operational function. One method of redundancy is to have one system in operation and the other on standby. If failure of the operating system takes place then automatic change-over to the standby system should immediately take place. Another method of redundancy is to have two systems operating "parallel". These principles of redundancy which can be easily realized in electronic systems are, however, not so easily realized in tribo-mechanical systems. An instructive example concerning the redundant design of bearing assemblies is shown in Figure 6.19 (Ref. 6.42).

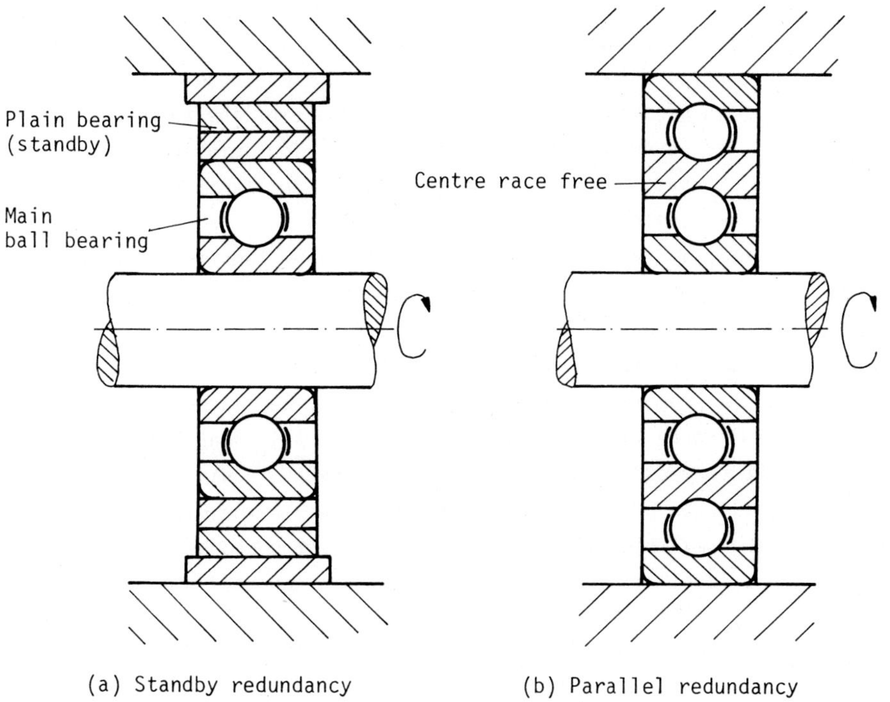

Figure 6.19 Redundant design of a bearing assembly.

In the standby case, a combination of a plain bearing and a ball bearing allows one to fail but for operation to continue on the other type of bearing. This is an interesting example because it also brings in the idea of diversity which is often used to overcome a common fault condition. In the case of the parallel redundant bearing assembly, the load can be carried by a single race hence surviving under certain conditions a failure on the other race. In this case, checks should be made at maintenance, whereas in the standby arrangement it would be useful to have a monitoring system to sense when the plain bearing is working.

6.6.3 Proper choice of operating variables

In order to obtain a proper functional behaviour of a tribo-mechanical system of a given design, the operating conditions (load F_N or pressure p, velocity v, operating temperature T, operating duration t) should not exceed certain limits. See also Haddon's abnormal energy exchange theory (Ref. 6.13), discussed in Section 6.4.1. These limits of operating variables may be found from the condition that the action of the operating variables may not lead to intolerable changes of the structure of the system. Although there is no general theory of the limits of operating conditions, for mechanical systems of simple geometry, the procedure of the so-called "zero wear model" (Ref. 6.43) may be a convenient starting point for estimating the limits of operational conditions for a quasi-constant structure of a tribo-mechanical system.

"Zero wear" is taken to be wear of such a magnitude that the surface roughness in the wear track is not significantly different from the initial surface finish in the unworn part of the surface. To quantify this statement, the zero wear limit is assumed when the depth of the wear track equals one-half of the peak-to-valley surface roughness. The model states that wear can be controlled by limiting the maximum (Hertzian) shear stress τ_{max} acting in the contact region. Defining one "pass" as a distance l of sliding equal to the length of the geometric contact area in the direction of sliding, then as an example, for 2000 passes surface topography changes should be of a "zero level" if the operating conditions lead to

$$\tau_{max} \leqslant \gamma_R \cdot \tau_y \qquad \tau_y : \text{ yield stress in shear}$$

The value of γ_R depends on the materials in contact and the degree of lubrication.

For quasi-hydrodynamic lubrication it is postulated that

$$0.54 \leqslant \gamma_R \leqslant 1$$

For boundary lubrication or dry sliding the following values are selected:

$$\gamma_R = 0.54 \quad \text{if the system has a low susceptibility for transfer}$$

or

$$\gamma_R = 0.20 \quad \text{if the system has a high susceptibility for transfer}$$

For i > 2000 passes the "zero wear condition" is given by

$$\tau_{max} \leqslant \left[\frac{2000}{i}\right]^{1/9} \gamma_R \tau_y$$

From this relation, for a given tribo-mechanical system of known values of γ_R and τ_y, the allowable number of sliding passes (for given operating shear stresses τ_{max}) or the allowable operating shear stresses (for a given number of passes i) may be estimated. Since this model is based almost entirely on stress considerations it may be not applicable for wear process including a considerable adhesive wear component (Ref. 6.44). Indeed, the form of the above equation, i.e., the one-ninth-power dependence, implies that the model may be applicable only for fatigue wear situations (see Section 4.4.2).

For hydrodynamically lubricated journal bearings, Vogelpohl's method may be applied in characterizing the allowable limits of operating variables (Ref. 6.45). Vogelpohl found that to a reasonable degree of accuracy the following relation characterize the conditions at the transition from full fluid film lubrication to mixed lubrication at the minimum of the Stribeck curve:

$$\frac{\bar{p}}{\eta \cdot v} = \frac{3}{2} \quad (m^{-1})$$

where $\bar{p}$: mean bearing pressure ($10^5 N/m^2$)

η : viscosity (10^{-3}Pas = cP)

v : surface velocity (m/s)

From this condition, for two given values of the set ($\bar{p}, \eta$, v) the critical transition value of the third one may be estimated.

Finally, for a lubricated system operating in the low speed, high load part of the Stribeck curve, under conditions of mixed or boundary lubrication, the operational limits of the system may be characterized by determining experimentally the "failure surface" for the given system (see Section 4.5.6, Figure 4.62).

6.6.4 MONITORING THE FUNCTION OF THE SYSTEM

The primary method of relating deterioration of a system to consequence is the monitoring of the performance of the system. The simplest form of such monitoring is the regular recording of relevant parameters in a log-book. Regular logging or automatic registering of important parameters and their derivatives, in comparison with reference information, provides the basis for performance-trend monitoring. In trying to monitor the performance of a system, it is necessary to decide which relevant parameters to monitor. From a systems point of view two classes of parameters may be distinguished:

(I) Input and output variables of the system.

(II) Relevant parameters of the system structure.

In recent years, several techniques have been developed which can be used for the on-line monitoring of operating tribo-mechanical systems. In a feed-back loop, the signals from the monitoring device may then be used to influence or control the system's input operating variables in order to obtain a proper functional behaviour of the system. The main techniques applicable to the monitoring of tribo-mechanical system are to be discussed in detail in Section 7.5.

This chapter on the functional behaviour of tribo-mechanical systems may be best concluded by citing some interesting remarks of the great mathematician John von Neumann concerning the function and malfunction of engineering and living systems. In a lecture at the Hixon Symposium, September, 1948, in Pasadena, California, entitled "The general and logical theory of automata" von Neumann made the following remarks (Ref. 6.46):

"In living organisms malfunctions of components occur. The organism obviously has a way to detect them and render them harmless. The system must, therefore, contain the necessary arrangements to diagnose errors as they occur, to readjust the organisms so as to minimize the effects of the errors, and finally to correct or to block permanently the faulty components. Our modus procedendi with respect to malfunctions in our artifical automata is entirely different. Every effort is made to detect every error as soon as it occurs. Then an attempt is made to isolate the component that caused the error as rapidly as feasible. This may be done partly automatically, but in any case a significant part of this diagnosis must be effected by intervention from the outside. Once the faulty component has been identified, it is immediately corrected or replaced. Note the differences in these two attitudes. The basic principle of dealing with malfunctions in nature is to make their effect as unimportant as possible and to apply correctives, if they are necessary at all, at leisure. In our dealings with artificial automata, on the other hand, we require an immediate diagnosis. Therefore, we are trying to arrange the automata in such a manner that errors will become as conspicious as possible, and intervention and correction follow immediately. The rationale of this difference is not far to seek. Natural organisms are sufficiently well conceived to be able to operate even when malfunctions have set in. They can operate in spite of malfunctions, and their subsequent tendency is to remove these malfunctions. An artificial automaton could certainly be designed so as to be able to operate normally in spite of a limited number of malfunctions in certain limited areas. Any malfunction, however, represents a considerable risk that some generally degenerating process has already set in within the machine. It is, therefore, necessary to intervene immediately, because a machine which has begun to malfunction has only rarely a tendency to restore itself, and will more probably go from bad to worse."

7 Tribometry: test, simulation, and control methods

7.1 Introduction

It is obvious that in a multidisciplinary subject like tribology, the experimental techniques of investigating and testing play an important role. In following the terminology of Coulomb, who described his friction-measuring device as a "tribometer", the whole field of experimental study in tribology is called "tribometry". This topic is very broad, ranging from friction and wear measurements, the testing of lubricants and evaluation of tribo-technical components to machinery condition monitoring and reliability testing. Although at first glance the experimental determination of the tribological behaviour of materials or lubricants seems easy, inasmuch as friction and wear are apparently not difficult to measure, the difficulties lie in the interpretation of the test results in connection with the various tribo-technical applications of materials and lubricants (Ref. 7.1).

During the first half of this century a number of tribometers came into existence with some features in common:

(i) simple geometry,
(ii) cheap test pieces,
(iii) used for "accelerated" testing,
(iv) at best of moderate accuracy,
(v) yielding results that can not be interpreted in a simple way.

The term "hybrid testers" was proposed, since service conditions were not accurately reproduced nor were physical parameters well defined (Ref. 7.2).

Today the same equipment is still popular in Industrial Laboratories, because it is convenient for acceptance testing and production control. However, if the results are interpreted as models of real tribo-engineering events, grave quantitative errors may be made (Ref. 7.2). Recently research laboratories, taking advantage of modern instrumentation, have developed various general-purpose tribometers. Equipped with sensitive recorders and analyzers, such sophisticated instruments permit the study of a great many variables (Ref. 7.3).

These few points indicate the great variety of tribometric techniques and imply that their application and the interpretation of results requires great care. In the field of tribometry it is again found that the systems approach can be used as a convenient guide-line for the description of the various tribometric methods and their possibilities. It has been emphasized in the forgoing chapters that both structural and functional aspects of a tribological system must be studied, and this leads to the following classification of tribometric techniques:

(I) Structural aspects

- (a) elements
- (b) properties of elements
- (c) interaction of elements

(a), (b) → testing of system component properties

(c) → investigation of tribological processes

(II) Functional aspects

Functional behaviour of the system

→ simulative tribo-testing

→ machinery condition monitoring

Consideration of the various aspects of tribometry from a systems point of view shows that the tribometric techniques can be classified into four basically different groups. It should be noted again that each group has its own (and limited) technical purpose and predictive value. Great care is needed for a mutual comparison of the results obtained by methods belonging to different groups. In the following, the four main groups of tribometry will be studied by discussing the main features of these techniques and reviewing their requirements and possibilities. The treatment starts with laboratory methods, i.e., the tribometric techniques for the investigation of friction and wear processes and simulative tribo-testing, and then turns to the more industrial aspects of tribo-testing, namely tribo-technical component testing and machinery-condition monitoring.

7.2 Friction and Wear Test Methods

One of the main aspects of tribometry is the investigation of friction and wear processes by means of laboratory tribo-testing devices. Reviewing the results of wear tests published in the international periodical WEAR over a period of 13 years, it has been found that approximately 50% of the reported wear tests were obtained with laboratory test rigs and that only 17% of the wear tests were performed with actual machine elements (Ref. 7.4) These data clearly illustrate the great importance of laboratory tribo-testing devices. It was indicated previously that in recent years accurate general-purpose tribometers equipped with sensitive recorders and analyzers have been developed in research laboratories. In combining such tribometers with the new powerful surface analytical tools, it is possible to study the elementary friction and wear processes in detail. It should be emphasized, however, that the application of these tribometers is useful mainly for investigations of the system structure - as for instance the changes in the properties of the system components due to wear processes. As a consequence of the concentration on the structural aspects, from a systems point of view the engineering "use-function" is frequently degenerate in tribometric models.

In Figure 7.1 the main parameters relevant to tribo-testing are compiled (Ref. 7.5). In the central part of the figure the elements, i.e., the material components of the test system are sketched. In most cases the test system consists of two solid specimens (1), (2) together with the lubricant (3) and the atmosphere (4). The inputs of the test-system are given by the operating variables compiled on the left-hand side of Figure 7.1. Through the action of the operating variables on the test-system elements, friction and wear processes occur. For the description of these tribological processes the tribometric characteristics compiled on the right of Figure 7.1 have to be measured. Supplementing these tribometric characteristics, surface characteristics of the specimens (1) and (2) have to be determined. In the following sections the main features of the parameters and characteristics shown in Figure 7.1 will be discussed briefly. (The measuring techniques discussed in this section in connection with laboratory test rigs may be adequately used also in the testing of tribo-technical components, see Section 7.4, and in machinery condition monitoring, see Section 7.5.)

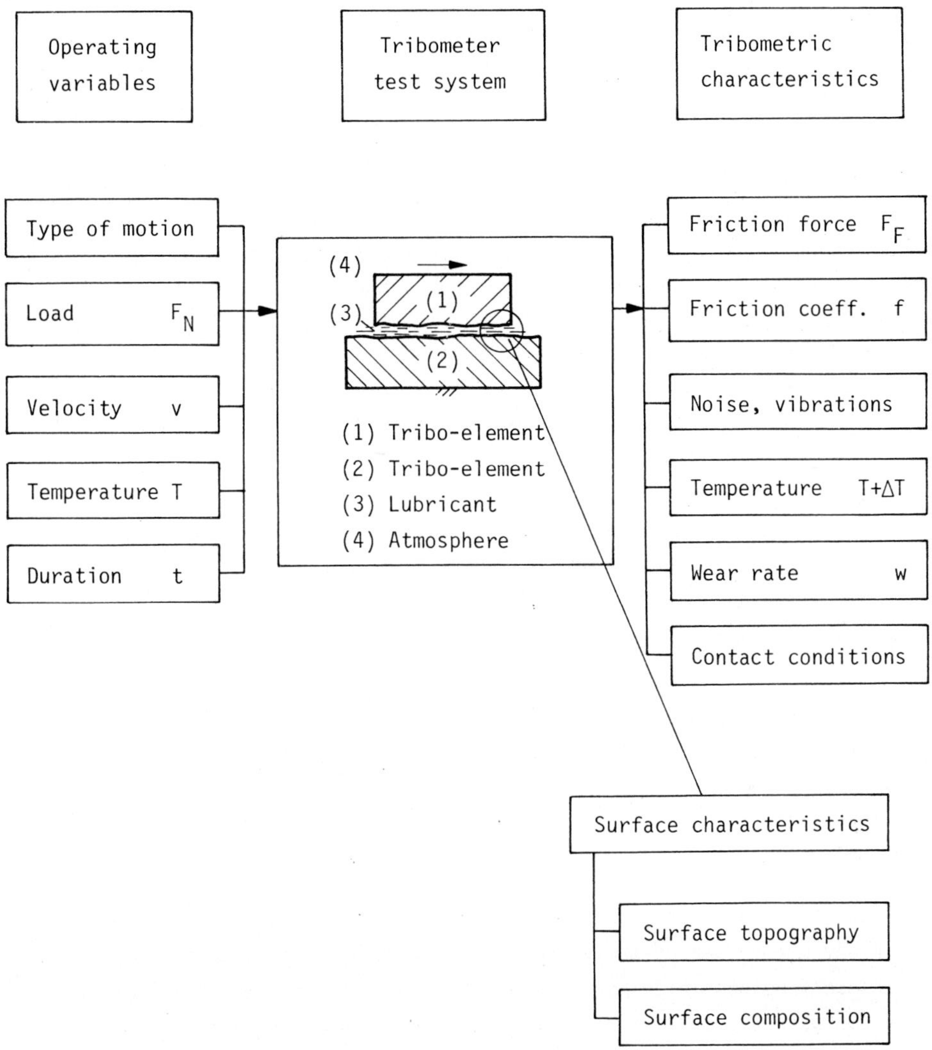

Figure 7.1 Characteristics and parameters relevant to tribo-testing.

7.2.1 TRIBOMETER TEST SYSTEM TYPES

As explained above, in most tribometers used in the laboratory, test specimens of simple geometry are utilized in order to start the tests with test conditions as well defined as possible. Usually one of the moving specimens consists of a rotating cylinder or disc against which another solid specimen of different curvature is pressed thus leading to nominal point, line or flat contact.In a compilation of friction and wear devices edited by the American Society of Lubrication Engineers (ASLE) which includes 234 test rigs (Ref. 7.6), the various test devices are classified according to their geometry into the following groups:

1. Multiple sphere
2. Crossed cylinders
3. Pin on flat (reciprocating or linear motion)
   (a) Moving pin
   (b) Moving flat
   (c) Multiple contact
4. Flat on flat (reciprocating or linear motion)
5. Rotating pins on disc (face loaded)
6. Pin on rotating disc (face loaded)
7. Cylinder on cylinder (face loaded)
8. Cylinder or pin on rotating cylinder (edge loaded)
9. Rectangular flat on rotating cylinder (edge loaded)
10. Disc on disc (edge loaded)
11. Multiple specimens
12. Miscellaneous

Besides the characterization of the test systems geometry, the other relevant properties of the elements of the test system must be specified as closely as possible. For this purpose, Part III of the "Tribological Systems Data Sheet" developed in Section 8.2 can be conveniently used (see Figure 8.5). Table 7.1 contains the relevant parameters to be specified in order to characterize the structure of the test system. In addition, the cleaning and surface preparation technique applied to the test specimens (1) and (2) should be specified.

<table>
<tr><th colspan="6">Structure of the tribo-system</th></tr>
<tr><td colspan="2">Properties of elements (initial/final)</td><td>Tribo-element (1)</td><td>Tribo-element (2)</td><td>Lubricant (3)</td><td>Atmosphere (4)</td></tr>
<tr><td colspan="2">Designation of element and material</td><td></td><td></td><td></td><td></td></tr>
<tr><td rowspan="3">Volume properties</td><td>Geometry/Dimensions/ Volume</td><td></td><td></td><td></td><td></td></tr>
<tr><td>Chemical composition</td><td></td><td></td><td></td><td></td></tr>
<tr><td>Phys.-mech. data:
Hardness
Viscosity $\eta(T,p)$
other</td><td></td><td></td><td></td><td></td></tr>
<tr><td rowspan="2">Surface properties</td><td>Topography descriptors (c.l.a.,etc)</td><td></td><td></td><td colspan="2" rowspan="2">Other data:</td></tr>
<tr><td>Surface layer data (if different from volume)</td><td></td><td></td></tr>
<tr><td colspan="2">Contact area A[ ]</td><td colspan="2"></td><td colspan="2" rowspan="3">Tribological interactions:
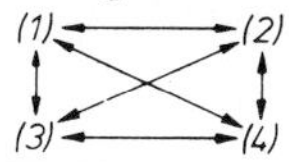
</td></tr>
<tr><td colspan="2">Ratio: $\frac{\text{contact area}}{\text{total wear track}}$ $\varepsilon[\%]$</td><td>$\varepsilon(1)$</td><td>$\varepsilon(2)$</td></tr>
<tr><td colspan="4">App. lubrication mode:</td></tr>
</table>

Table 7.1 Parameters to be specified to characterize the structure of a tribo-testing system.

A very important aspect of the description of the structure of the test system concerns the "interaction characteristics", including for instance

(i) the initial contact area and its changes during the test (Ref. 7.7),

(ii) the initial contact pressure and its changes during the test (Ref. 7.7),

(iii) the ratio of the area of contact to the whole wear track (Ref. 7.8),

(iv) the relative volumes and surface areas of the test specimens (Ref. 7.9).

These points are important for the transmission, translation and dissipation of mechanical work introduced to the test system. For example, as explained in Chapter 3 and 4 the transmission of mechanical work occurs via the contact area whereas for the emission of friction-induced acoustic or thermal energy into the environment, the surfaces not actually in contact play an important role. These points are discussed in Section 7.3 in connection with simulative tribo-testing.

7.2.2 Control of operating variables

The operating variables of a friction and wear test should be selected to be consistent with the nature of the investigation. As shown on the left of Figure 7.1, the operating variables of a tribometer are given by the following quantities:

- type of motion
- load F_N
- velocity v
- (ambient) temperature T
- test duration t

The type of motion is connected with the geometry of the test system and is given by one of the four basic motion types, namely:

- sliding
- rolling
- spin
- impact

or by a superposition of these basic types (see Section 8.2.2).

(a) LOAD

The load can be applied by various means to the test system, e.g. as a mass (dead load), by a spring, by hydraulic means or by electromagnetic means.

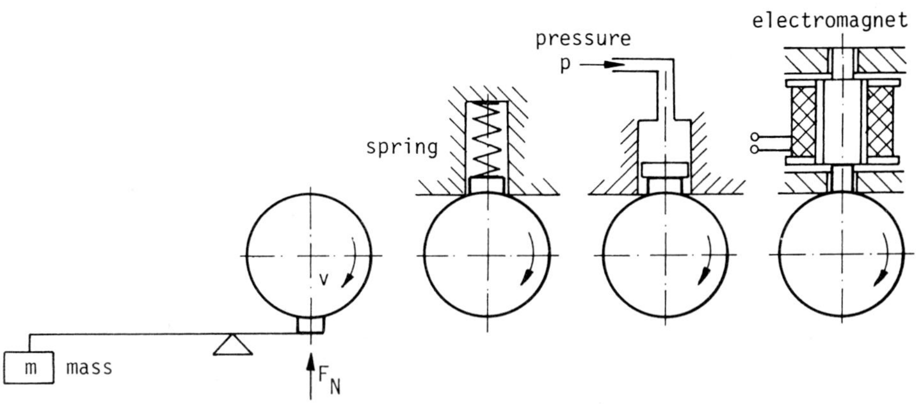

Figure 7.2 Load application methods.

In Figure 7.2 the different possibilities for the application of the load are shown schematically for the example of a pin-on-cylinder test system.

In static loading the different principles of load application lead to the same value of the load force F_N. In dynamic loading, however, owing to the different mass-spring-damper combinations inevitably involved in the different loading principles, different load-time behaviour may result. Therefore the dynamic $F_N(t)$ behaviour may be different from the initial static load F_N, (Ref. 7.10).

For the measurement of the load, force transducers based on strain gauges, inductive elements or pressure-sensitive detectors evaporated on the surface of the specimen in the contact zone are available. The measurement of the load by means of an electro-mechanical force transducer has the great advantage that the F_N-signal together with the signal of the friction force F_F can be fed to an electronic divider which calculates directly the actual value of the friction coefficient $f = F_F/F_N$ during the test. This aspect is discussed in more detail in the next section.

The load F_N applied to the test system determines, in connection with the area of contact A, the contact pressure $p = F_N/A$. Since the value of A may change due to wear during the test the value of p may also undergo some changes as function of time, as indicated in the foregoing section (Ref. 7.7).

(b) VELOCITY

For the characterization and control of the velocity between the test specimens of a tribotesting system, the same considerations as in the case of the load relate to the dynamic behaviour. For the control of the velocity, electronic or opto-electronic transducers are commercially available. The velocity is an important parameter in tribotesting in as much as it determines the lubrication regime within the Stribeck curve (see Section 4.5.1). It is also the independent variable in the product $F_F v$ which gives the friction-induced power loss of a tribo-testing system.

(c) TEMPERATURE

The (ambient) temperature of the test system is another important operating variable since it determines the thermal state of the system - at least at the initial stage of the test. (The problem of the friction-induced temperature is discussed in the next section.) For the control of the temperature, the following two techniques are utilized most often:

(i) thermocouple techniques

(ii) infrared-pyrometer techniques

The characteristics of these techniques are discussed in the next section.

(d) TEST DURATION

The set of the operating variables (load F_N, velocity v, temperature T) acts on the tribotesting system as function of time or duration t. As for the set (F_N, v, t) the test duration t also should be selected to be consistent with the nature of the investigation. Attempts are frequently made to shorten the test duration in so-called "accelerated tests" by increasing the severity of the operating conditions of F_N, v, T. This procedure, however, is only permissible if the nature of the tribological processes is not changed. Consider, for example, a certain test system running under given operating variables (F_N, v, T) with tribological processes occurring within the "outer" surface layers (see Figure 4.5) of the test specimens, thus giving a very low wear rate. Consequently, in order to obtain sufficient removal of material for an accurate wear measurement, a long test duration is needed. If, in order to shorten the test duration, the severity of the test is now increased, e.g., by increasing the load F_N, a higher wear rate may indeed result. However, the tribological processes may now act within the "inner" surface layers thus changing entirely the friction and wear mechanisms and increasing the possibility of catastrophic failure due to scuffing or seizure. This simple example indicates the importance of the proper choice of operating conditions of F_N, v, T in connection with the test duration t for a given test situation. Instead of increasing the severity of the operating conditions it may be more appropriate to increase, where possible, the sensitivity of the detectors and analyzers in order to obtain the necessary quantities within a certain test duration.

The tribo-testing operating variables discussed hitherto refer to test systems of a "closed structure", given for example by test systems in which the elements (1), (2), (3), (4) of the tribo-testing system are entirely within a closed specimen chamber as indicated in Figure 7.1. If, on the other hand, the test system structure is "open" with respect to lubricant (3) and environmental atmosphere (4), the flow rates of (3) and (4) are also important operating variables. Where, for instance, a liquid lubricant circulating system is used, the lubricant flow rate and the residence time of the lubricant film on the test specimens should be controlled, in addition to the operating variables discussed above. Another group of systems with an open system's structure concerns test systems related to industrial processes as in agriculture, mining, earth moving, drilling, dredging, etc. The operating "input" of these systems is variable due to the nature of soils, rocks, sledges, etc. Special care is needed in the testing of such systems with an "intrinsically open system structure" (Ref. 7.11).

7.2.3 TRIBOMETRIC CHARACTERISTICS

The tribometric characteristics to be measured in tribo-testing should allow proper characterization of the friction and wear processes. According to the general scheme developed in Chapter 3 these characteristics can be classified into three groups:

(i) Mechanical quantities : Friction force,
friction coefficient,
noise and vibration.

(ii) Thermal quantities : Friction- induced temperatures

(iii) Material quantities : Wear rates.

Besides these tribometric characteristics, information on the contact conditions, such as the lubricant film thickness or the amount of metallic contact, should be obtained.

(a) FRICTION FORCE AND FRICTION COEFFICIENT

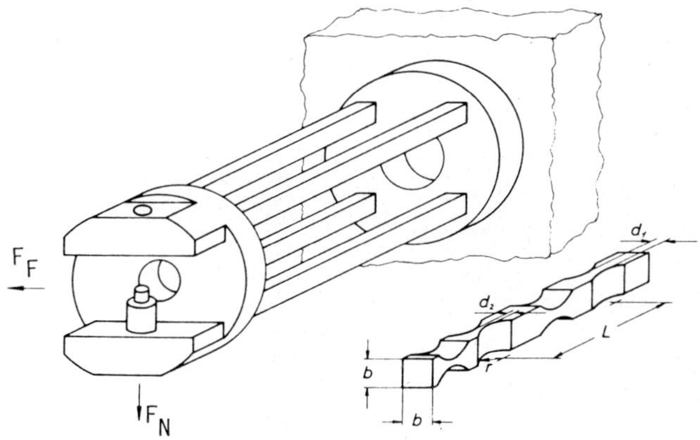

Figure 7.3 Schematic diagram of a two-component force transducer, to the right details of the design of the four rods are shown.

For the measurement of the friction force, electro-mechanical force transducers based mainly on strain-gauge techniques or inductive effects are available commercially. As indicated in the foregoing section, it is advantageous to measure both friction force F_F and load F_N and to calculate the friction coefficient directly during the test. A force transducer for the simultaneous measurement of F_N and F_F independently from each other is shown in Figure 7.3.

The force transducer is a device containing four bars with notches perpendicular to each other (Ref. 7.12). In the centre of symmetry of the four bars one of the specimens of a tribometer - for instance the pin of a pin-on-disc tribometer - is inserted. In each notch a strain gauge is fixed, detecting the strains as functions of F_N and F_F. The sensitivity and the eigenfrequency of the force transducer can be varied through the dimensions of the bars and the locations and dimensions of the notches. The block diagram of the determination of the friction coefficient from the signals of F_F and F_N by means of an electronic divider is shown in Figure 7.4. With the help of this measuring device it is possible to obtain all necessary information on the mechanics of a tribo-testing system as a function of time. Moreover, with the use of a discriminator it is possible to stop the test if certain (adjustable) levels of the friction coefficient are passed.

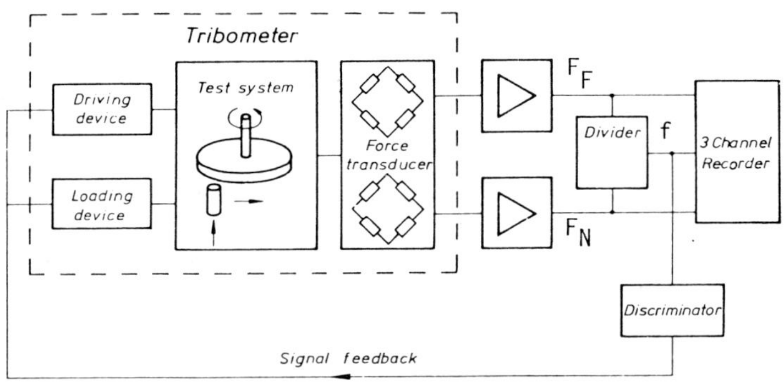

Figure 7.4 Block diagram of signal flow.

(b) NOISE AND VIBRATION

As other "mechanical" capabilities of a tribo-testing device, measurements of noise and vibrations should be mentioned (Ref. 7.13). Although important information on the contact situation of the test system may be deduced from such measurements, they have not usually been reported in connection with laboratory friction and wear test rigs. These techniques are, however, often used in monitoring the functional behaviour of complete tribo-engineering systems, such as ball bearings or gear trains (see Section 7.5).

(c) TEMPERATURE

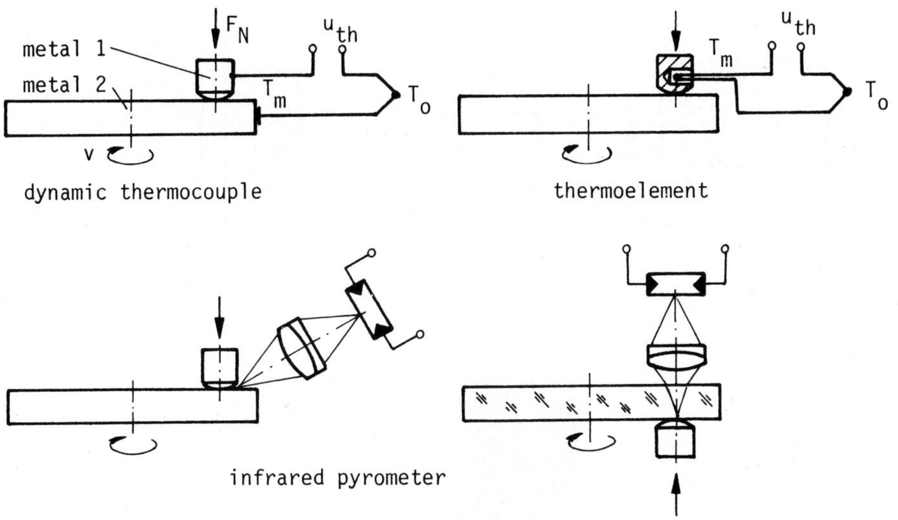

Figure 7.5 Temperature measuring methods.

The main techniques for the measurement of friction-induced temperatures are illustrated in Figure 7.5. As can be seen, in most cases thermocouples of different types or optical infrared pyrometer techniques are utilized. The techniques may be difficult to handle or the results difficult to interpret, so that it is still not easy to perform accurate and meaningful measurements of friction-induced temperatures. The thermoelements commercially available have time reponses from 10 sec to 10 msec, depending on the size and the thermocouple materials. In using thermocouples two methods may be applied:

(i) one (or both) of the actual test specimens forms part of the thermocouple ("natural thermocouple"),

(ii) one or more complete thermocouples are inserted in one or both of the test specimens.

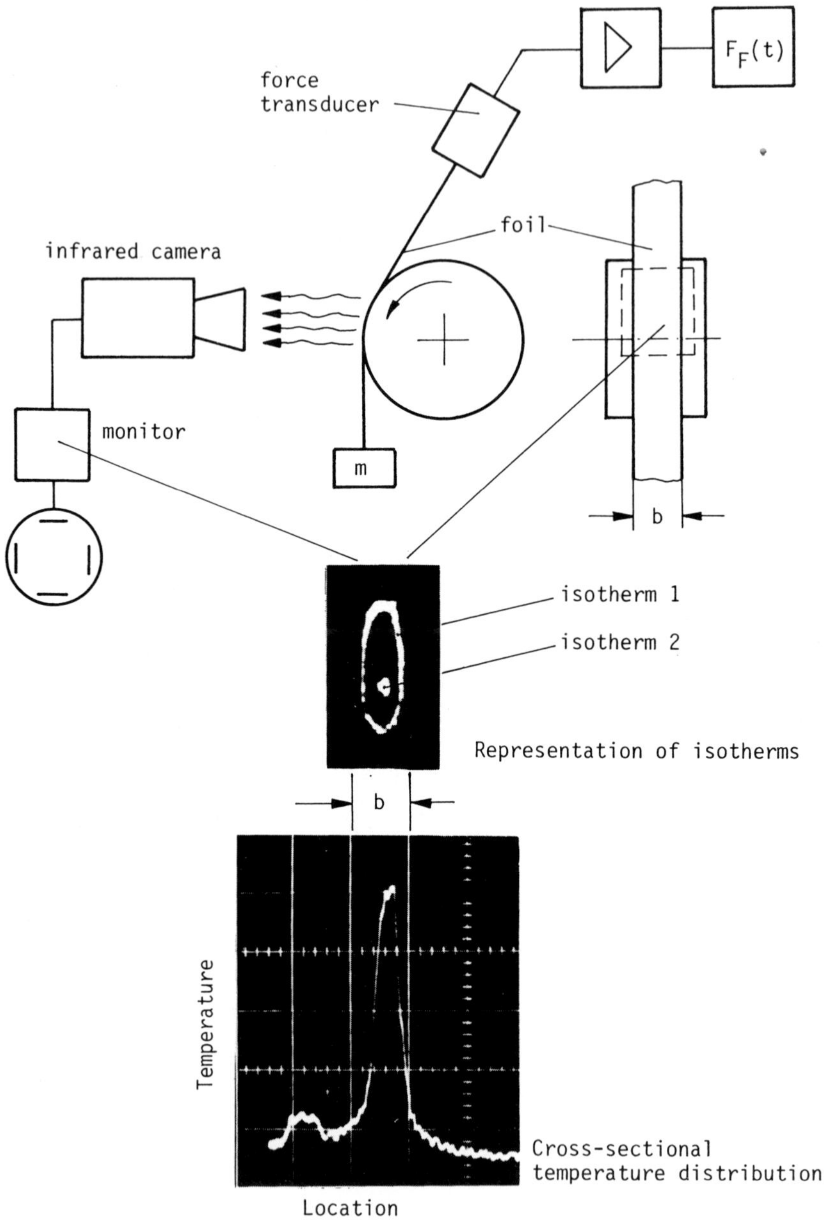

Figure 7.6 Measurement of friction-induced temperature with infrared camera.

In case (i) it is possible, in principle, to obtain information directly from the frictional interface. This information, however, may be influenced by interfacial electromotive forces resulting, for example, from the action of lubricant additives. Also circulating currents in multiple contacts may be of influence. In case (ii) it is not possible to obtain the temperature information directly from the frictional interface. Therefore, attempts have sometimes been made to determine temperature gradients by inserting several thermocouples at different distances from the interface and extrapolating from these measurements to an estimated value at the frictional interface. (For a discussion of the theories of friction-induced temperatures see Section 4.3.4).

With pyrometric techniques, either the temperature radiation emitted from the surrounding of the frictional interface of optically non-transparent specimen is detected or one partner has to be made optically transparent in the infrared region. In both instances the optical emissivity as a function of temperature and wavelength of the radiation is the crucial point. An experimental example of the application of the IR-technique to the determination of friction-induced temperatures by measuring the radiation from the back of a thin foil ($d \approx 70$ μm) sliding against a rotating shaft is illustrated in Figure 7.6 (Ref. 7.14). In a similar way, by using an optically transparent partner Winer and co-workers determined the actual temperatures within a sliding EHD contact (see Figure 4.51). In comparing these experimentally determined data with temperature calculations from the Blok-theory a good correlation between the experimental and the theoretical values has been found (Ref. 7.15).

(d) WEAR

Also for the determination of wear, different measuring techniques may be utilized (Ref. 7.16, 7.17). According to the German standard DIN 50321, "Wear-quantities", wear may be detected by measuring "direct" wear-quantities, such as

(a) changes of geometry of the specimens:
- (i) changes in linear dimensions
- (ii) changes in cross-sections
- (iii) changes in volume

(b) changes of the mass of the specimens

(c) the amount of worn material-loss

or "relative" wear-quantities, i.e., wear rates such as

(d) the wear-time-ratio (wear velocity)

(e) the wear-distance-ratio

or the wear coefficient, defined as

$$(f)\quad K' = \frac{\text{wear volume}}{\text{load}\cdot\text{distance}} \quad \left(\frac{mm^3}{Nm}\right)$$

or the dimensionless Archard's wear coefficient

$$(g)\quad K = \frac{\text{wear volume}\cdot\text{hardness}}{\text{load}\cdot\text{distance}}$$

For each type of wear measurement (a), (b), (c), appropriate detectors are commercially available. Some of the basic wear-measuring techniques are discussed in relation to the condition-monitoring of machines, see Section 7.5. The type of the measuring technique should be chosen to be consistent with the nature of the investigation, and the wear data obtained should be reported in terms that are useful for engineering applications (Ref. 7.18).

In this connection it should again be emphasized that the wear-quantities must not be misinterpreted as intrinsic material constants but must be treated as system-dependent quantities which depend on both the operating variables and the structure of the wearing system under consideration. A check-list to assist in the compilation of relevant data in connection with wear measurements is developed in Chapter 8.

Supplementing the measurement of the mechanical, thermal and materials characteristics, information on the contact condition within a tribometer may be useful. Here two groups of characteristics are of interest:

(i) the extent of real solid-solid contact (including the real area of contact)

(ii) the lubricant film thickness

For the measurements (i), the techniques of optical microscopy, described in Section 7.6, and the electrical contact resistance (ECR) measuring techniques have been applied (Ref. 7.19). The principle of the ECR technique is illustrated in the block diagram in Figure 7.7 (Ref. 7.20). A small electrical voltage (10-100 mV) is applied to the specimens and the time dependence of the voltage drop on a series resistance is detected. Usually the voltage drop fluctuates over several orders of magnitude ($< 0.1\,\Omega$ up to $> 100\ k\Omega$) with impulse durations down to the μsec range as illustrated in Figure 7.8. By rapid electronic means (discriminators, amplifiers, counters)

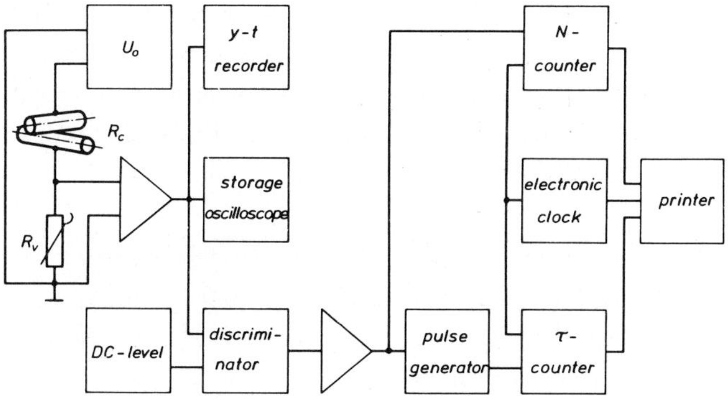

Figure 7.7 Block diagram of ECR signal processing.

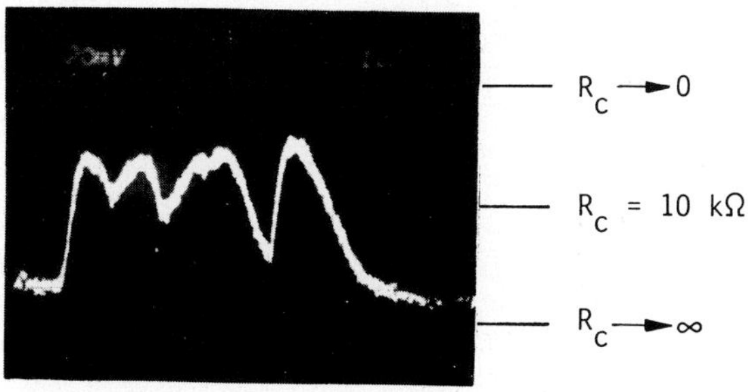

Figure 7.8 Typical ECR oscillogram of system (a), Figure 7.9.

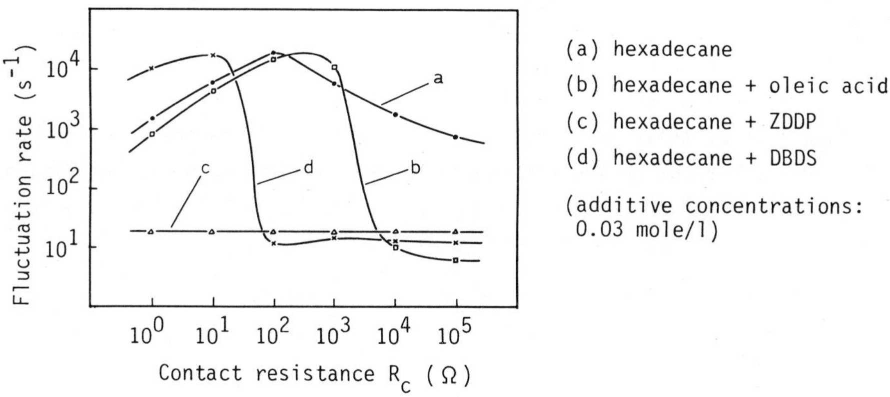

Figure 7.9 ECR fluctuation rates (lubricated sliding AISI 52100 steel cylinders, r=2.5mm, p_{oH}=0.67GN/m^2, v=1.5cm/s, T=25°C).

it is possible to calculate characteristics such as the "no-contact-time fraction" (see Figure 4.58) or impulsive fluctuation rates as shown in Figure 7.9. Although these results permit conclusions on the extent of metallic contact or on the action of lubricant additives to be drawn, great care is needed in their interpretation because of the effects of multiple contact or interfacial electromotive forces.

Lubricant film thickness (ii) may also be studied by means of optical microscopy (see Section 7.6), by the capacitance method, or by an X-ray technique. A detailed discussion of the various methods for measuring lubricant film thickness is provided in Ref. 7.21.

Having discussed the various measuring techniques relevant to friction and wear testing it is now possible to supplement the compilation of the relevant test parameters listed in Figure 7.1 by the corresponding measuring techniques. Although in tribotesting applications it may not be possible to measure all of the parameters listed in Figure 7.1 it should be emphasized once more that every parameter can influence markedly the results of a friction and wear test.

7.3 Simulative Tribo-Testing

In simulative tribo-testing an attempt is made to simulate the tribological behaviour of a practical tribo-engineering system, or the behaviour of some part of it, by means of a laboratory test system. Obviously, this field of tribometry is extremely difficult due to the complexity of tribological processes, its numerous influencing factors and the additional requirement of an appropriate "simulation criterion" (Ref. 7.22, 7.23).

The starting point in simulative tribo-testing should be the collection of available data on the practical system (PS) and the test system (TS). For the collection of these data and "casehistories", the data sheet developed in Section 8.2 can be conveniently used. Based on these data, the conditions for simulative testing may then be specified. From a formal systems point of view, the procedure of simulative tribo-testing requires:

(a) A similarity of the functions of PS and TS, i.e.:
  (i) similarity of the inputs and the outputs
  (ii) similarity of the functional input-output relations

(b) A similarity of the structures of PS and TS, i.e.:
  (i) similarity of system elements

(ii) similarity of system element properties

(iii) similarity of tribological interactions

In trying to fulfil these conditions, first the materials, the lubricants and the atmospheric environment should be chosen to be identical in both PS and TS. Then the geometrical and contact conditions of the test system should be adjusted to those of the practical system, taking into account the problem of appropriate scale factors. Finally, the operating variables should be adjusted in order to obtain the same tribological interactions in the test system as in the practical system. Clearly, simulative tribotesting as outlined in these few sentences is an extremely difficult task and a generally accepted procedure is not yet available. In order to obtain at least a kind of guideline, some of the main aspects of simulative tribotesting, will be discussed briefly here, considering as a starting point a pin-on-disc tribometer test system, shown in Figure 7.10.

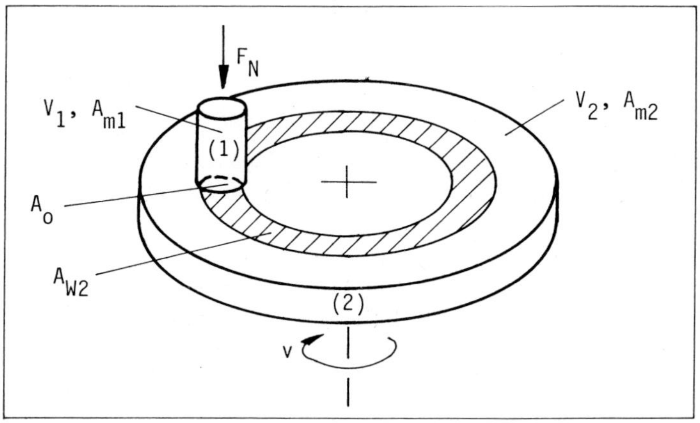

Figure 7.10 Geometrical characteristics of a tribotesting model system.

For the task of simulative tribo-testing, the test system shown may be applicable for pure continuous sliding with the materials (1), (2) identical with those of the practical system. The conditions of dry or lubricated sliding should be realized with the atmosphere and lubricant identical to those in the practical system. Then the important "system-independent" properties of the elements, including chemical composition, elastic modulus, hardness, viscosity, etc. will inherently be identical in PS and TS. After the choice of the elements of the system and their system-independent properties, the geometrical and contact conditions of the test system should be compared with those of the practical system. In this connection the quantities listed in Table 7.2 are important.

Geometric or contact properties	System elements	
	(1)	(2)
Volume	V_1	V_2
Mantle surface	A_{m1}	A_{m2}
Contact area (geom.)	A_o	
Contact area (real)	A_r	
No-contact area	$A_{m1}-A_o$	$A_{m2}-A_o$
Wear track area	A_{W1}	A_{W2}

Table 7.2 Geometric characteristics of test specimens.

The characteristics compiled in Table 7.2 are not only important for the contact pressures $p_o = F_N/A_o$ and $p_r = F_N/A_r$ they are also relevant to the processes of the transmission and dissipation of mechanical work in the system, as explained in Table 7.3.

Tribological process	Relevant geometrical quantity
Generation of frictional heat	Contact area (A_o, A_r)
Absorption of frictional heat	Volume (V_1, V_2)
Emission of frictional heat to environment	No-contact area ($A_{m1}-A_o$), ($A_{m2}-A_o$)

Table 7.3 Relevance of geometrical quantities for tribological processes.

Besides these geometric characteristics, relative quantities may also be important, for example the ratio of the heat-emitting area to the heat-absorbing volume of each of the elements (1) and (2):

$$o = \text{ratio} : \frac{\text{heat emitting area}}{\text{heat absorbing volume}}$$

$$o_{1,2} = \frac{A_{m1,2} - A_o}{V_{1,2}}$$

Another important relative quantity is given by:

$$\varepsilon = \text{ratio} : \frac{\text{contact area}}{\text{wear track area}}$$

$$\varepsilon_{1,2} = \frac{A_o}{A_{W1,2}}$$

For the example of the pin-on-disc tribometer, it follows that

for the pin : $\varepsilon_1 = 1$

for the disc : $\varepsilon_2 < 1$

Characteristics related to the value of ε are compiled in Table 7.4.

Characteristics of elements for which $\varepsilon = 1$	Characteristics of elements for which $\varepsilon < 1$
permanent contact	intermittent contact
permanent friction heating	intermittent friction heating
no direct influence of atmospheric environment on wear area	influence of atmospheric environment on the area $A_W - A_o$
no macroscopic cyclic stressing	cyclic stressing

Table 7.4 Dependence of tribological processes on the contact area to wear-track-area ratio ε.

The discussion of the geometric and contact conditions show that the problem of appropriate "scale factors" for the simulation of these quantities should be paid great attention. An example of the determination of scale factors based on similarity theory and dimensional analysis for the simulative testing of a drum brake was reported by Pogosian (Ref. 7.22).

An important point in simulative tribo-testing is the realization of the same tribological processes as in the practical system. For this purpose, the operating variables should be adjusted in connection with the geometric and contact conditions in order to obtain the same friction and wear processes in TS and PS. A simulation procedure for the design of high-performance dry bearings, suggested by Play and Godet (Ref. 7.24), is based on the assumption that identical coefficients of friction and even wear rates in the TS and the PS are obtained for equal average pressures, velocities, contact temperatures and identical contact types. In an iterative procedure, convergence between two temperatures is sought in this method. First an experimental simulation of pressure and velocities found in the mechanism to be designed is set up and temperatures are measured. The thermal path of the simulator can be modified by different degrees of cooling. Second, temperatures in the actual mechanism are calculated by finite-element techniques for the coefficient of friction found experimentally. The theoretical and experimental temperatures are compared and the simulator cooling conditions are modified until both temperatures are equal. The problem is considered to be solved when pressures, velocities and temperatures are correctly simulated.

Another procedure for simulative testing of the tribological behaviour of bearing materials to be used in oil lubricated bearings has been proposed by de Gee (Ref. 7.25). This procedure starts from a consideration of the friction and lubrication state in the Stribeck curves of both the practical system (journal bearing) and the test system (pin-on-ring tribometer).

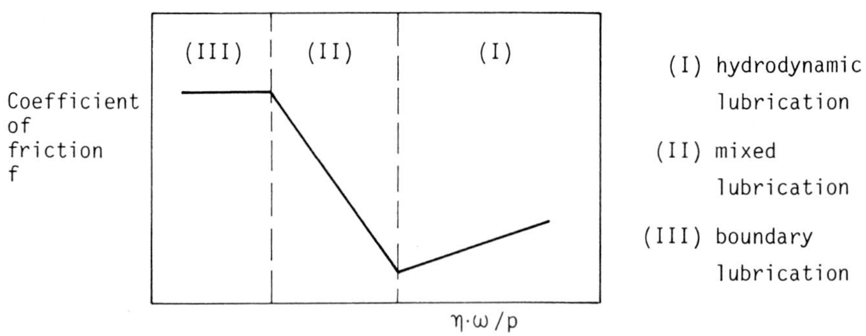

Figure 7.11 Simplified form of the curve of friction coefficient f vs bearing characteristic number $\eta\omega/p$.

In Figure 7.11, a simplified form of the curve of coefficient of friction f versus bearing characteristic number $\frac{\eta\omega}{p}$ is shown, i.e., the Stribeck curve, discussed in Section 4.5.1, (η is the dynamic viscosity of the lubricant, ω the angular velocity and p the load per unit projected area). In the simulative tribotesting procedure suggested by de Gee, the same lubrication state in the test system as in the practical system is established, as far as possible, by adjusting the operating conditions in the test system until f becomes independent of the parameter $\frac{\eta\omega}{p}$. This implies boundary lubrication only and is done in the following steps:

(i) Use the same materials, lubricants and atmospheres in TS and PS.

(ii) Choose an appropriate surface roughness of the test specimens. To a first approximation it may be assumed that the R_a value of a machined surface increases with increasing radius of curvature of the journal, r_{PS}, according to

$$R_a \approx 1.2 \left[r_{PS}\right]^{1/2} \qquad R_a \text{ in } \mu m \text{ and } r_{PS} \text{ in m}$$

(iii) Apply a realistic normal load $F_{N_{TS}}$ to the test system. The most widely acceptable compromise is to make the force per unit projected pin area in the TS equal to the force per unit projected bearing area of the PS.

(iv) Choose a temperature T_{TS} of the lubricant bath of the TS corresponding to that of the PS. Because of friction heating the selection of a realistic temperature may be difficult. In such cases it may be wise to introduce the test temperature as an independent parameter, performing tests at different temperatures and presenting separate results for each temperature level.

(v) Establish for all materials or material-lubricant combinations to be tested, the lower limit of the angular velocity $\bar{\omega}$ for which the friction coefficient f becomes independent of ω. If the lowest $\bar{\omega}$ value thus established is $\bar{\omega}_{min}$, subsequent experiments are performed at a constant ω value, $\omega_{test} < \bar{\omega}_{min}$.

(vi) At the conditions of

load F_N = $F_{N_{TS}}$
angular velocity ω = ω_{TS}
temperature T = T_{TS}

perform experiments with all material-lubricant combinations to be tested. Measure friction during the entire test period and measure wear of the stationary specimen during or after termination of the test. The test duration t should be long enough to allow for determination of the total amount of wear by weight-loss measurement.

By applying this simulative tribo-testing procedure, de Gee obtained a final ranking of candidate bearing alloys to be used in lubricated journal bearings.

In the simulative tribo-testing procedure proposed by de Gee, the friction and lubrication state of the system (specified as location within the Stribeck curve) served as "simulation criterion". As other "simulation criteria" which indicate a similarity of tribological processes within the test system and the practical system the appearance of worn surfaces or vibration characteristics may be used.

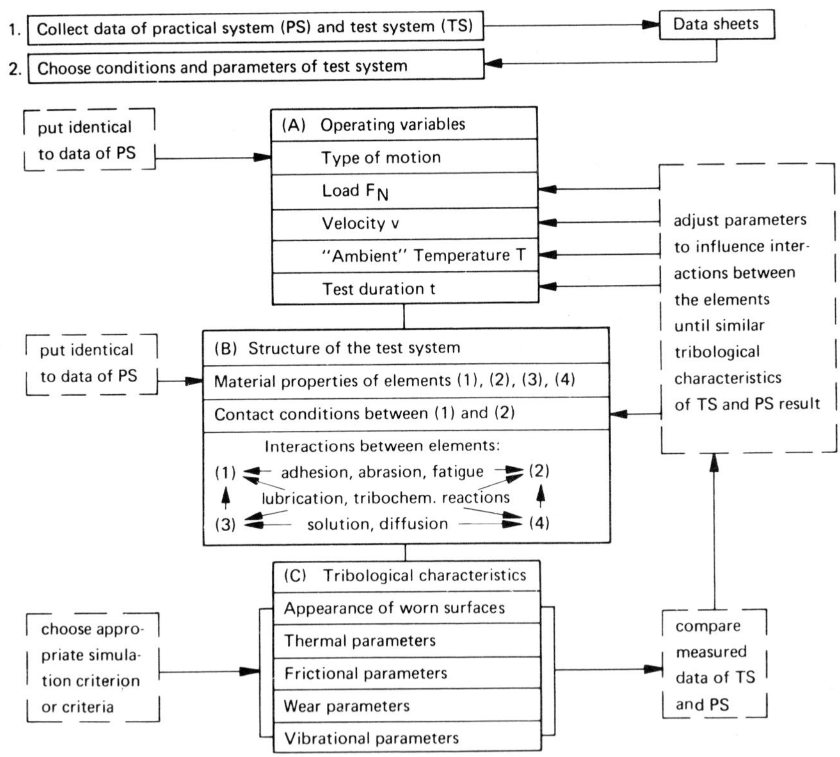

Figure 7.12 Schematic outline for simulative tribo-testing.

In summarizing the various aspects of simulative tribo-testing, from a systems point of view, a guide-line as outlined in Figure 7.12 may be recommended. According to Figure 7.12 we may proceed through the following steps in choosing the conditions of a test system in an attempt at simulative testing:

(i) make the type of motion and the material properties of the elements (1), (2), (3), (4) identical in both TS and PS.

(ii) choose an appropriate simulation criterion or simulation criteria,

(iii) adjust the operating variables (F_N, v, T, t) as well as the contact conditions until similar tribological characteristics of TS and PS are observed.

Although it is clear, that a procedure as indicated in Figure 7.12 can only be regarded as a very rough outline, successful simulative tribotesting has, in fact, been based on principles similar to the guide-line of Figure 7.12, see Section 8.5. For example, Heinke has reported a simulation of the tribological behaviour of a vane pump (PS) by means of a pin-on-disc apparatus (TS) (Ref. 7.26). In this instance, the appearance of the worn surfaces of PS and TS, studied by means of a scanning electron microscope (SEM), served as a simulation criterion. If Heinke's study is represented in the manner of Figure 7.12, a schematic diagram for his procedure appears as shown in Figure 7.13.

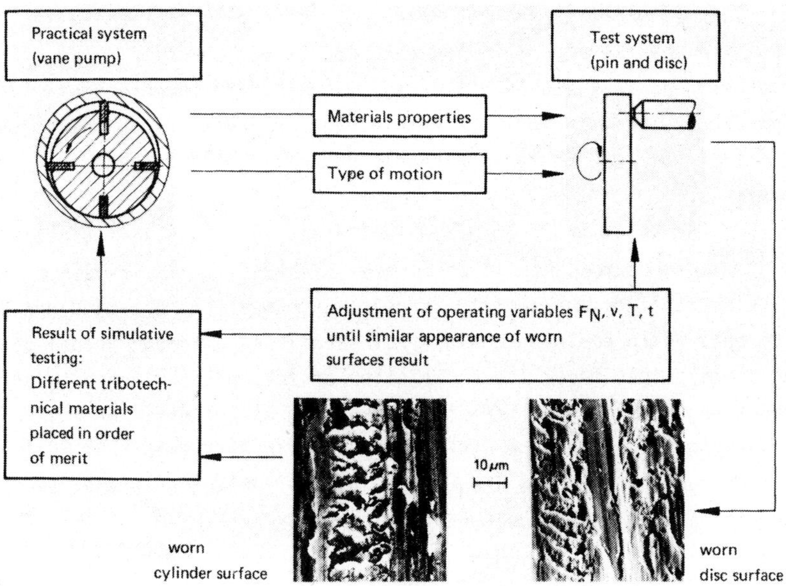

Figure 7.13 An example of simulative tribo-testing (Ref. 7.26).

It can be seen that, as a first step, the materials and the type of motion were made identical. Then, the values of the operating variables (F_N, v, T, t) were adjusted until the SEM pictures of the worn surfaces in the TS and the PS were the same. The result of the simulative testing in this case was the placing of different tribo-technical materials in an order of merit.

7.4 Testing of Tribo-Technical Components

The characterization of the performance of a given tribo-engineering system requires its practice-oriented testing. These tests should be performed with the actual components of the tribo-engineering system. Due to the great variety of tribo-mechanical systems (see Appendix A) and their components (see Appendix B) the test procedures may vary widely in scope and performance. Therefore, in this section only some general remarks will be made.

The testing of tribo-technical components can be broadly classified into (a) the testing of tribo-technical machine elements and (b) the testing of lubricants.

(a) TESTING OF TRIBO-TECHNICAL MACHINE ELEMENTS

In the testing of actual tribo-technical machine elements some of the friction and wear measuring techniques described in Section 7.2 may be applicable. For the measurement of mechanical and thermal quantities (such as contact pressures, friction forces or friction-induced temperatures) force-detecting transducers, including strain gauges or pressure- or temperature-sensitive coatings, may be fixed directly on the machine element to be investigated.

In some cases, in order to avoid full-scale testing a semi-practical testing of machine components is performed. Consider, as an example of this type of tribo-testing, the testing of gears. A number of test machines have been developed which employ gears as test specimens. Although actual gears are used as test specimens, the correlation of test performance with service performance requires careful control of the test parameters. For instance, if the gear-teeth-ratio of the gear test system is different from that of the practical gear system the frequencies of the contact of a gear tooth of one gear with a given gear tooth of the other gear may be different, thus leading to different running-in behaviour of the two systems.

Machine name	IAE	Thornton	FZG	Shell	Ryder	Western gear
Centre dist. (mm)	82.6	127	91.4	76.2	88.9	50.8
Gear face (mm)	4.8	10.8	20.1	6.4	6.4	9.5
Number teeth	15-16	24-26	16-24	17-19	28-28	32-32
Pitch diam. (mm)	121.1	127	127	152.4	254	406.4
H.P. of drive	4.5	30	30	40	40	40
Max speed (rpm)	6000	12000	12000	30000	15000	30000
Max load (kN per mm width)	2.29	0.89	0.75	1.26	0.89	1.07
Load applied while running	no	no	no	yes	yes	yes

Table 7.5 Characteristics of some spur gear rigs (Ref. 7.27).

The most widely used gear test machines employ spur gears. It is not possible here to go into the details of the various types, but the characteristics of six common spur-gear test machines are presented in Table 7.5 as compiled by Brown (Ref. 7.27). These test-machines all employ the power-circulating or "four-square" principle. The application of this principle to gear testing was suggested first by Kutzbach in 1926 (Ref. 7.28). He also showed that it has the advantage that power is needed only to overcome the friction in the test and slave gears and their support bearings, see Figure 7.14.

With gear test machines, the following main tribometric characteristics are determined in order to describe the tribological behaviour of a gear set:

(a) friction losses
(b) wear resistance
(c) surface fatigue
(d) scuff load (load-carrying capacity)

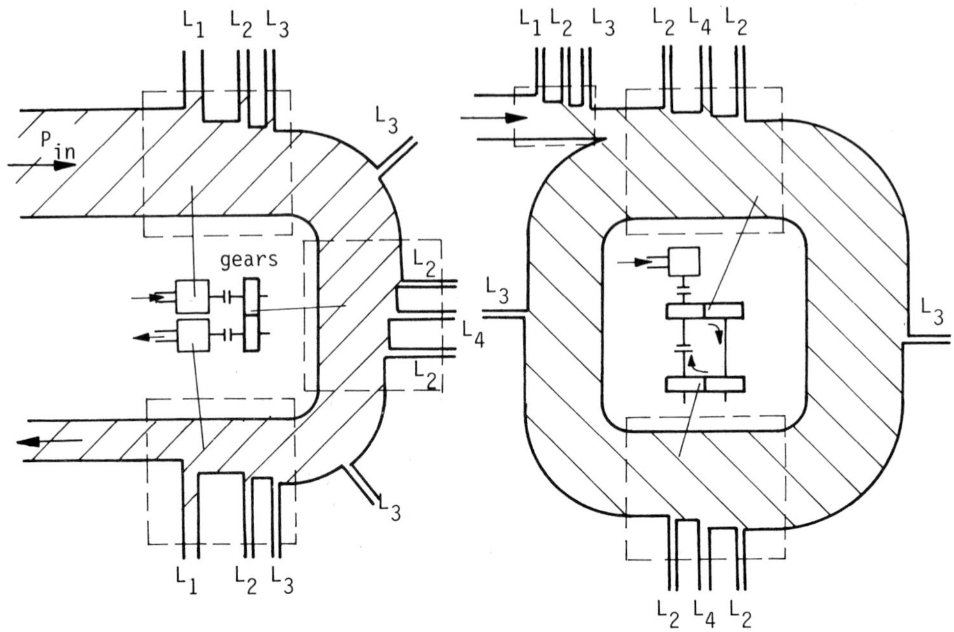

Losses: L_1 electrical; L_2 bearing-friction; L_3 air-friction; L_4 gear-friction

Figure 7.14 Power transmission and power circulating principles (Ref. 7.28).

These tribometric characteristics are in most cases also determined in the testing of other tribo-technical components, for example ball and roller bearings. As mentioned above, these tribometric characteristics concentrate, however, on the loss-output function and the wear-induced change of the system's structure. In addition to this type of testing of tribo-technical components, for the characterization of the performance of actual machinery the control and monitoring of the use-function of the tribo-engineering system is needed. This aspects of tribometry is discussed in the next section.

(b) TESTING OF LUBRICANTS

Also for the testing of the other group of tribo-technical components, namely the lubricants, several aspects must be considered. As in the characterization of every component of a tribological system, the testing and characterization of system-independent and system-dependent properties of lubricants should be distinguished. The system-independent lubricant characteristics describe the properties "per se" of the lubricant, independently of the tribo-engineering application of the lubricant. The main lubricant characteristics falling into this category are:

1. chemical composition (including content of S, P, Cl, H_2O)
2. dirt content (including ash content)
3. density and specific gravity
4. specific heat and thermal conductivity
5. dielectric constant
6. acidity, basicity, neutralization number
7. aniline point
8. flash and fire point
9. cloud, pour and flock points
10. hydrolytic stability
11. oxidation characteristics
12. evaporation and volatility
13. interfacial tension
14. emulsion characteristic
15. foaming
16. viscosity
17. viscosity-temperature characteristic
18. viscosity-pressure characteristic

For the testing and specification of these system-independent lubricant properties and characteristics, well-known tests have been worked out and standardized. This has been done, for instance, in the USA by the American Society for Testing and Materials, Committee on Standards on Petrolem Products and Lubricants (ASTM-D2), in England by the Institute of Petroleum (IP) and in the Federal Republic of Germany by the Fachausschuß Mineralöl- und Brennstoffnormung im Deutschen Institut für Normung (FAM-DIN). The details of the various tests can be found in the official publications of these institutions; for general compilations of these tests the reader is referred to Refs. 7.29 to 7.31.

In contrast to the standardized tests for the system-independent physical, chemical and technological properties of lubricants, the testing of the functional behaviour of lubricants should be performed in connection with the technical function of the actual tribo-engineering system in which the lubricant is used. These tests, assess predominantly the overall ability of a lubricant to permit rubbing surfaces to operate without scuffing seizing or other manifestation of material destruction. They can be broadly classified in three groups (Ref. 7.32):

(i) Testing of lubricants in simplified bench tests.

In these tests, the performance of lubricants is tested with simplified test geometries leading to point, line or flat contact. The more familiar tests of this type are the Four Ball, Timken, SAE or Amsler, Falex and Almen. Most of these tests were devised to differentiate between EP and non-EP oils, and their accuracy is sometimes not good enough to grade different levels of EP activity. Erratic results can occur if operating variables (e.g. the temperature of the lubricant) are not closely controlled. As mentioned above, the term "hybrid testers" was proposed as neither service conditions nor physically well defined parameters were reproduced (Ref. 7.2). Predicting the performance of lubricants on the basis of these tests alone is almost impossible. On the other hand, they are convenient for acceptance testing, production control and they are good indicators of batch variations of lubricants.

(ii) Testing of lubricants with tribo-technical components.

Because of the above shortcomings, a different type of lubricant testing is required to permit control of as many variables as possible while simulating actual performance requirements. A convenient way of doing this is to test lubricants in the laboratory, where operating conditions can be controlled, the parts under test being those used in the complete tribo-engineering unit. The best known examples of this type of lubricant testing are the laboratory tests performed with gear-test rigs as compiled in Table 7.5. Consider, as a further example of this type of testing, the testing of hypoid gear lubricants (Ref. 7.33). In these tests, full-scale rear axles are set up on laboratory test beds and driven by engines and transmissions while load (F_N) is applied through absorption dynamometers on the driven axles at certain velocities (v) and oil temperatures (T) for a given period of time (t). After the test program is completed, gears are inspected for signs of scuffing, wear or other distress and the lubricant rated as either pass or fail. The test procedure is based on correlation with field tests on similar axles in full-scale vehicles, which also showed differences between the lubricants.

(iii) Testing of lubricants in full-scale tests

There is general agreement that the only satisfactory means of evaluating the performance characteristics of lubricants is by full-scale tests of its actual use in tribo-engineering systems. Although these tests vary broadly in scope and procedure, they may be classified according to the broad categories which account for the majority of lubricants consumed, namely

(a) General machinery lubricants
(b) Hydraulic fluids and turbine oils
(c) Internal combustion engine oils
(d) Steam cylinder oils
(e) Compressor oils
(f) Gear oils

Test procedures for the full-scale testing of these main groups of lubricants can be found in Refs. 7.34 and 7.35.

Although the practical value of full-scale testing of lubricants is agreed, the cost of field or proving-ground tests is considerable, so that this type of tribo-testing is generally used only as final proof of the decisions made while developing the design of an actual tribo-engineering system.

7.5 Machinery Condition Monitoring

With the increase of sophistication of contemporary machinery any disturbance or failure of the proper functioning of complex machines will tend to have serious consequences. To illustrate the costs which can be involved, it has been reported that a simple bearing failure in a fully integrated steel mill can lead to shut-down which at full output rate may cost ¢ 150 - ¢ 300 per minute. A similar bearing failure in a modern generator set could involve a loss of ¢1 - ¢20 per minute until the set was again operational (Ref. 7.36). In order to prevent such failures of tribo-engineering systems, increasing attention is now being given to tribometric machinery condition or "health" monitoring techniques.

In recent years, sensitive detectors and techniques for the diagnosis of malfunctions in machinery have been developed (Ref. 7.37). Emphasis is being given to "on-line monitoring techniques" as a means of detecting deterioration so that remedial action can be taken before the breakdown point is reached. In addition, there is growing recognition that future progress cannot rely upon diagnosis alone. It is becoming increasingly important to monitor the "state of health" of machinery on either an intermittent or continuous basis. A prognostic approach can then be adopted and remedial action can be taken at a convenient time in the normal cycle of operation of the equipment (Ref. 7.38).

The systems approach as applied throughout this volume can be conveniently used for the classification and description of the tribometric machinery condition monitoring techniques. Since the characterization of a tribo-engineering system requires the description of the system function as well as its structure, the machinery condition monitoring techniques can be broadly classified in (a) tribometric techniques for the monitoring of the system function, and (b) tribometric techniques for the monitoring of the system structure.

(a) MONITORING OF THE FUNCTION OF TRIBO-MECHANICAL SYSTEMS

The monitoring of the system's use-function would require the control of operating input variables and its corresponding use-outputs, namely:

- the type of motion,
- the input and output forces, or contact pressures,
- the input and output velocities,
- the temperatures at the input and the output.

Although the use-function of tribo-engineering systems is connected with the above quantities, information on the proper function of a given tribo-engineering system is often indirectly obtained by monitoring the loss-outputs corresponding to the use-input and use-output quantities (Ref. 7.39). For the movement from diagnosis towards prognosis the signals from the loss-outputs indicating the ailments of machinery may be fed back by means of servo-control equipments or microprocessors to the input, thus influencing the operational inputs in order to establish a proper functioning of the whole system.

According to the general scheme of the transmission and transformation of mechanical work within a tribo-mechanical system as illustrated in Figure 3.4, the following loss-output quantities should be monitored:

(i) Friction force and frictional changes

An increase or decrease in the friction of tribo-engineering systems like bearings, gears etc. can be a pointer towards decreased performance and incipient failure (see Section 4.5.6).

(ii) Noise and vibration
Changes in the noise spectrum associated with particular tribo-engineering components operating under normal conditions in known environments can indicate incipient failure. Two types of acoustic signals may be distinguished:

- noise fields due to vibrations, impacts or aerodynamic processes emitted in a relatively low frequency range (10 to 20 kHz),
- impuls-like acoustic signals of low amplitude due to microstructural changes, like micro-cracking, emitted in a frequency range of about 50 kHz to 1.5 MHz (acoustic emission) (Ref. 7.40).

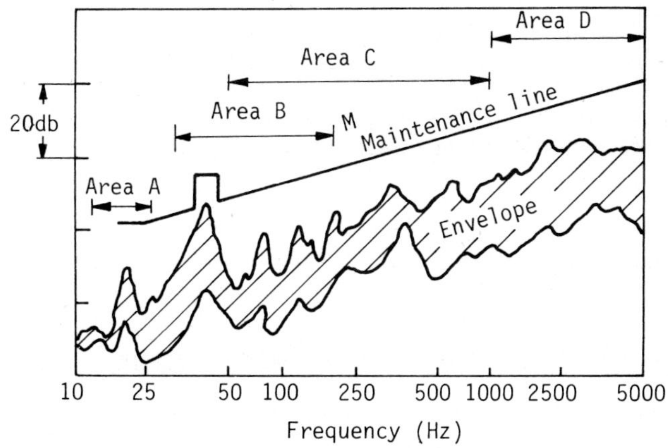

Figure 7.15 Noise diagram of submarine engine (Ref. 7.41).

As a practical example of the noise-analysis of tribo-engineering systems, in Figure 7.15 a noise diagram is shown which is used as maintenance indicator for the monitoring of submarine engines. The different frequency ranges A to D correspond to different types of mechanical faults (Ref. 7.41).

(iii) Heat emission and temperature changes
Tribo-induced heat and temperature rise may lead to thermal distortion and thermal stresses and may both adversely affect the mechanical strength of the machine components and the properties of the lubricant, thus influencing the performance and safety of the machine. In the condition monitoring of rolling element bearings and sliding bearings, the temperature levels compiled in Table 7.6 are recommended as thresholds for the alarm or the stoppage of the bearing (Ref. 7.42).

Location	Temperature level (^{o}C)			
	Rolling element bearing		Sliding bearing	
	alarm	stop	alarm	stop
1. housing	70	90		
2. rings	90	120		
3. bearing bushing (2mm below minimum oil gap)			90	120
4. lubricant within bearing	80	100	80	100
5. lubricant outlet	80	100		

Table 7.6 Temperate data relevant to the monitoring of rolling element bearings and sliding bearings.

For the monitoring of characteristics (i), (ii), (iii) a host of sensing devices is available today, based on the measuring techniques described in detail in Section 7.2.

(b) MONITORING OF THE STRUCTURE OF TRIBO-MECHANICAL SYSTEMS

Besides the monitoring of the input and output parameters of a tribo-engineering system indicating the (momentary) functional behaviour of the system, the monitoring of parameters characterizing the state of the system's structure is required. In this connection attention should be drawn to the important difference between the information and the conclusions that follow from the monitoring of functional and structural parameters. These differences can be easily explained in considering the conceptual tribo-process diagram of a given tribo-mechanical system, see Figure 3.4. If only the functional parameters on the conceptual functional plane of a given tribo-engineering system are monitored, the system structure may undergo unperceived gradual changes, for example on the conceptual material planes, leading after a certain operating duration to an unexpected failure although during the whole time the functional parameters remained in their proper range. If, on the other hand, the monitoring of the relevant parameters of the structure of the system indicate a constancy of the system structure and if, in addition, the functional inputs are within their proper range, the occurrence of a sudden failure is rather unlikely. Although the on-line monitoring of the characteristics of the system structure is, in general, more difficult than the monitoring of the characteristics of the functional inputs and outputs, convenient detectors have been developed for this purpose also in recent years.

The following characteristics may be monitored as an indication of the actual state of the structure of a tribo-engineering system:

(i) Lubricant supply and lubrication mode

If the volume of lubricant to the component is inadequate or if the physical and chemical properties of the lubricant change in service, the lubrication mode may change leading to a deterioration of the performance of the whole system. Methods of monitoring oil supply of a machine are well known and have long been practiced. The techniques involved range from the visual checking of the oil level in the sump or oil tank using a sight glass, to the insertion of oil pressure gauges and oil flow meters. These detectors are often connected to an automatic alarm system so that the oil flow is monitored continuously. Further, the lubricant quality should be monitored by taking samples of the used oils at intervals and subjecting them to laboratory tests to determine whether relevant lubricant properties have changed.

(ii) Surface conditions of machine components

Changes in surface roughness or the appearance of pits or cracks are a powerful indicator of incipient failure, particularly in rolling elements and gears. The techniques for the investigation of surfaces are discussed in some detail in Section 7.6. The application of these techniques in machinery condition monitoring is hindered by the fact that moving surfaces have to be investigated. To overcome these difficulties attempts have been made to use lasers for the in situ detection of surface roughness characteristics. Another possibility is to make a replica of the surface to be monitored. It has been reported, for example, that the surface examination of a hypoid gear may be carried out by such methods without removing it from its housing (Ref. 7.43).

(iii) Wear behaviour of the system

As characteristics of the wear-induced changes of the structure of tribomechanical systems, (a) geometrical changes of the moving components or (b) worn material loss rates may be monitored. As an example of class (a) in Figure 7.16 a method for piston ring condition monitoring is shown.

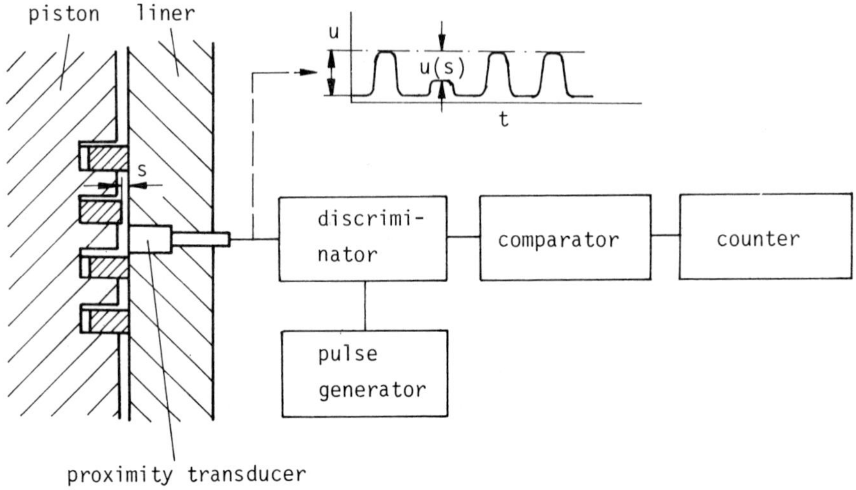

Figure 7.16 Schematic representation of piston-ring-condition monitoring technique.

The piston ring condition is sensed by means of proximity transducers, which are mounted flush with the running surface of the cylinders. The transducer detects rings not in proper contact with the cylinder, as well as broken piston rings. From the electronic signal-processing equipment shown in Figure 7.16, information can be obtained about the piston ring condition.

Various techniques are also available for the monitoring of wear output rates. Careful examination of debris or lubricant contaminants can indicate their origin and allow conclusion to be drawn about their formation, and hence the condition of inaccessible moving parts. For example, it is not possible to examine in situ the working parts of a jet engine, but each drop of lubricant which circulates through the moving parts carries with it evidence of its experience in passage. Careful examination of the oil and any particles it may contain allows interpretation of the conditions. The wear debris contained in the lubricants in operating mechanisms may be monitored by the following methods:

(α) Spectrographic oil analysis procedure (SOAP)

Very small concentrations of metallic wear products (1-2 part/million) suspended in used lubricating oil can be identified by spectrographic analysis (Ref. 7.44). Using the information thus obtained together with a knowledge of chemical composition of oil-wetted parts, it is possible to determine which components are wearing. For example, abnormal levels of iron in the sump oil of a Diesel engine can indicate excessive cylinder bore wear.

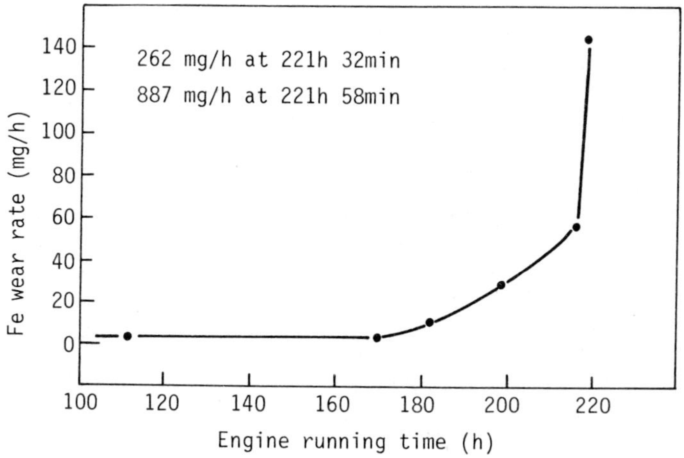

Figure 7.17 Spectrometric analysis - Olympus engine 59332 in Concorde prototype aircraft (Ref. 7.45).

As an example of the application of the SOAP technique in Figure 7.17 a typical plot of the Fe-rate in an Olympus engine installed in the concorde prototype aircraft is shown (Ref. 7.45). The rapid increase at 220 hours was found to arise from heavy fretting wear on the bevel spline location on the layshaft driven by the port gear box.

(β) Magnetic chip detectors

The use of a magnet in a lubrication system provides a simple and effective method for monitoring the contamination of liquids. A strong magnet will attract particles in an oil stream, as for instance metal flakes such as arise from fatigue fragmentation. Use of this detector involves removal of the magnetic probe at regular intervals and the insertion of a fresh probe while the original is retained for the assessment of the particles adhering to it.

(γ) Radioactive tracer methods

The use of radioisotopes, artifically produced by neutron irradiation offers a particularly convenient method for following the movement of material during deformation, transfer, or the formation of wear debris. In recent years, a great reduction in background radiation by implanting radioactive ions instead of activate the whole sample has been obtained. In using a thin-layer activation technique it is possible to differentiate between the wear from different parts of moving machine elements.

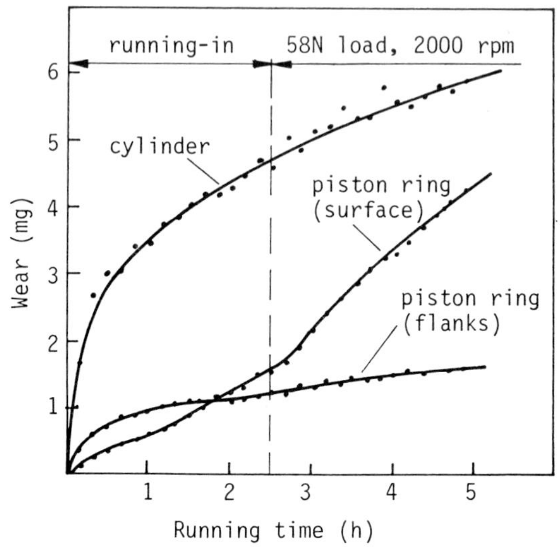

Figure 7.18 Three-component wear measurement with radioisotopes (Ref. 7.46).

As an example (Ref. 7.46), in Figure 7.18 wear curves from the different parts of a Diesel engine, namely the cylinder liner, the running surface and the flanks of the piston rings are shown. The possibilities of the application of radioactive tracer in machinery condition monitoring have been reviewed in detail by Gervé (Ref. 7.46).

(δ) Ferrography

Ferrography is a technique developed to separate wear debris from the lubricant and spread it according to size on a transparent substrate for examination in an optical or scanning electron microscope (Ref. 7.47). The Ferrograph analyzer consists of a pump to deliver a sample at low rate, a magnet to provide a high-gradient magnetic field near its poles and a treated transparent substrate (Ferrogram slide) on which the particles are deposited. The oil sample, diluted with a special solvent to promote the

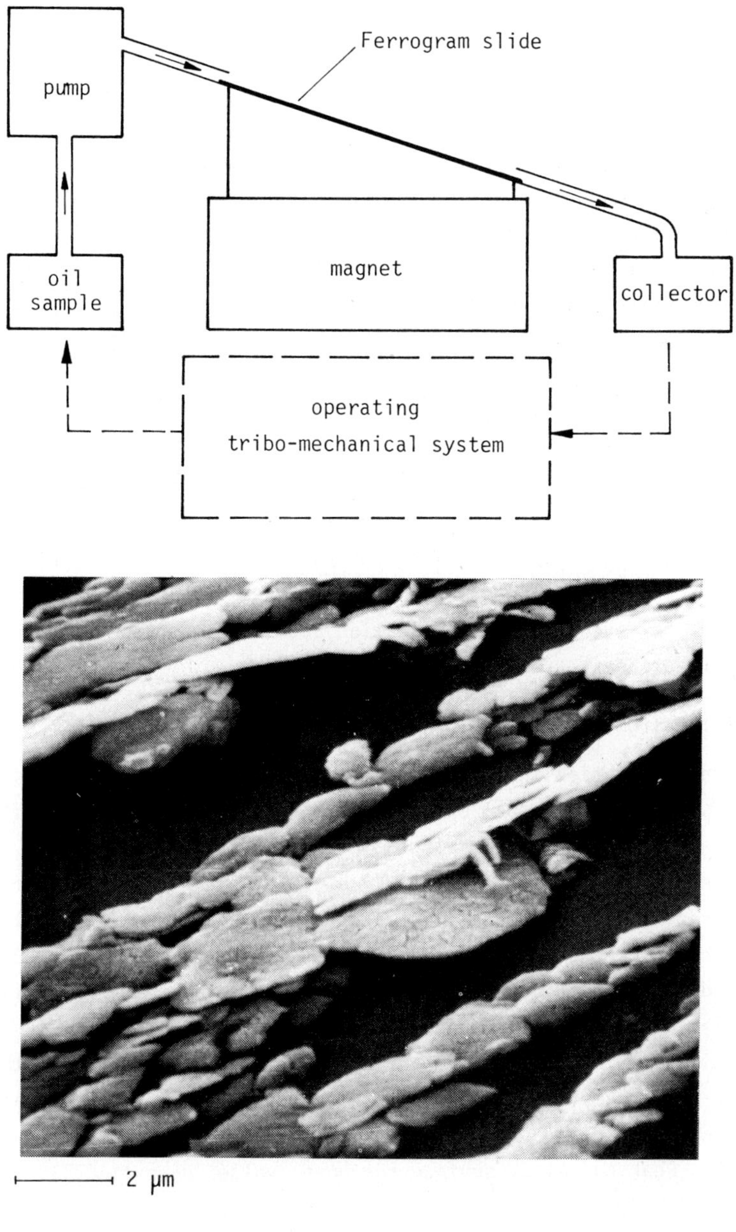

(a)

(b)

Figure 7.19 Illustration of the Ferrography technique: (a) schematic representation of Ferrograph analyzer, (b) SEM micrograph of deposited wear particles (Ref. 7.48).

precipitation of the wear particles, is pumped across a transparent substrate which is mounted at a slight incline, the magnetic particles adhere to the substrate, distributed approximately according to size. The quantity of wear particles and their size distribution can be determined by optical density measurement. In Figure 7.19 the application of the Ferrograph technique (Ref. 7.48) is illustrated.

The monitoring techniques described in this section are applied in various types of machinery as for instance in the failure early warning system for aircraft engines. It may be concluded that machinery condition-monitoring techniques, in addition to enabling a prognostic approach to failure prevention can be an aid to the design of advanced machinery for immediate maintenance and failure free service without costly repair work.

7.6 Surface Investigation Techniques

Since friction and wear processes occur primarily at surfaces, surface investigation techniques play an important role for laboratory tribo-testing as well as for the practice-oriented testing of tribo-engineering equipment (Ref. 7.49). For a complete investigation both the topography and the composition of surfaces have to be studied.

(a) INVESTIGATION OF SURFACE TOPOGRAPHY

Surfaces can be studied in a qualitative manner by microscopic techniques. For plane surfaces, the usual optical microscopy can be utilized. With the use of Nomarski lenses it becomes possible to study surfaces with a resolution in depth of about 50 Å (Ref. 7.50). For example with this technique it is possible to detect slip bands around wear tracks as well as to estimate the thickness of thin polymer films (thickness less than 200 Å) transferred to metallic counterfaces in the sliding of polymer-metal couples (Ref. 7.51), as shown in Figures 7.20 and 7.21.

Quantitative measurements of surface irregularities, such as steps, scratches, or grooves, can be made on smooth surfaces with an optical interference microscope or in applying multiple beam interference microscopy (Ref. 7.52). Peak-to-valley surface roughness values down to 0.015 µm can be measured (7.53). The technique can be improved further by applying the method of equidensities, which splits up every interference line into two lines, thus improving the sensitivity and providing a 3-dimensional picture

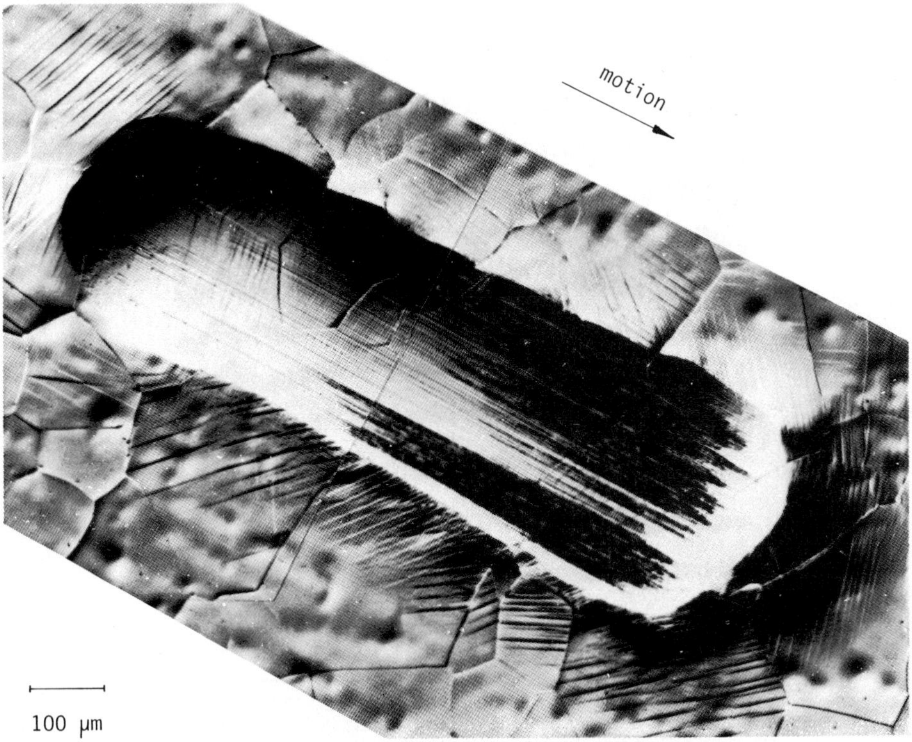

Figure 7.20 Nomarski optical micrograph of an Al surface after sliding contact with a glass ball.

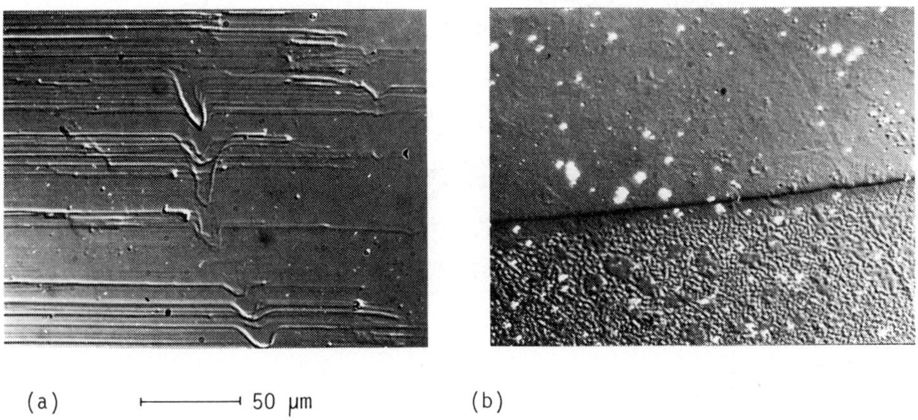

Figure 7.21 Nomarski optical micrographs: (a) PTFE transfer on glass, (b) step of evaporated gold for comparison, h=230 Å.

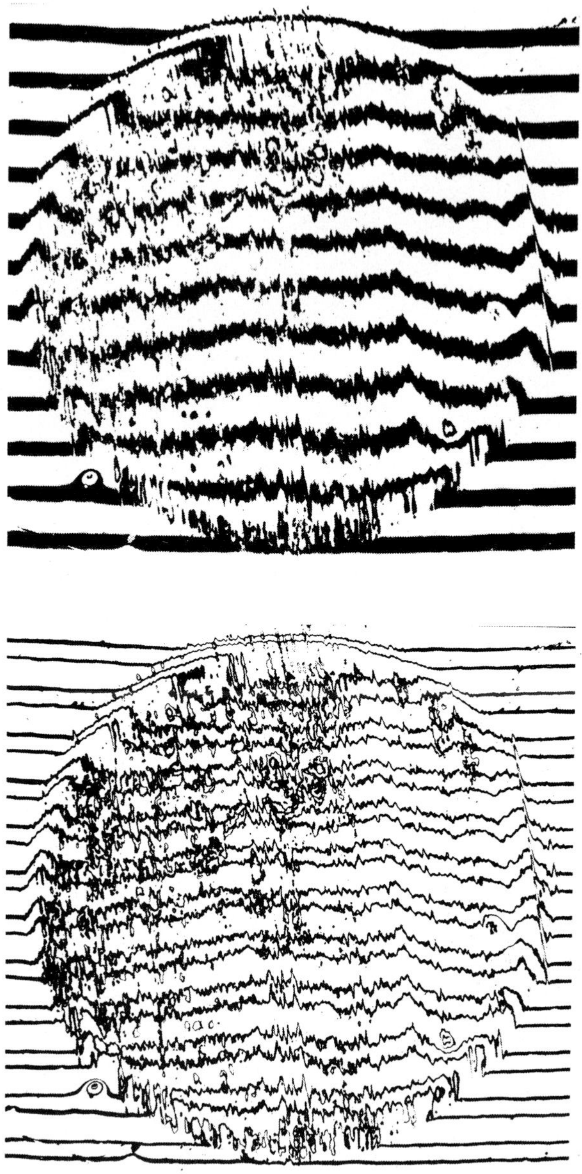

(a)

(b)

Figure 7.22 Interferogram (a) and corresponding equidensities (b) of a circular wear scar on a plane (spacing of interference lines: $\lambda/2$ = 0.27 μm).

of the surface roughness, see Figure 7.22. By following an individual interference line with a planimeter, surface roughness data, such as the R_a (center-line-average) value can be determined for the various cross-sections of the profile.

The application of optical microscopic techniques to investigations of worn surfaces is limited to the study of relatively plane and smooth surfaces because of the very poor depth of focus of optical techniques, being only 0.1 µm at a magnification of say 500 times. Therefore the introduction of the scanning electron microscope (SEM) to the study of worn surfaces has been very advantageous (Ref. 7.54). Due to the extremely short wavelength of the electron beam, both high magnification and high depth of focus can be obtained. For example, at a magnification of 5000 times the depth of focus is still higher than 10 µm. Therefore, this technique can be conveniently used in the study of relatively rough surfaces as illustrated by the examples shown in Section 4.4, demonstrating the different appearance of worn surfaces. The SEM can also be used to detect the state of deformation below the surface with the "electron channelling pattern (ECP)" technique. In tilting the electron beam with respect to the surface spot under study, the diffraction of the electron beam at the different crystallographic planes leads to characteristic ECP lines (Ref. 7.55). Under optimum experimental conditions, with the ECP technique, which is analogous to the Kikuchi technique of the transmission electron microscopy, information on dislocation densities can be estimated from the width, w, of the ECP lines (Ref. 7.56). In Figure 7.23 examples of ECP diagrams are shown (Ref. 7.57).

Profilometer methods are preferred for the quantitative determination of the surface topography. A small diamond needle with a tip radius of about 5 µm is moved across the surface and the deration of the needle is amplified by electronic means. The vertical resolution has a value of about 0.01 µm whereas the horizontal resolution is determined by the radius of the tip of the needle. Feeding the electronic output of the profilemeter to a computer it is possible to calculate the various surface roughness measures described in Section 4.2.1.

Moreover, characteristics such as the Fourier analysis, the autocorrelation function and related quantities can be obtained. With the help of a specimen stage (x-y-motion in steps) surface profile maps can be obtained. The application of surface profilometry to the study of tribo-induced surface topography changes can be seen from the examples discussed in Section 5.2.1. A surface contour map of an Al surface before and after a contact deformation process is shown in Figure 7.24 (Ref. 7.58).

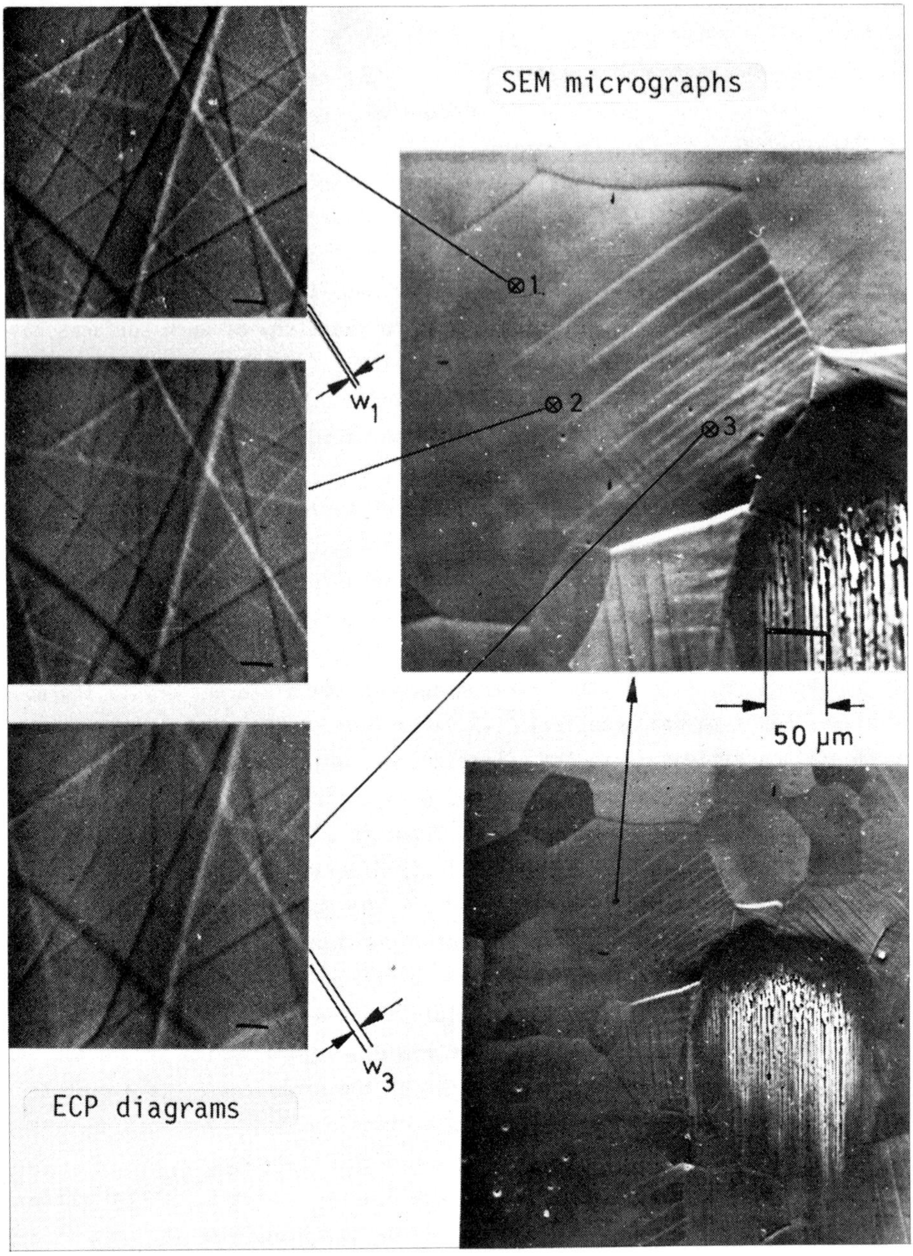

Figure 7.23 ECP diagrams near the end of a wear track on an Al-surface, indicating an increase of sub-surface dislocation density of about two orders of magnitude from location 1 to location 3.

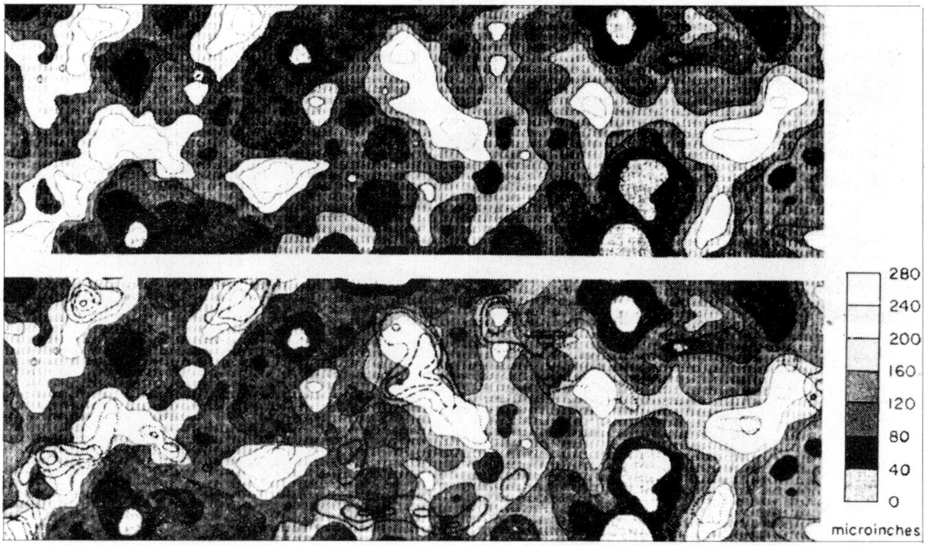

Figure 7.24 Surface contour map before and after contact deformation (Ref. 7.58).

(b) INVESTIGATION OF SURFACE COMPOSITION

For the investigation of the composition of surfaces several powerful tools are available today.

In the electron probe microanalyzer (EPMA) an electron beam is scanned across the surface and the X-ray radiation emitted is analyzed by means of a spectrometer. The concentration of the chemical elements belonging to the characteristic X-ray radiation is registered on a screen. Since the X-ray emission originates from the topmost surface atoms as well as from atoms some µm below the surface one resolved spot on the screen is related to a specimen volume of about 1 μm^3 (Ref. 7.59). Therefore, the EPMA can be used only for the study of the surface composition of relatively thick surface layers. In Figure 7.25 the identification of the chemical nature of a wear particle on a copper pin which has been slid against an iron disc is shown. The comparison of the positive images from the Cu-K_α radiation and the Fe-K_α radiation indicate that the wear particle consists mainly of iron.

Surface-investigation techniques for the analysis of the topmost surface layers have been developed only in recent years. The main principle of these techniques is that the surface is bombarded with photons, electrons,

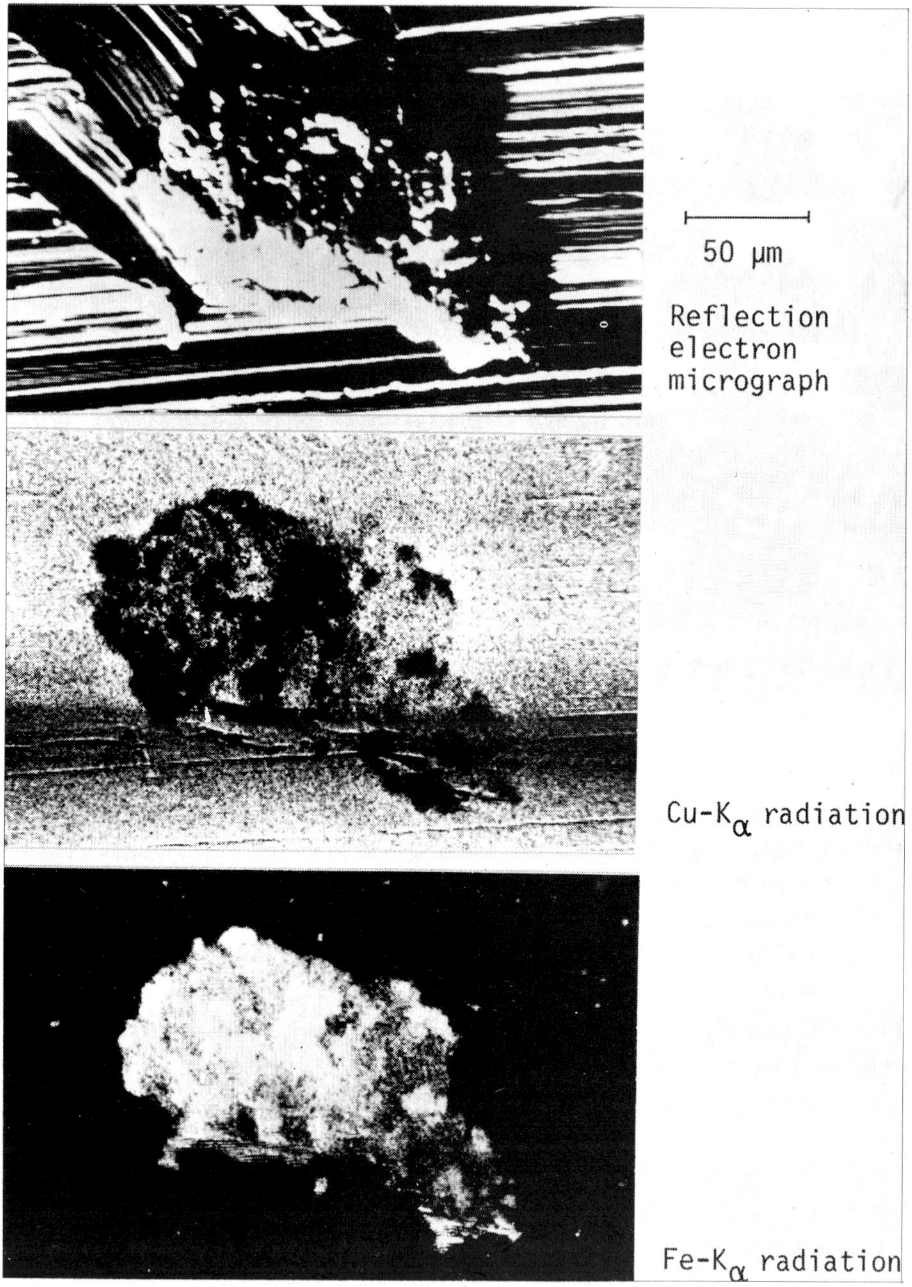

Figure 7.25 Example of an EPMA identification of the chemical nature of a wear particle.

ions, etc. and the emitted particles which carry some information about their history, are analyzed. In Figure 7.26 a general scheme of the various physical processes utilized in surface analyzing techniques is shown (Ref. 7.60).

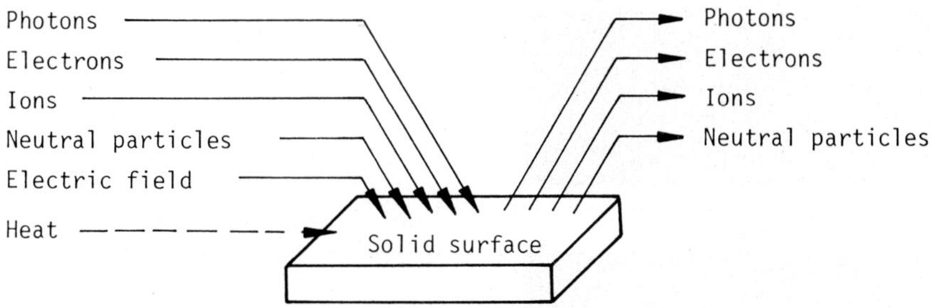

in / out	Photons hν	Electrons e	Ions i	Neutrals N	Heat kT	Electric field E
Photons hν						
Electrons e	ESCA	AES,LEED				
Ions i			SIMS			FIM
Neutrals N						

Figure 7.26 Methods for surface analysis in the monolayer range.

The main techniques which have been applied already in tribological studies are:

$(e \rightarrow e_A)$: AES : Auger electron spectroscopy

$(e \rightarrow e_B)$: LEED : Low energy electron diffraction

$(h\nu \rightarrow e)$: ESCA : Electron scattering for chemical analysis

$(i \rightarrow i_s)$: SIMS : Secondary ion mass spectroscopy

$(E \rightarrow i)$: FIM : Field ion microscopy

The application of these new powerful surface-investigation techniques to tribological problems is still in its beginnings. In the following, brief descriptions of the basic physical principles of these techniques together with reference to some examples of their application will be given. For

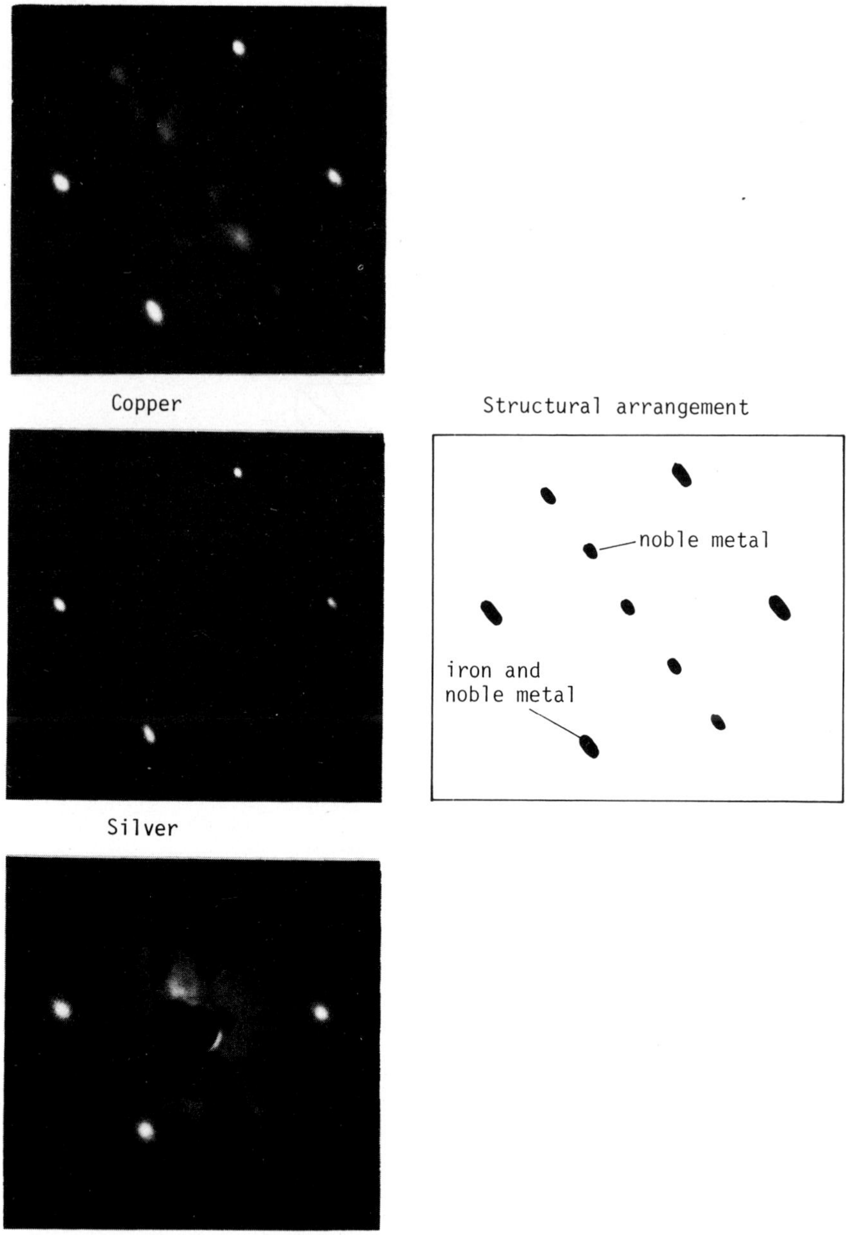

Figure 7.27 LEED photographs of iron (011) surface after adhesion of noble metals (Ref. 7.61).

more details the reader is referred to the review articles, Ref. 7.61, 7.62, 7.63.

In the AES technique, the surface to be investigated is bombarded by a monoenergetic electron beam (E ≈ 3 keV) and the energy distribution of the emitted secondary electrons (Auger-electrons) is measured. The energy of the Auger electrons is a function of the surface material. In this way, the chemical elements at the surface can be detected, with the exception of light chemical elements (hydrogen to nitrogen). With AES a semiquantitative analysis with a resolution of 1/1000 of a monolayer is possible. An example of the application of the AES technique is shown in Figure 5.7.

In the LEED technique, an electron beam of low energy (E≈20...2000eV) is diffracted at a single crystal surface. Measured quantities are: location, amplitude, shape and intensity of diffraction spots given by the elastically diffracted electrons. From this technique, information can be obtained on the geometric configuration of the topmost surface atoms, but not on their chemical nature. Figure 7.27 presents LEED patterns for the iron (011) surface after adhesive contact with noble metals. The structural arrangement of the adherent noble metal on the iron after fracture of adhesive contact is essentially the same (Ref. 7.61).

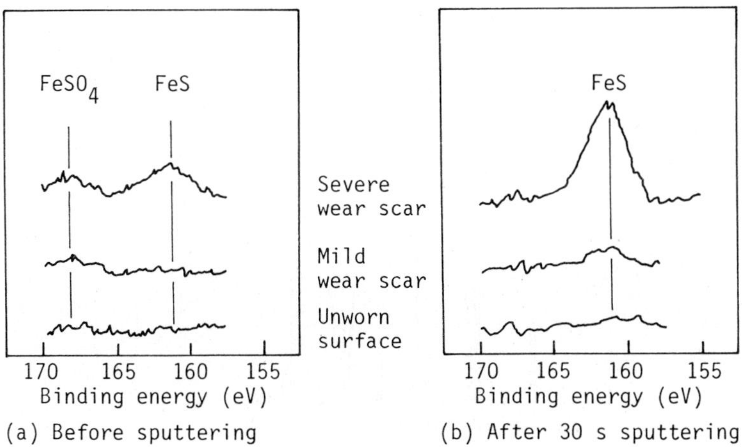

Figure 7.28 Sulfur (2p) XPS features from wear scars and unworn surface (Ref. 7.64).

The ESCA technique utilizes the effect of photoelectric emission of surfaces. Depending on the energy of the incident radiation the techniques of ultraviolet photoelectron spectroscopy (UPS), extreme UPS (XUPS) and X-ray photoelectron spectroscopy (XPS) can be distinguished. The spectrum of the photoelectrons gives a picture of the electronic structure of the surface layers thus allowing conclusions about their chemical composition. With this technique it is also possible to determine the state of chemical bonding of the surface atoms. An example of the application of the ESCA technique for the study of the action of the lubricant additive dibenzyl-disulphide under mild and severe wear conditions is shown in Figure 7.28 (Ref. 7.64) (see also Section 5.2.2).

In the SIMS technique an emission of ions from the surface is obtained through the bombardement with argon ions ($E \approx 3$ keV). The ions are then analyzed by means of mass spectroscopy. With this technique all chemical elements can be detected. The resolution of the technique has been reported to be better than 10^{-6} of a monolayer (Ref. 7.65). The SIMS technique was used to study the composition of surface layers formed on the running surfaces of both a pin-on-disc tribometer and an internal combustion engine lubricated with a commercial engine oil (Ref. 7.66). In this study, the spectra of positive and negative secondary ions were measured before and after the test runs and the ratio of the measured ion intensities after and before the test runs were determined.

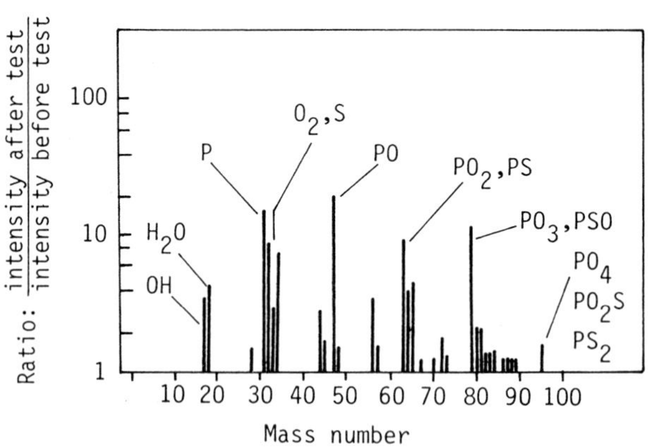

Figure 7.29 Spectra of negative ions detected with the SIMS technique (Ref. 7.66).

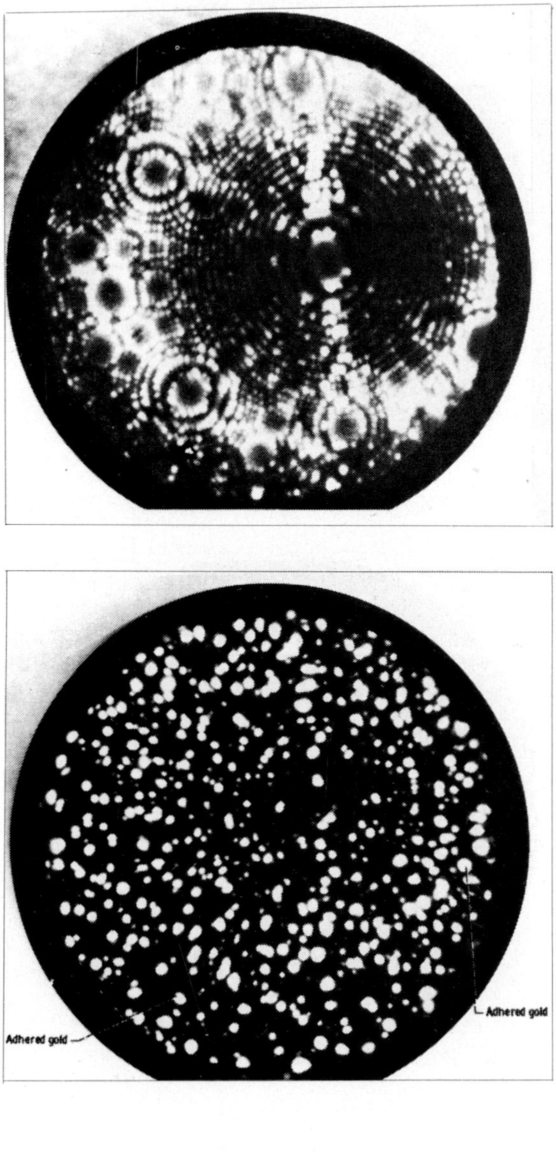

Figure 7.30 FIM micrographs of tungsten prior and after contact with gold (Ref. 7.61).

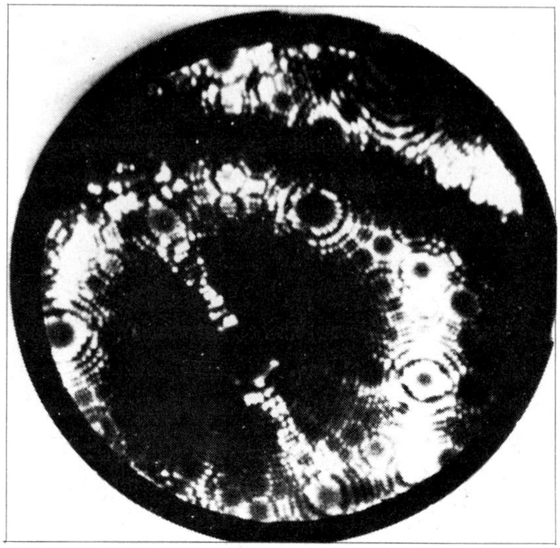

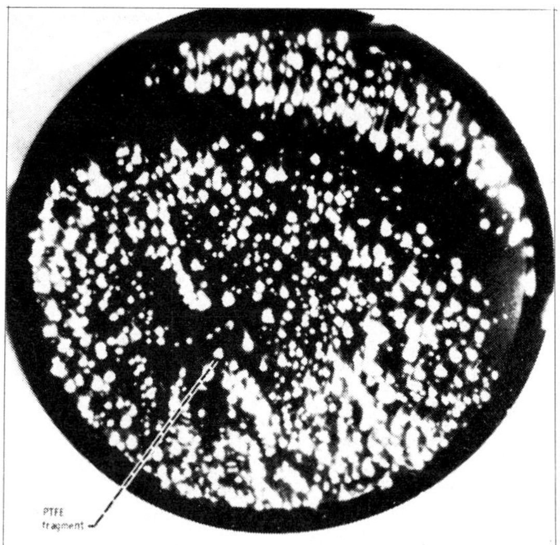

Figure 7.31 FIM micrographs of tungsten prior and after contact with PTFE (Ref.7.61).

With the FIM technique it is possible to get a picture of the atomic arrangement of surfaces (Ref. 7.67). As specimen an extremely thin tip of a metal with high melting point (W, Ni, Pt) is used. This is subjected to an electric field of high strength, as a result of which particles are emitted and become visible on a fluorescent screen. This microscope is truly unique in that it is the only device available today which will allow the examination of individual atom sites and surface structures in atomic detail. It has an atomic resolution of 2.5 Å. In Figures 7.30 and 7.31, FIM micrographs of a tungsten surface before and after contact with the metal gold, and the polymer polytetrafluorethylene are shown (Ref. 7.61). It can be seen that in all cases material from the counterpartner adheres on the tungsten surface. This then is the direct proof of the occurrence of adhesion on an atomic scale!

The techniques and application examples discussed in this section indicate that very powerful tools for surface investigations are available today. In combining these techniques with the other tribometric techniques and the theoretical triboanalysis methods and checklists described in the next chapter some further progress in the difficult field of tribotesting may be anticipated.

8 Practical systems methodology

8.1 Introduction

The discussions in the preceding chapters have amply illustrated the multidisciplinary nature and the tremendous inherent complexity of tribology. It has been shown, among other things, that friction and wear of material pairs are not intrinsic material properties but system-dependent interaction characteristics. It follows that:

(i) It is not possible to deal with the subject of tribology in simple terms.

(ii) Any proper analysis and solution of tribological problems require the consideration of numerous parameters and influencing factors (Ref. 8.1).

(iii) Consequently, "monistic" simplified approaches to tribological problems may lead to grave errors in that important parameters or influencing factors may be neglected or overlooked (Ref. 7.2).

Although most workers in the field of tribology are well aware of these facts, it is a common experience that in publications dealing with friction, lubrication and wear important parameters are often omitted or are reported in a manner or in units unsuitable for comparison with other results or for further applications. If, for instance, wear rates are given in mass units it is often impossible to calculate from these data wear rates in terms of length units required for engineering applications. Other typical omissions concern the characterization of important characteristics of system components, such as surface roughness data, or the description of the atmospheric or environmental conditions. These omissions and shortcomings are, of course, not intentional but are consequences of the multidisciplinary nature of tribology. It is clear that a physicist or chemist, who publishes experimental results of investigations of friction or wear processes, is mainly interested in describing the physical or chemical content of his investigation rather than in thinking of possible links of his findings with engineering applications. On the other hand, it is often not possible to draw basic generalized conclusions from the results of engineering investigations of tribo-technical components of machinery. This in turn necessitates the repetition of studies already performed but incompletely described in the literature and partly explains the vast number of publications produced every year in the field of tribology. (As mentioned in Section 1.3, about 8000 tribological papers are currently published every year.) From all this it follows that both the lack of a convenient framework for the description of tribological results and the lack of a convenient methodology for the solution of tribological problems form severe obstacles to progress. Discussion of the multifarious aspects of tribology from a systems point of view as presented in the foregoing chapters has provided the basis for a systems approach to these problems. For this purpose, the general theoretical framework as presented in the tribo-process diagram, Figure 3.4, should be brought into a form suitable for engineering applications. In the following, the development of a "data sheet" suitable for the analysis and solution of tribological problems will be described and examples of the application of systems methodology will be given (Ref. 8.2).

8.2 A Tribological Systems Data Sheet

As a starting point for an engineering systems approach to the analysis of tribological problems, consider a typical tribo-technical system, namely the gear-box as shown in Figure 8.1.

As described in Chapters 2 and 3, the two partners which form the "tribologically interacting surfaces", i.e., gear 1 and gear 2, can be hypothetically separated from their environment by the proper choice of a system's envelope. All components of the system are within this envelope and are part of the structure of the system. The structure consists of the elements of the system, their relevant properties and their interrelations, described formally by the set $S = \{ A, P, R \}$.

The inputs of the operating variables are transformed through the structure of the system into outputs which are used technically: the use-outputs. Simultaneously, as a consequence of the tribological interactions between the elements, loss-outputs occur, denoted in summary by the terms friction and wear losses. The way in which the inputs are transformed into outputs determines the technical function of the system.

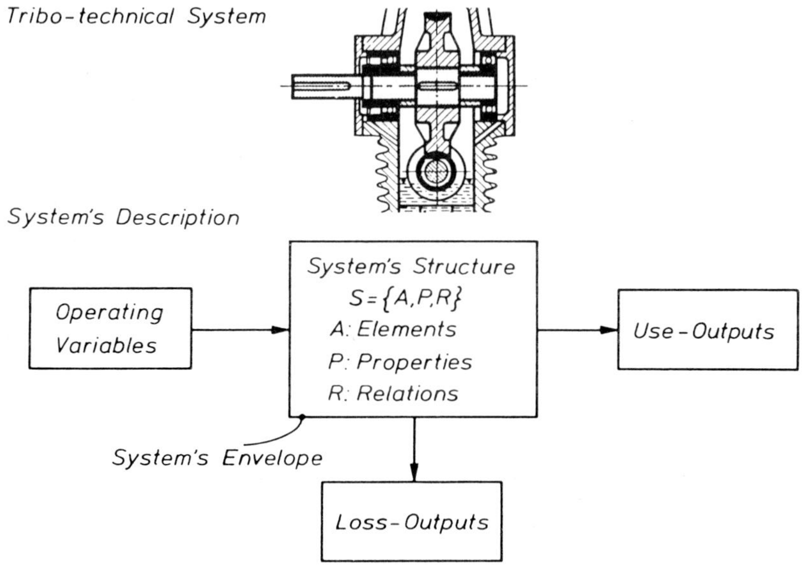

Figure 8.1 Analysis and description of a tribotechnical system.

From the discussion of the tribo-engineering system illustrated in Figure 8.1, as well as the theoretical considerations of Chapters 2 and 3, it follows that the relevant tribological parameters can be divided into the following four groups as compiled in Table 8.1.

In the next four sections, the main features of the relevant parameters of these groups will be discussed. Since most of the various systems aspects have been already treated in some detail in the preceding chapters, the following considerations will be summarizing in nature.

I Technical function of the tribo-system

Transmission of motion, information, energy, materials

II Operating variables

Type of motion
Load, Velocity, Temperature
Time
Materials flow

III Structure of the tribo-system

a) Elements

(1) Tribo-element (3) Lubricant
(2) Tribo-element (4) Atmosphere

b) Properties of elements

Relevant geometrical and materials properties of (1), (2), (3), (4)

c) Relations between elements

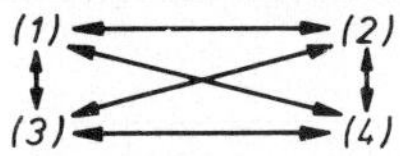

IV Tribological characteristics

a) Tribo-induced changes of system's structure

b) Tribo-induced energy losses

c) Tribo-induced materials losses

Table 8.1 Fundamentals of a systems analysis check list for tribo-mechanical systems.

8.2.1 TECHNICAL FUNCTION OF THE SYSTEM

In Section 3.2 it was explained that the technical function of tribological systems is connected with the transmission or transformation of the basic quantities

- motion
- information
- work
- materials

In using these four basic quantities or related quantities, the technical function of the various tribological systems may be classified in terms of the input-output relations of these quantities. The principles of the method of classifying tribo-mechanical systems according to their function were illustrated in Table 3.2. A more detailed classification is given in the Appendix (A).

8.2.2 Operating variables

The most characteristic operating variable of a tribo-mechanical system is the type of relative motion between tribo-element (1) and tribo-element (2). The basic types of motion are

- sliding
- rolling
- spin
- impact

as illustrated in Figure 8.2 for the simple case of the relative motion of a ball (1) against a plane (2).

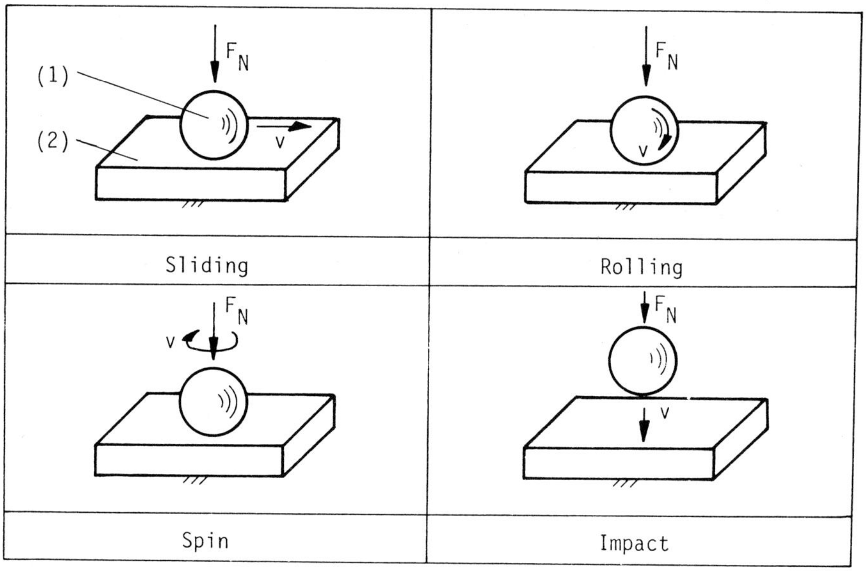

Figure 8.2 Basic types of relative motion.

It can be shown that every type of relative motion between two solid bodies can be expressed as a superposition of these four basic types of motion. In addition to the characterization of the type of motion, its dependence on time should be specified, being for example:

- continuous
- oscillating

- reciprocating
- intermittent

The other operating variables are the following quantities:

- load F_N
- velocity v
- temperature T
- distance of motion s
- operating duration t

For some tribological systems, these physical operating variables are accompanied by material inputs, e.g., the flow rate of the lubricant. Some disturbing inputs may also be present, e.g., vibration and radiation.

8.2.3 STRUCTURE OF THE SYSTEM

As compiled in Table 8.1, the structure of a system is given by the system's elements, their relevant properties and their interrelations, described formally by the set $S = \{A, P, R\}$.

(a) ELEMENTS OF THE SYSTEM, $A = \{a_i\}$

If the system's envelope (for definition, see Section 2.2) is located as closely as possible around the "interacting surfaces in relative motion", it appears that in most tribological systems four different basic elements are involved in the friction and wear processes.

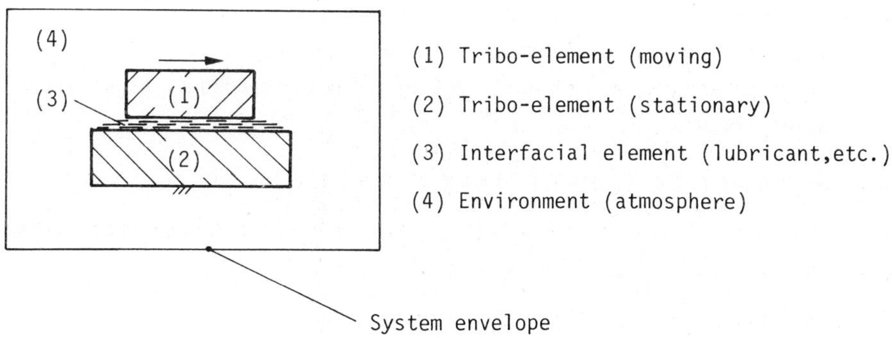

Figure 8.3 Basic elements of a tribo-mechanical system.

As illustrated in Figure 8.3, the components that form the pair of the "interacting surfaces in relative motion" are named "tribo-element (1)" and "tribo-element (2)". The other two basic elements are the lubricant (3) and the environment (4).

(b) PROPERTIES OF THE ELEMENTS, $P = \{ P(a_i) \}$

Any tribological process is influenced by many properties of the basic elements (1), (2), (3), (4). Owing to the great variety of tribo-mechanical systems and tribological processes, it is very difficult to provide a comprehensive general compilation of the tribologically relevant properties of the systems elements. The following properties of the elements have been found to be of primary concern for the tribological behaviour of the system:

(i) Properties of the tribo-elements (1) and (2)
The tribologically relevant properties of the elements (1) and (2) can be sub-divided into "volume" properties and "surface" properties. The study of the tribological processes, as discussed in Chapter 4, has shown that the following main properties of the elements (1) and (2) are relevant for the tribological behaviour of the system:
- Volume properties: geometry, chemical composition and metallurgical structure; materials data, including elastic modulus, hardness, density, thermal conductivity.
- Surface properties: surface roughness and surface composition.

(ii) Properties of the lubricant (3)
The main relevant properties of the lubricant are the viscosity and its dependence upon temperature and pressure, together with the chemical composition of the lubricant.

(iii) Properties of the environmental atmosphere (4)
The main relevant properties of the atmosphere are its chemical composition and the amount and pressure of its components, especially water vapour.

(c) INTERRELATIONS BETWEEN THE ELEMENTS, $R = \{ R(a_i, a_j) \}$

The tribological interactions between the elements of a tribological system were studied in detail in Chapter 4 under the headings of

- contact processes
- friction processes
- wear processes
- lubrication modes

In Figure 8.4 the basic tribological processes are expressed in the form of schematic diagrams for systems of increasing complexity, i.e., increasing number of interacting system's elements (Ref. 7.16).

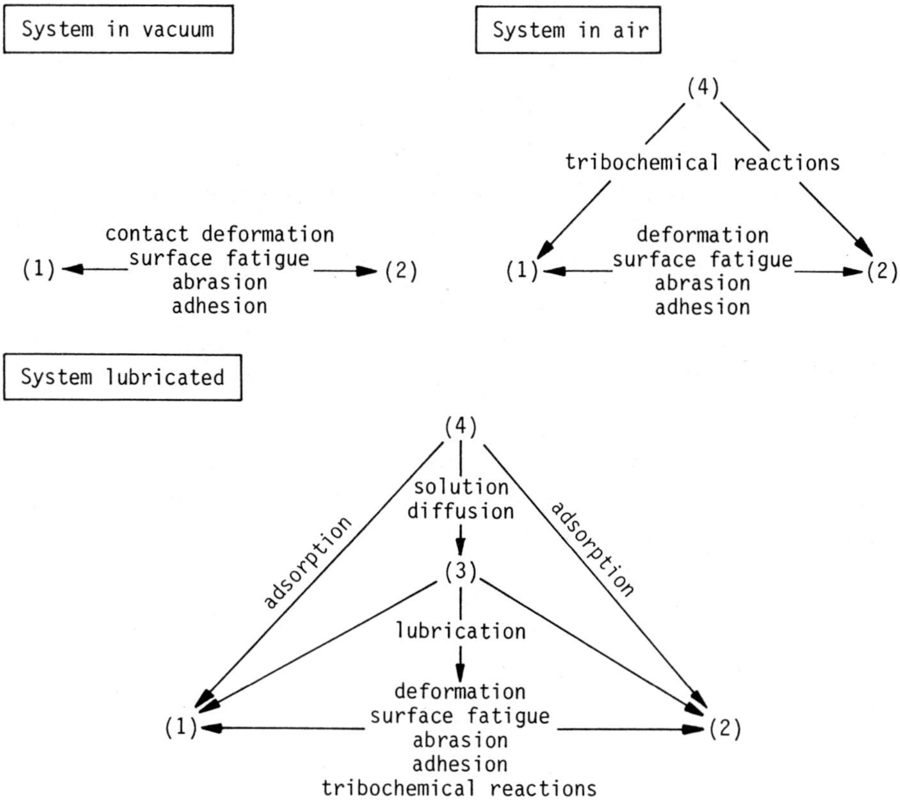

Figure 8.4 Schematic representation of tribological interactions.

In this connection it should be emphasized that besides the friction and wear processes between the system's elements (1) and (2) the processes of diffusion and effusion between the elements (3), the lubricant, and (4), the environment, are of great importance for the behaviour of tribological systems. (The influence of diffusion processes (4)⟶(3) is illustrated in Figure 8.8; for an instructive example of effusion processes (3)⟶(4) see Ref. 8.3.)

8.2.4 TRIBOLOGICAL CHARACTERISTICS

The tribological characteristics may be divided into the following three groups:

(a) Tribo-induced changes of system's structure
(b) Tribo-induced energy losses
(c) Tribo-induced material losses

According to Chapter 5, tribo-induced changes of a system's structure (a) may concern:

(i) the destruction of elements or the creation of new elements in a tribo-system,
(ii) changes in the properties of the elements, for instance, changes in contact topography and surface composition,
(iii) changes in the interrelations between the elements, as for instance, changes of the wear mechanisms under the action of the operating variables.

The other two main tribological characteristics (b) and (c) listed in Table 8.1, i.e., the friction-induced energy losses and the wear-induced materials losses, may be expressed formally as:

friction losses = f(operating variables; system's structure)
wear losses = f(operating variables; system's structure)

Consequently, friction coefficient, f, and wear rate, w, of a tribo-mechanical system may be expressed formally as:

$$\boxed{f = f\,(X;\,S)} \qquad \boxed{w = f\,(X;\,S)}$$

where X : Operating variables
S : {A, P, R} : System's structure

Although the parameter groups X, S, are not independent variables since they are connected with each other through the tribological interrelations R, the above symbolic representation of friction and wear characteristics can be conveniently used as starting point for the practical application of the systems methodology, see Section 8.5.

Summarizing the various parameters of tribo-mechanical systems, a data sheet results as shown in Figure 8.5. On the basis of the general explanations given in the last sections, the data sheet should be self-explanatory.

I Technical function of the tribo-system

II Operating variables

Type of motion[1]:		Duration of operation t[ ]:
Load[2] $F_N(t)$ [ ]	Velocity[3] $v(t)$ [ ]	Temperature[4] $T(t)$ [ ]
Other op. variables:		Location:

III Structure of the tribo-system

Properties of elements (initial/final)		Tribo-element (1)	Tribo-element (2)	Lubricant (3)	Atmosphere (4)
Designation of element and material					
Volume properties	Geometry/Dimensions/Volume				
	Chemical composition				
	Phys.-mech. data: Hardness, Viscosity $\eta(T,p)$, other[5]				
Surface properties	Topography descriptors (c.l.a., etc)			Other data:	
	Surface layer data (if different from volume)				

Contact area A[ ]			Tribological interactions:
Ratio: $\frac{\text{contact area}}{\text{total wear track}}$ ε [%]	ε(1)	ε(2)	
App. lubrication mode:			

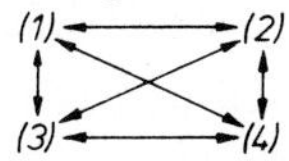

IV Tribological characteristics

Changes in properties of the elements[6]	Friction data (vs time t or distance s)	Wear data (vs time t or distance s)
Other characteristics (e.g.: contact resistance, vibrations, noise, etc):		Appearance of worn surfaces:

1) e.g.: sliding, rolling, oscillating, reciprocating, etc.
2) the contact pressure p is given by $p = F_N/A$
3) velocity of tribo-element (1) relative to tribo-element (2)
4) temperature at stated location
5) e.g.: density, thermal conductivity, Young's modulus, etc.
6) e.g.: changes in hardness of (1),(2), changes in viscosity of (3), changes in humidity of (4), etc.

Figure 8.5 Tribological systc s data sheet.

8.3 THE DESCRIPTION OF TRIBO-ENGINEERING SYSTEMS

With the help of the tribological systems data sheet, the relevant parameters of tribo-engineering systems can be easily compiled in a comprehensive manner. As a typical application example, in Figure 8.6 the characteristics of a journal bearing operating at the minimum of the Stribeck curve are compiled with data taken from the literature (Ref. 8.4). The main reason for choosing this example was the fact that most of the important parameters of this tribo-engineering system have been reported by the investigator of Ref. 8.4. In discussing this example, only the procedure of compiling the data sheet is really relevant here, not the applied measuring techniques or result per se of the quoted investigation. Therefore, only some main features of the example are pointed out.

The data sheet clearly indicates the main features of this most typical tribological system:

The technical function (I) of the system can be described as "guidance of motion".

The operating variables (II) are given by the type of motion, namely "continuous sliding" of a duration of 40 min together with the given values of load F_N, velocity v, and bulk oil temperature T.

The structure (III) of the system is described by the elements of the system, their relevant properties and interrelations. In this case, tribo-element (1) is given by a bearing bushing consisting of a lead-tin-bronze and tribo-element (2) is a steel shaft. As lubricant (3) a mineral oil of the type SAE 20W-20 is used and the atmosphere (4) is laboratory air. Together with the designation of the elements and materials the data sheet contains all important properties of the elements. The data sheet also indicates the tribological interactions between the elements of the system which can be characterized as a "running-in" period leading to topographical changes which are within the initial roughness of the bearing bush element (2).

As tribological characteristics (IV) emphasis is laid in the characterization of wear rates. Although the original wear rates are given in mass units, since the dimensions of the bearing are given, it is possible to estimate wear-time rates and wear-distance rates. Because the wear measurements have been performed by means of radio-tracer techniques, it is possible to distinguish between the wear rates of tribo-element (1), the bearing bushing, and tribo-element (2), the bearing shaft.

I Technical function of the tribo-system			
Journal bearing: guidance of motion			

II Operating variables		
Type of motion[1]: continuous sliding	Duration of operation t: 40 min	
Load[2] $F_N(t)$ [N]	Velocity[3] $v(t)$ [m/s]	Temperature[4] $T(t)$ [°C]
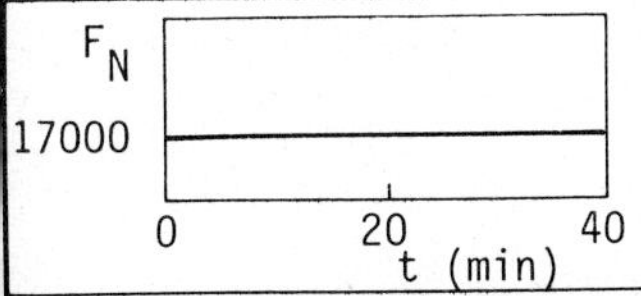	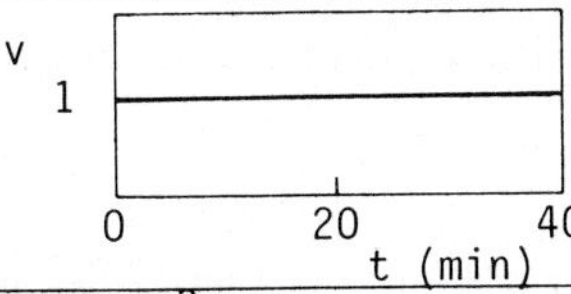	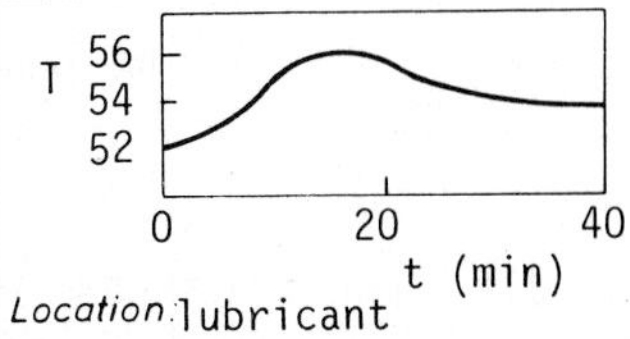
Other op. variables: oil flow rate: 500 cm^3/min		Location: lubricant

III Structure of the tribo-system					
Properties of elements (initial/~~final~~)		Tribo-element (1)	Tribo-element (2)	Lubricant (3)	Atmosphere (4)
Designation of element and material		bushing bronze	shaft steel	mineral oil SAE 20 W 20	air
Volume properties	Geometry/Dimensions/Volume	l/d = 0.4	d = 65 mm	——	——
	Chemical composition	15 Pb; 7 Sn 78 Cu	0.2 C; 0.8 Mo 0.2 Cr; 0.4 V	——	——
	Phys.-mech. data: Hardness (N/mm^2) Viscosity $\eta(T,p)$ other[5] thermal conductivity	H=700 45 kcal/mh°C	H=2630 3 kcal/mh°C	η =130 cP/20°C η =17.5cP/80°C density: 0.9 g/ml	——
Surface properties	Topography descriptors (c.l.a.,etc) R_t (μm)	2.5 - 3.0	0.5	Other data: ratio: $\frac{\text{radial clearance}}{\text{shaft radius}}$ ψ_r = 0.2 %	
	Surface layer data (if different from volume)	——	——		
Contact area A		1xd = 17 cm^2		Tribological interactions: 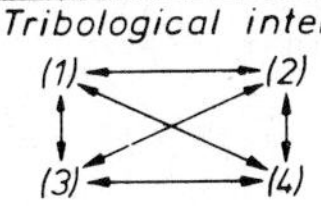 running-in process	
Ratio: $\frac{\text{contact area}}{\text{total wear track}}$ ε [%]		$\varepsilon(1)$ 100	$\varepsilon(2) \approx 20$		
App. lubrication mode: quasi-hydrodynamic					

IV Tribological characteristics		
Changes in properties of the elements[6]	Friction data (vs time t or distance s)	Wear data (vs time t or distance s)
(1): $R_t < 2$ μm	Minimum of Stribeck curve	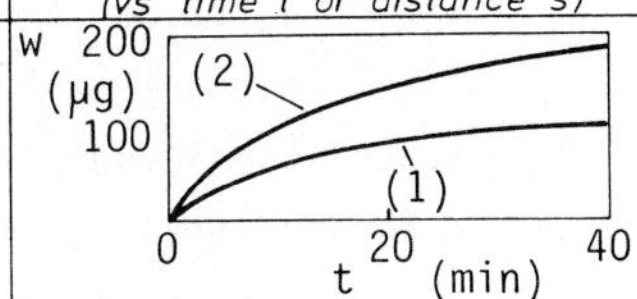
Other characteristics (e.g.: contact resistance, vibrations, noise, etc): ——		Appearance of worn surfaces: (1): run-in (2): slightly roughened

Figure 8.6 Description of a journal bearing.

In a similar manner as the example described in Figure 8.6, the characteristics of other tribo-engineering systems can be described with the help of the tribological systems data sheet. For an adequate description of the various tribo-engineering systems, the data sheet may be individually modified, i.e. shortened, extended or grouped in another order. However, in all applications, the total of the four groups of parameters as compiled in Table 8.1 should be taken into account in any case.

8.4 The Presentation of Friction and Wear Research Data

The introductory remarks of this chapter have amply demonstrated the need for a useful framework for the representation of friction and wear research data. It is important that results be presented in a form suitable for subsequent independent evaluation and, if possible, correlation with results obtained from other sources. The value of all work, whether basic or applied, would be greatly increased if some basic standards of reporting experimental variables would be accepted. In the following, the application of the tribological systems data sheet to the presentation of friction and wear research data will be illustrated in discussing two examples.

In Figure 8.7, the relevant parameters and data of an investigation of a sliding cobalt-cobalt couple under conditions of high vacuum are listed (Ref. 8.5). In this case, the tribo-system, a pin-on-disc apparatus, consists of only the two elements (1) and (2). The tribological systems data sheet clearly shows the main features of this tribo-system, namely:

(i) the increase of the system's temperature with increasing sliding velocity,
(ii) the differences in the ratio of contact area to total wear track, ε, for the tribo-element (1) and tribo-element (2) leading to differences in the thermal situations of (1) and (2),
(iii) the transformation of the crystal structure of the pin at a certain temperature,
(iv) the changes of friction and wear behaviour with increasing sliding velocity as a consequence of (i), (ii), (iii).

In another example, shown in Figure 8.8, the data of a research study with a ball-on-cylinder tribo-testing system are compiled (Ref. 8.6). By contrast with the foregoing example, in this example the tribological system consists of four elements, i.e., a sliding steel-steel couple (1), (2), together with

I. Technical function of the tribo-system

Friction and wear test with pin-on-disc apparatus

II Operating variables

Type of motion[1]: continuous sliding	Duration of operation t[h]: 7 h run-in time

Load[2] $F_N(t)$ [N]	Velocity[3] $v(t)$ [m/s]	Temperature[4] $T(t)$ [°C]
F_N=10 N = const.	v = 0.51-10.2 m/s (10 steps)	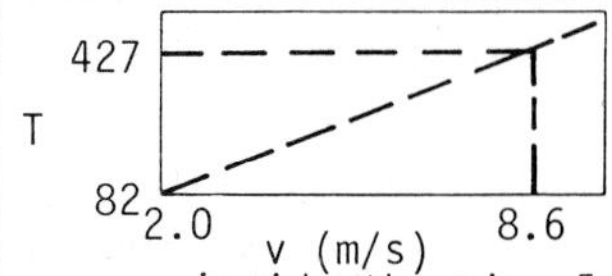
Other op. variables: ———		Location: inside the pin, 5mm from contact surface

III Structure of the tribo-system

Properties of elements (initial/~~final~~)	Tribo-element (1)	Tribo-element (2)	Lubricant (3)	Atmosphere (4)
Designation of element and material	pin Co (polycryst)	disc Co (polycryst)	———	———
Volume properties: Geometry/Dimensions/Volume	r=4.76 mm V=1.45 cm³	d=63.5 mm V=29.4 cm³	———	———
Volume properties: Chemical composition	99,88Co; 0.026 O_2; 0.02 Zr; 0.02 Ni; 0.01 Fe; 0.03 Cu		———	———
Volume properties: Phys.-mech. data: Hardness, Viscosity η(T,p), other[5]: thermal conduction	165 (cm/s/cm²/°C)	165 (cm/s/cm²/°C)	———	———
Surface properties: Topography descriptors (c.l.a., etc) R (μm)	0.1 - 0.2	0.1 - 0.2	Other data: vacuum 10^{-9} Torr (130 nPa)	
Surface properties: Surface layer data (if different from volume)	———	———		

Contact area A [mm²]	$A_{Hertz} \approx 0.05$	
Ratio: contact area / total wear track ε [%]	ε(1) 100	ε(2) < 1
App. lubrication mode: unlubricated		

Tribological interactions: (1) ⟷ (2)

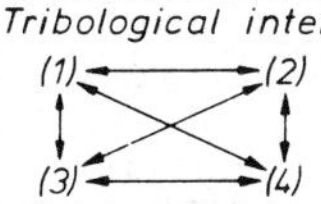

adhesion, plastic deformation, metal transfer

IV Tribological characteristics

Changes in properties of the elements[6]: crystal structure	Friction data ~~(vs time t or distance s)~~	Wear data ~~(vs time t or distance s)~~
disc: hexagonal; pin: hex. → fcc, T=417 °C	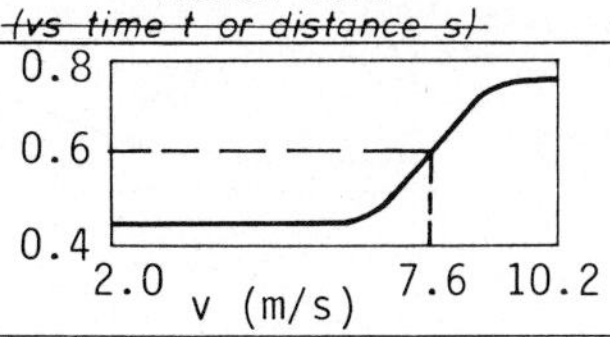	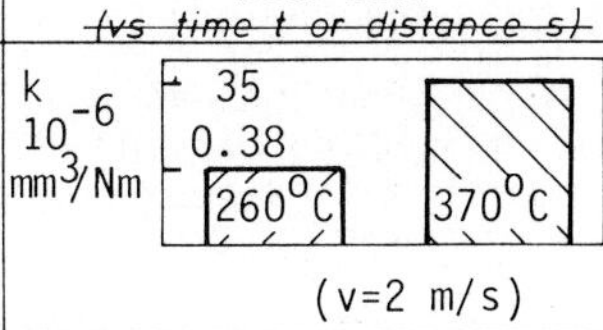

Other characteristics (e.g.: contact resistance, vibrations, noise, etc):	Appearance of worn surfaces:
———	plastic flow (T=260 °C) metal transfer (T=370 °C)

Figure 8.7 Example of the representation of friction and wear research data.

I Technical function of the tribo-system

Friction and wear test with ball-on-cylinder apparatus

II Operating variables

Type of motion[1]: continuous sliding | *Duration of operation t* 100 min

Load[2] $F_N(t)$ [N]	*Velocity*[3] $v(t)$ [m/s]	*Temperature*[4] $T(t)$ [°C]
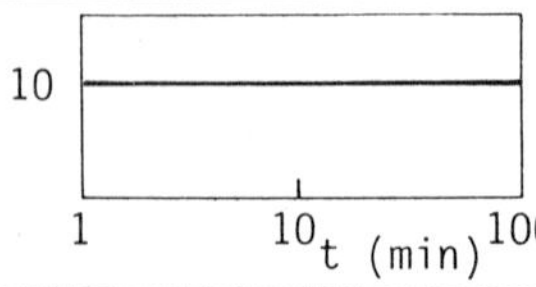	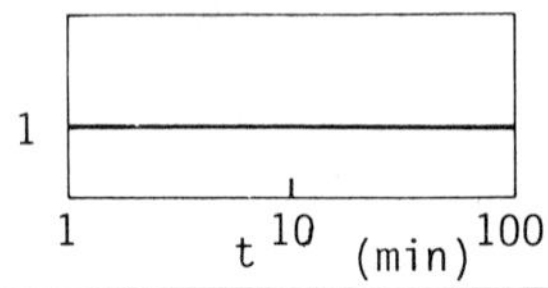	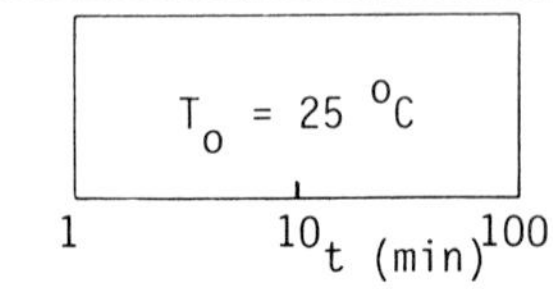

Other op. variables: ——— | *Location:* ambient

III Structure of the tribo-system

Properties of elements (initial/~~final~~)		Tribo-element (1)	Tribo-element (2)	Lubricant (3)	Atmosphere (4)
Designation of element and material		ball 52100 steel	cylinder 52100 steel	base oil (Bayol 35)	(A): Ar, dry ¦ (B): air dry
Volume properties	Geometry/Dimensions/Volume	d=12.7 mm	d=44.4 mm	———	———
	Chemical composition	0.98 C 1.50 Cr	0.98 C 1.50 Cr	hydrocarbon (highly ref.)	100 Ar ¦ 80 N_2 20 O_2
	Phys.-mech. data: Hardness (Rockwell), Viscosity $\eta(T,p)$, other[5]	R_c = 61	R_c = 20	η /25°C=2.5 cP	———
Surface properties	Topography descriptors (c.l.a., etc) cla (μm)	0.02 - 0.05	0.25 - 0.33	*Other data:* impurities in (4): (A): 5 ppm water 20 ppm O_2 (B): 20 ppm water	
	Surface layer data (if different from volume)	———	———		

Contact area A [mm²]: $A_{Hertz} \approx 0.01$

Ratio: $\frac{\text{contact area}}{\text{total wear track}}$ ε [%]: ε(1) 100 | ε(2) < 1

App. lubrication mode: mixed lubrication

Tribological interactions:

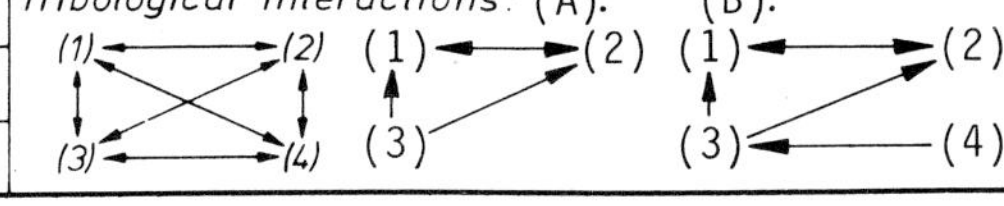

IV Tribological characteristics

Changes in properties of the elements[6] (1) and (2)	Friction data (vs time t or distance s)	Wear data (vs time t or distance s)
Formation and rubbing away of metal oxides	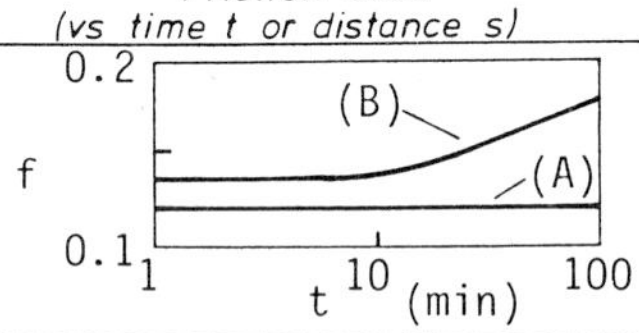	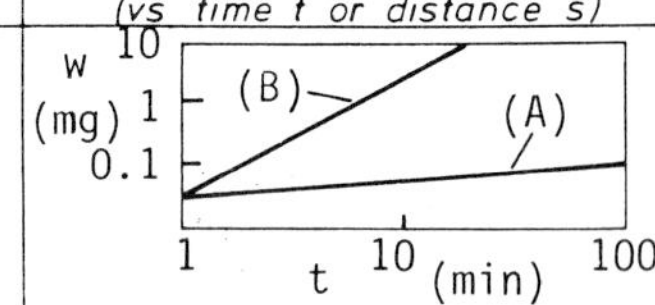

Other characteristics (e.g.: contact resistance, vibrations, noise, etc): ———

Appearance of worn surfaces: Corrosive wear phenomena

Figure 8.8 Example of the representation of friction and wear research data.

a lubricant (3) and the atmosphere (4). In this investigation, the influence of different atmospheric environments on the tribological behaviour of lubricated steel-steel contacts was studied. Also in this case, the compilation of the experimental data in the form of a tribological data sheet has the advantage that all relevant parameters are registered. The main features of this friction and wear test, which was run under a constant set of operating variables (F_N, v, T), are given by the differences in the tribological interactions between the four elements (1), (2), (3), (4). If dry argon is used as atmosphere, the element (4) is an "inert" one. If, in contrast, dry air is used as atmosphere (4), the diffusion of atmospheric oxygen into the lubricant (4)⟶(3) connected with tribo-chemical processes (4), (3)⟶(1), (2) distinctly influences the friction and wear mechanisms. This leads to remarkable differences in the friction and wear data of the systems (A) and (B).

These examples - which could easily be supplemented by examples from other studies - clearly illustrate the advantages of a compilation of the relevant tribological parameters in the form of tribological systems data sheets. The publications quoted here are exceptionally detailed. Nevertheless, two items on the data sheets had to be estimated. However, tribologists will be familiar with the problem of comparing the work of different authors. As mentioned above, it is usually found that some important parameters, required for comparative studies or for the application of the study to practice, are missing. If, in future publications, authors would add such a tribological systems data sheet no relevant parameter would be omitted and the usefulness of the publications would be much enhanced (Ref. 8.7). Another major practical aspect is that data banks are being compiled for many aspects of technology. A first step for tribology would be the use of standard data checklists as developed here.

8.5 Application of Systems Methodology to the Solution of Tribological Problems: Some Case Studies

In concluding this treatise on tribology, the application of systems techniques to the solution of practical tribological problems should be outlined. As emphasized throughout this volume, in attempts to reach a systematic solution of tribological problems, the whole set of the four basic groups of systems parameters discussed in Section 8.2 and their interdependencies must be taken into consideration under the headings:

I Technical function of the system
II Operating variables
III Structure of the system
IV Tribological characteristics

Again, the essential point is that only by systematic consideration of all main variables of the four basic groups of parameters can oversight of important influencing factors be avoided. Once this "systems thinking" is adopted, the application of systems methodology to the solution of tribological problems appears to be a quite straightforward application of "technological common sense" on a broad scale.

The case studies discussed in brief in the following are chosen to illustrate systematic approaches to the solution of tribological problems from different classes of tribo-engineering systems, components and processes as well as to illustrate material-selection procedures, wear-reduction attempts and failure analyses. In most cases, the solution of the problem has been obtained by a combination of systems methodology and laboratory tribo-testing (see Sections 7.2 and 7.3). While the tribological data sheet developed in Section 8.2 served as a convenient "check-list" for the consideration of pertinent parameters, for the sake of brevity, in the case studies only the main relevant parameters are individually compiled and only the main steps of the solution of the problem are outlined. The examples presented in the following indicate that proper solutions of tribological problems can be obtained by different means, for example change of design (see Section 8.5.5), appropriate material and lubricant selection (see Section 8.5.2, 8.5.3 and 8.5.4) appropriate interfacial and environmental conditions (see Section 8.5.6 and 8.5.8), appropriate operational ranges of load, velocity or temperature (see Section 8.5.1 and 8.5.7) in connection with the other systems parameters of the problem under discussion.

8.5.1 INVESTIGATION OF THE VALIDITY OF "COULOMB-FRICTION" FOR POLYMER-STEEL SLIDING PAIRS

(a) PROBLEM

Polymer-steel pairs to be used in a precision engineering application under dry sliding conditions should exhibit a (low) friction coefficient which must be constant under varying operating conditions. The conditions for constant "Coulomb-friction" of suited polymer-steel pairs are to be specified.

(b) APPROACH

The "Amontons-Coulomb law" of dry sliding friction states that the friction coefficient f is independent of load F_N and geometrical area of contact A_0 and thus independent of the nominal contact pressure $p_0 = F_N/A_0$ (see Section 4.3.2). In addition, for low sliding velocities v, it is often assumed that the friction coefficient is also independent of the sliding velocity v. In contrast, many "deviations from Coulomb-friction" have been reported in the literature, so that a systematic study of the dependence of the friction coefficient on the various system parameters has to be performed.

Theoretically, from a systems view point, the friction of a given sliding pair depends on both the set of operating variables and the system's structure. As described in Section 8.2.4 this can be expressed formally as:

friction coefficient = f(operating variables; system's structure)

or $f = f\ (X;\ S)$

Therefore, the following parameters, at least, must be considered as potential influencing variables:

Operating variables, $\{X\}$:

- F_N : load, or contact pressure
- v : sliding velocity
- T : temperature
- s : sliding distance

System's structure $S = \{A, P, R\}$:

- $\{A\}$: elements of the system
  - (1) polymer (to be specified)
  - (2) steel (AISI 52100)
  - (3) air
- $\{P\}$: relevant properties of (1), (2), (3)
- $\{R\}$: frictional interrelations between (1), (2), (3)

Clearly, the influences of the many potential parameters can be studied only in carefully controlled laboratory tests. A literature search shows (see for example Ref. 8.8 and 8.9) that from the materials commercially available at present, the following polymers may exhibit a low coefficient of friction ($f \approx 0.1$) if sliding against steel:

polytetrafluorethylene (PTFE)

high density polyethylene (HDPE)

In using these materials in connection with the ranges of operating variables given by the engineering application conditions, friction tests were performed with a pin-on-disc tribometer utilizing the two-component force transducer described in Section 7.2.3. (For further details of the tribometer, see Ref. 8.10.) In Table 8.2, the test conditions are listed.

<table>
<tr><th colspan="5">Operating variables</th></tr>
<tr><td colspan="2">Type of motion</td><td colspan="3">sliding</td></tr>
<tr><td colspan="2">Load F_N(N)</td><td colspan="3">2 - 20</td></tr>
<tr><td colspan="2">Velocity v (m/min)</td><td colspan="3">10^{-5} - 10</td></tr>
<tr><td colspan="2">Temperature (ambient) T(°C)</td><td colspan="3">20 - 70</td></tr>
<tr><td colspan="2">Sliding distance s (cm)</td><td colspan="3">< 10</td></tr>
<tr><td colspan="2">Contact pressure p (N/mm²)</td><td colspan="3">0.6 - 6</td></tr>
<tr><td colspan="2">Contact area A_o (mm²)</td><td colspan="3">3.1</td></tr>
<tr><th colspan="5">System's structure</th></tr>
<tr><td>Elements / Properties</td><td colspan="2">Tribo-element (1) (stationary pin)</td><td>Tribo-element (2) (rotating disc)</td><td>Surrounding atmosphere (3)</td></tr>
<tr><td>Material</td><td>PTFE</td><td>HDPE</td><td>AISI 52100 steel</td><td>dry synthetic air</td></tr>
<tr><td>Dimensions (mm) (diameter)</td><td>2</td><td>2</td><td>42</td><td>—</td></tr>
<tr><td>Chemical Composition</td><td>$(C\text{-}F_2)_n$</td><td>$(C\text{-}H_2)_n$</td><td>0.98 C
1.5 C</td><td>20 O_2
80 N_2</td></tr>
<tr><td>Tensile strength (MN/m²)</td><td>9</td><td>12</td><td>—</td><td>—</td></tr>
<tr><td>Hardness H_V (N/mm²)</td><td>—</td><td>—</td><td>8000</td><td>—</td></tr>
<tr><td>Thermal conductivity (W/m/°C)</td><td>0.25</td><td>0.33</td><td>70</td><td>—</td></tr>
<tr><td>Surface roughness R_t (μm)</td><td colspan="2">1 - 2</td><td>< 0.1</td><td>—</td></tr>
<tr><td>Ratio ε (%)</td><td colspan="2">100</td><td>≈ 1</td><td>—</td></tr>
</table>

Table 8.2 Conditions of friction tests.

(c) RESULTS

In Figure 8.9 the variation of friction coefficient of the PTFE-steel and HDPE-steel sliding pairs as a function of both contact pressure p and sliding velocity v and an ambient temperature of $T = 23^{o}C$ is shown. In addition, for PTFE-steel the f(p, v) curves for an ambient temperature of $T = 70^{o}C$ are also plotted (broken lines). The experimental results show that two friction regimes can be distinguished:

(i) a low-friction regime with an approximately constant coefficient of friction, i.e.,

$$f = 0.04 \pm 0.01 \quad \text{PTFE-steel}$$
$$f = 0.12 \pm 0.02 \quad \text{HDPE-steel}$$

for

$$p = 0.6 - 6 \text{ N/mm}^2$$
$$v = 0.01 - 1 \text{ mm/min}$$

(ii) a high-friction regime with an increasing coefficient of friction, i.e.,

$$f = 0.05 - 0.35 \quad \text{PTFE-steel}$$
$$f = 0.10 - 0.35 \quad \text{HDPE-steel}$$

for

$$p = 0.6 - 6 \text{ N/mm}^2$$
$$v = 1 \text{ mm/min} - 10 \text{ m/min}$$

In the two different friction regimes, different appearances of the sliding surfaces are also found with repeated sliding, as shown in Figure 8.10. In the low-friction regime (i) a thin oriented film, thickness $d \approx 20$ nm, is observed, whereas in the high-friction regime (ii) the polymer transferred to the sliding steel counterface consists of lumps of polymers of a thickness up to several μm. (For a discussion of the physical mechanisms of the different sliding behaviour in regimes (i) and (ii) see Ref. 8.10 and 8.11.)

(d) FOLLOW-UP

The results of this study confirm the well known fact that very low coefficients of friction for dry PTFE-steel sliding pairs can be obtained. However, "Coulomb-friction", i.e., a constant friction coefficient independent of contact pressure and sliding velocity is found approximately only under very limited conditions: in the general case, the dependence of the friction coefficient on both the set of operating variables (F_N, v, T, s) as well as on the system's structure $S = \{A, P, R\}$ must be taken into consideration. For the engineering application under question, the operating conditions

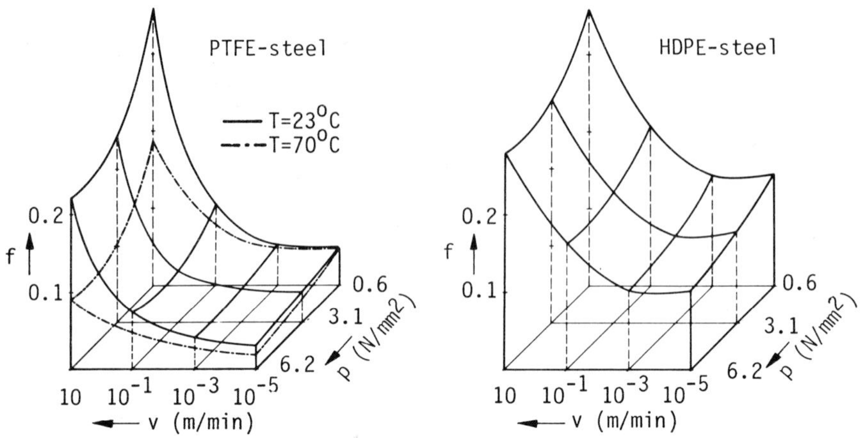

Figure 8.9 Friction coefficient of polymer-steel sliding pairs as a function of velocity v and pressure p.

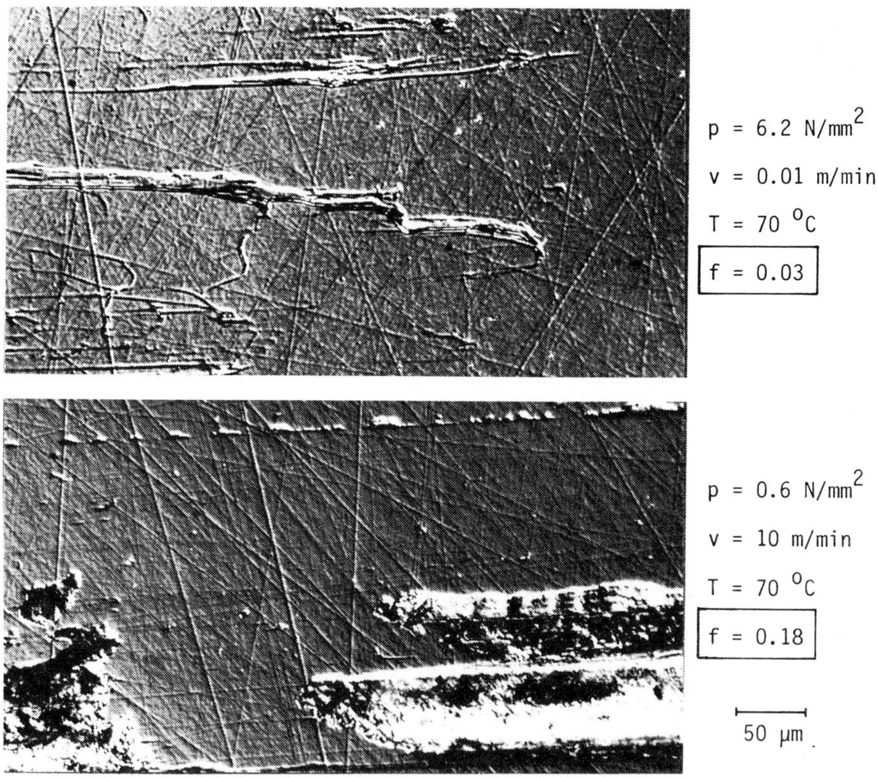

Figure 8.10 Appearance of steel surface after sliding against PTFE.

should be within the limits of the above regime (i) in order to obtain a low friction coefficient of an approximately constant value.

8.5.2 CHARACTERIZATION OF THE TRIBOLOGICAL BEHAVIOUR OF WEAR-RESISTANT DIFFUSION SURFACE COATINGS

A characterization of the wear resistance of newly developed surface treatments (boriding, vanadizing) in comparison with surface treatments already used in practical tribo-engineering applications (carburizing, nitriding) is to be made.

(b) APPROACH

Wear-resistant surface coatings are used in different tribo-engineering applications under broadly varying conditions. Since at the development stage of new surface treatments, suitable engineering application conditions are usually still not specified, a characterization of the wear behaviour of these coatings in general terms is necessary. In Section 4.4 it was emphasized that wear is not a single process but a variety of processes. To characterize (or control) wear, each of these processes must be isolated and studied independently. From a systems viewpoint, the wear of a given sliding pair depends - like friction - on both the set of operating variables and the system's structure. As described in Section 8.2.4, this can be expressed formally as:

$$\text{wear rate} = f(\text{Operating variables; System's structure})$$

$$w = f(X; S)$$

Therefore, to characterize the wear resistance of the surface coatings the following parameters must be considered:

Operating variables, $\{X\}$:

- F_N : load, or contact pressure p
- v : sliding velocity
- T : temperature
- s : sliding distance

System's structure, $S = \{A, P, R\}$:

- $\{A\}$: elements of the system (to be specified below)
- $\{P\}$: properties of the elements (to be specified below)

{R} : wear mechanisms:
- surface fatigue
- abrasion
- adhesion
- tribo-oxidation

It follows that in order to characterize the tribological behaviour of wear-resistant surface coatings, the materials should be studied successively under different conditions in which only one of the four basic wear mechanisms prevails. In the following, only the adhesive and the abrasive wear resistance of some diffusion surface coatings will be characterized. In Table 8.3 the materials, surface treatments and relevant material properties are listed.

Properties \ Materials	(A) C45	(B) 42CrMo4	(C) 42CrMo4	(D) C100
Surface treatment	case-hardened	nitrided	borided	vanadized
Surface structure	martensite	ε-Fe_xN	Fe_2B, FeB	VC
Surface hardness H_V (load 2N) (N/mm^2)	8320	6400	15600	22300
Thickness of surface layer (μm)	——	13	100	19
Surface roughness R_t (μm)	0.6	0.8	1.3	3.0
Hardness of base material H_V(load 100N) (N/mm^2)	6100	3480	2740	2780

Table 8.3 Properties of surface coatings.

The adhesive wear resistance of the surface-treated materials (A) - (D) sliding against themselves has been tested in vacuum where the formation of protective tribochemically formed interfacial layers is reduced and thus the adhesion component is the dominating wear process. The abrasive wear resistance has been studied with a disc-on-abrasive-paper technique where the abrasive "micro-cutting" process is the dominating wear process. (For further details see Ref. 8.12.)

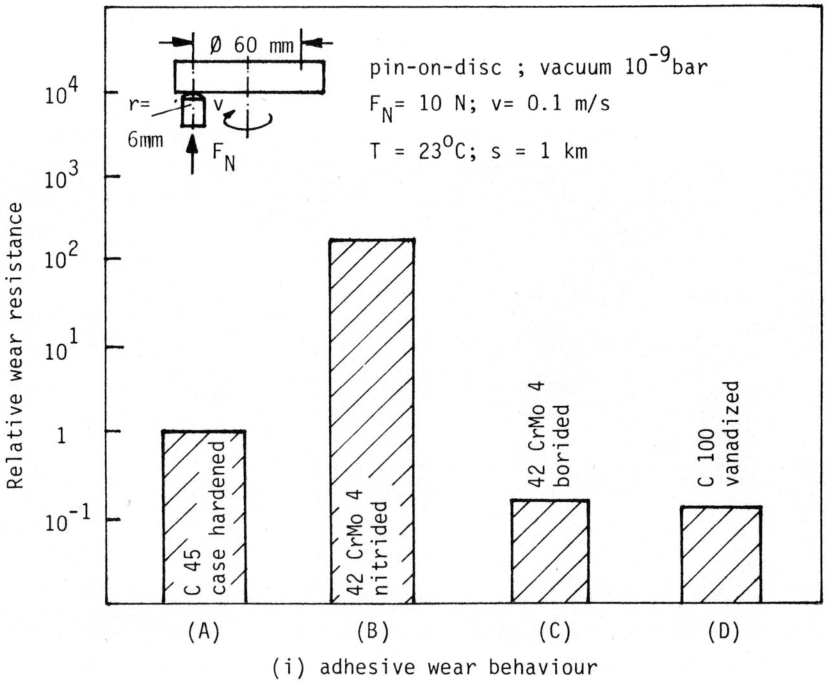

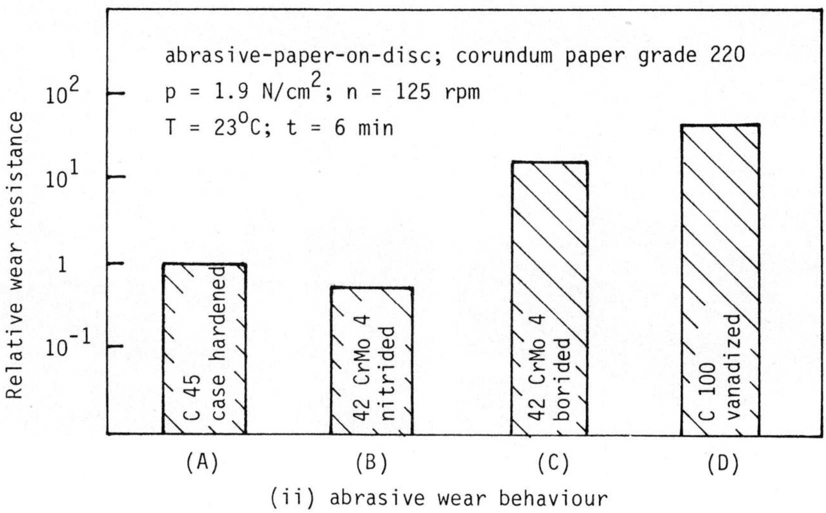

Figure 8.11 Wear resistance of surface coatings tested under (i) adhesive and (ii) abrasive wear conditions.

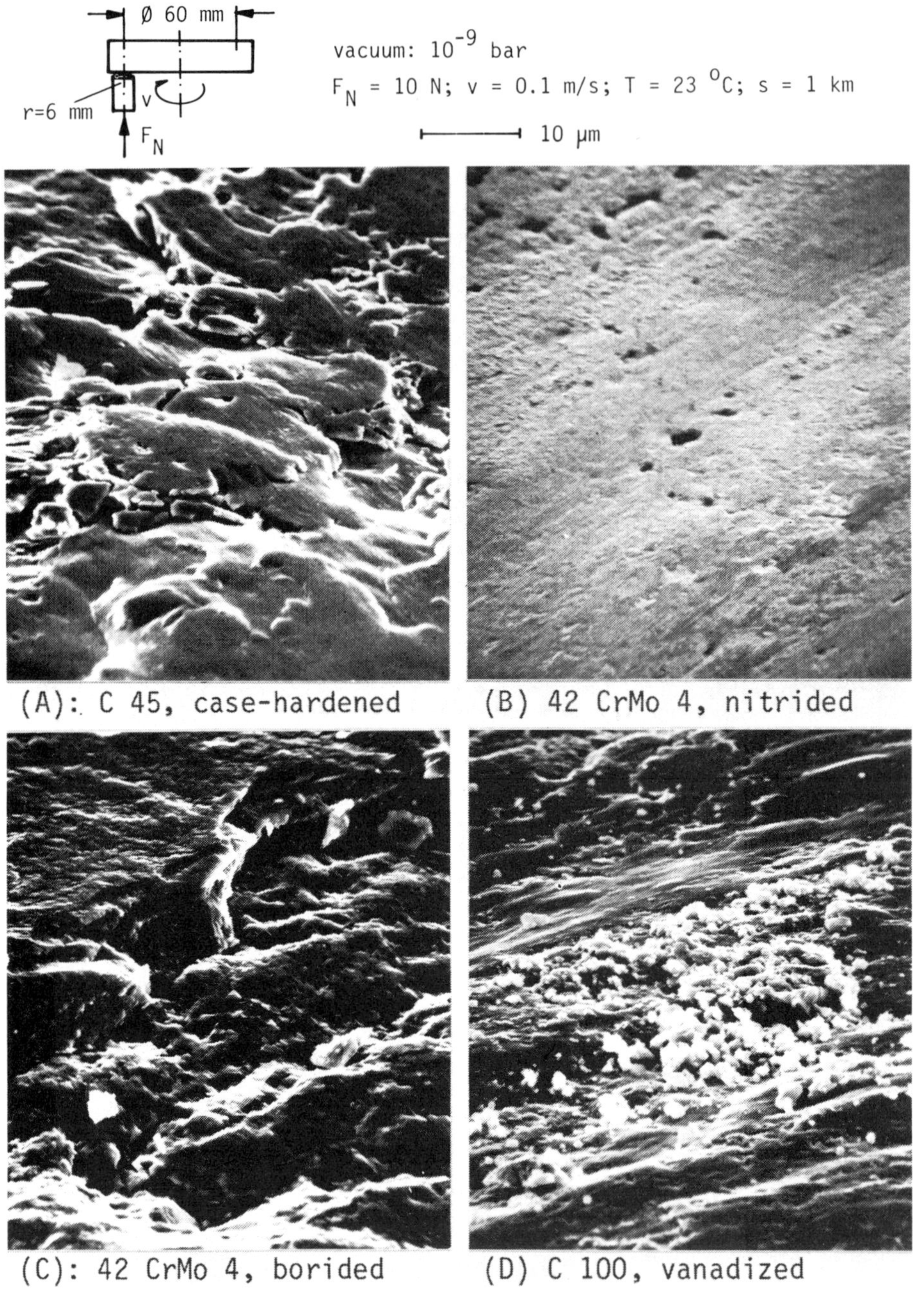

Figure 8.12 Appearance of wear surfaces of pins after testing in vacuum.

(d) RESULTS

In Figure 8.11 the relative wear resistance of the surface coatings under conditions of adhesive wear and abrasive wear together with the applied test conditions is shown with typical results.

In the adhesive-wear situation, the measured wear resistance and the corresponding appearance of worn surfaces, shown in Figure 8.12 indicate that the nitrided surface coatings exhibit the best wear-resistance behaviour. The poorest wear resistance has been found for the (very hard) vanadized steel. In this case, the surface-coating layer was penetrated so that the adhesive wear action extended to the (soft) base material. It should be mentioned that the adhesive wear resistance of the borided coating, which also shows a relatively poor behaviour according to Figure 8.11, is markedly improved if oxygen is available. In this case, the adhesive wear resistance of borided steel is clearly better than that of case-hardened steel.

In contrast, in the abrasive-wear situation, the best wear-resistant behaviour is found for the vanadized surface coating. As described in Section 4.4.3, this behaviour can be explained in terms of the hardness of the tested materials and the abrasive. Since corundum, which was used as an abrasive (hardness H_V = 17000 N/mm^2), is considerable harder than both the case-hardened steel (A) and the nitrided steel (B), both materials (A), (B) are in the "severe wear" regime, whereas the borided steel (C) and the vanadized steel (D) are still in the "mild wear" regime due to their high hardness. If the tests were performed with a silicon carbide abrasive paper (H_V = 27000 N/mm^2), the borided steel also changes to the severe wear regime, whereas the vanadized steel remains in the mild wear regime.

In summary, the studies on the adhesive and the abrasive wear resistance of some diffusion surface coatings show that under the choosen test conditions a quite different (in fact, a reverse) ranking of the surface coatings in order of merit is found in the two cases:

(i) adhesive wear resistance :
1. nitriding
2. case hardening
3. boriding
4. vanadizing

(ii) abrasive wear resistance :
1. vanadizing
2. boriding
3. case hardening
4. nitriding

(d) FOLLOW-UP

The results of this study confirm that for a pre-characterization of the wear behaviour of surface coatings besides the operating conditions the dominating wear mechanism must be known. It follows that different surface treatments may be suggested to be used in different tribo-engineering applications. For example, for the prevention of severe adhesive wear (or scuffing) of gears, nitriding may be suitable whereas for the mitigation of abrasive wear attack, as in dredge pumps, vanadizing may be best.

8.5.3 MATERIAL SELECTION FOR OFF-SHORE BEARING APPLICATION

(a) PROBLEM

A material selection for a journal bearing operating at 30 m below sea level (component of anchor-buoy) has to be made. According to the dimensions of the journal bearing, a linear wear rate of 0.6 µm/h of the bearing bushing corresponds to a specific wear rate of approximately $4 \cdot 10^{-6}$ mm^3/Nm. Because of strength considerations, the wear rate of the journal is to be smaller than $0.1 \cdot 10^{-6}$ mm^3/Nm. The objective was to find a material combination that meets these requirements. Moreover, attack (scratching) of the journal surface should be as mild as possible in order to minimize the chance of scratch-induced fatigue effects occurring.

(b) APPROACH

The tribological systems parameters which have to be taken into consideration in the material-selection process can be seen from the tribological data sheet of a journal bearing, Figure 8.6. For material selection a simulative tribo-testing procedure with a pin-and-ring test rig was applied. A diagrammatic presentation of this rig is shown in Figure 8.13. The specimen assembly consists of a pin (1) and a ring (2). The pin, made from the bearing material to be tested, is pressed under a known normal force against the curved surface of the rotating ring, made from the journal material to be tested. The contact surface of the pin is preshaped to fit the radius of curvature of the ring. Diameter ring: 50 mm; dimensions pin (rectangular cross section): 12.0 x 8.0 mm; overlap percentage $\varepsilon = 7.6\%$. The specimens were surrounded by a box, filled with artificial sea water of 21°C. The specimens were completely submerged.

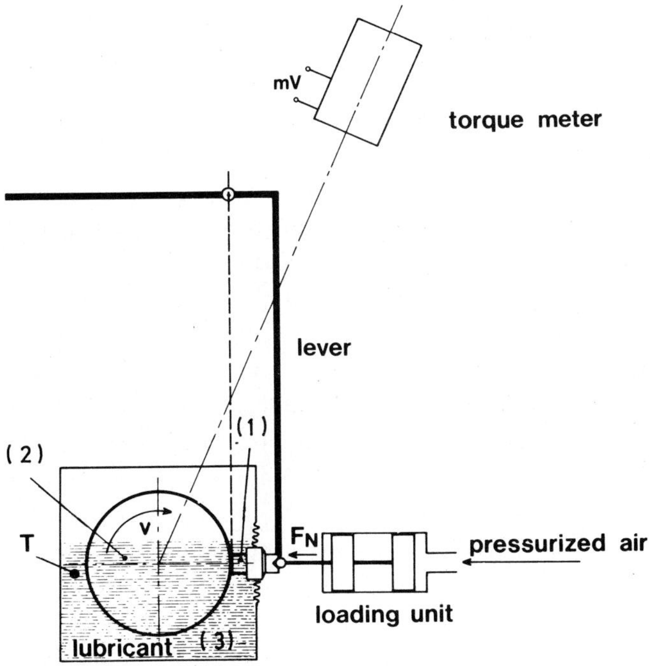

Figure 8.13 Diagrammatic presentation of pin-and-ring test rig.

Operating variables	Buoy bearing	Pin-and-ring test rig
Type of motion	sliding (reciprocating)	sliding (continuous)
Load F_N (N)	$1.5 \cdot 10^7$	$3.85 \cdot 10^3$
Pressure p (N/m^2)	$4.0 \cdot 10^7$	$4.0 \cdot 10^7$
Velocity v (m/s)	$v_{max} = 0.01$	v = 0.05
Total sliding distance s (km)	1000	18
Geometry and dimensions	Journal diameter 0.42 m l/d = 2.14	Ring diameter 50 mm Pin cross section 12 x 8 mm
Environment	Seawater (5-35°C)	Artificial seawater (21°C)

Table 8.4 Conditions in practice and in simulative tribo-testing.

(i) Operating variables:

All experiments were performed under a normal force F_N = 3850 N and at a sliding speed of v = 0.05 m/s (continuous rotation). At F_N = 3850 N, the projected pressure p = $4.0 \cdot 10^7$ N/m, i.e. equal to p in practice. The sliding speed v was set at 0.05 m/s (i.e. 5 times v_{max} in practice). For an experiment duration t = 100 hrs, the total sliding distance s in one experiment was 100 x 3600 x 0.05 x 0.001 = 18 km.

Table 8.4 compares the conditions in practice with those in simulative testing. It can be seen that there are some differences between the practical situation and the test situation, associated with F_N, v and s. As to F_N, the choice of a much lower value in laboratory testing was justified primarily by considering that by equalizing the pressure p, the amount of deformation involved in the laboratory test will be equal to that in practice. As to v, the reasoning was that the main effect of an increase in v is to increase the temperature in the conjunction area due to frictional heating. Although this will certainly be the case in the present situation, even at v = 0.05 m/s the above temperature rise was so low (order of magnitude: a few degrees K) that it could probably be neglected (note that the sea-water present in practice as well as in the laboratory test cools the rubbing surfaces quite effectively). As regards s, the acceptance of 18 km instead of 1000 km was based on the assumption that the wear process would rapidly reach a steady-state condition. Again, by virtue of the fact that thermal effects are virtually negligible, this assumption seemed realistic.

(ii) Materials

Materials for pin and ring were preselected on the basis of a number of non-tribological criteria, for instance, resistance to sea-water corrosion and price, as well as tribo-technical experience and handbook data. Eventually it was decided to make the test specimens from the materials described in Table 8.5. Because of restricted time, only a limited number of combinations of pin and ring materials was tested.

(c) RESULTS

A summary of results is given in Table 8.6. This shows that, from the two bronzes with MoS_2, type (a) did not come up to expectations. However type (b), combined with ring materials (C) and (E), met the requirements as regards wear rate. However, even with material (b) the rings were slightly damaged, see Figure 8.14. Combinations of different cobalt-type alloys, i.e., material combinations (cD), (cE), (cF) and (dF), were successful as far as the wear rate of the pin (bushing) material was concerned. However, in all cases an unacceptably high wear rate of the rings was observed, ranging from 0.5 units for combination (cF) to 1.5 units for combination (dF).

Material	Code	Type	Hardness level (10^7 N/m^2)
Pin (bushing) (1)	a b	cast bronze with MoS_2	150-250
	c d	cobalt-chromium alloys	500-600
	e f g h	phenolic resins + fillers + solid lubricant	30-40
Ring (journal) (2)	A	Cr-Ni alloy	200
	B	Cu-Ni alloy	200
	C	Fe-Ni alloy	250
	D E F	Co-Cr alloys	500-600

Table 8.5 Types of preselected materials.

Finally, all combinations involving resin-based pins were successful in that wear rate of the pins never exceeded 4 units. Minor damage of the rings of the type shown in Figure 8.14 occurred with material combinations (eA), (fB) and (hB). However, if used in combination with pin materials (e) and (g), rings of material (B) remained completely undamaged, see example in Figure 8.15.

(d) FOLLOW-UP

On the basis of the above results, application of materials (e), (g) and (B) seemed well justified. Further, it looked as if combination (bC) or (bE) might well be chosen as a possible back-up solution.

Since performing the laboratory tests, prototype testing has been performed with combinations (gB) and (bC) test duration: 0.5 year; total slid-

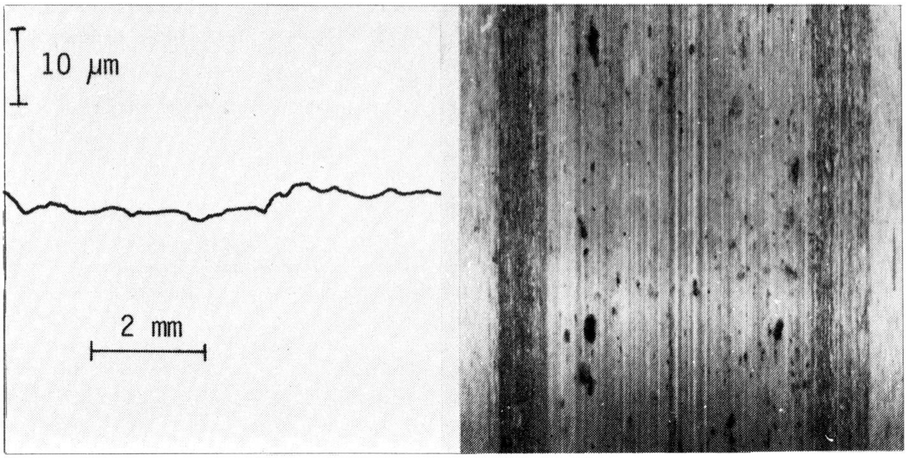

Figure 8.14 Profile diagram and photomicrograph of pin surface after test. Material combination: bC.
Damage classification: slightly damaged.

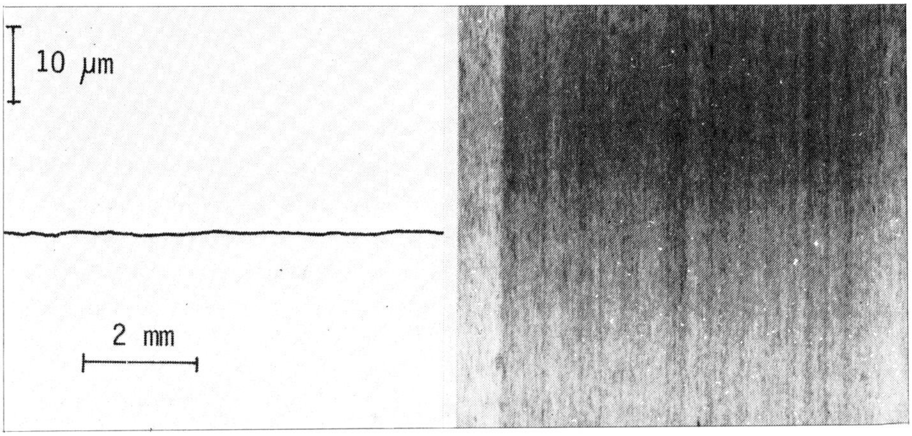

Figure 8.15 Profile diagram and photomicrograph of pin surface after test. Material combination: gB.
Damage classification: undamaged.

ing distance: 25 km). The results fully corroborate the pin and ring results and thus justify in retrospect the choice of the test conditions in the laboratory tests.

Material		Wear rate K' ($10^{-6} mm^3/Nm$)	
pin	ring	pin	ring
(a) cast bronze with MoS_2	(A) Cr-Ni alloy	10	< 0.1
	(B) Cu-Ni alloy	9	0.6
	(C) Fe-Ni alloy	7	0.3
	(D) Co-Cr alloy	6	0.1
	(E) Co-Cr alloy	10	< 0.1
	(F) Co-Cr alloy	6	< 0.1
(b) cast bronze with MoS_2	(C) Fe-Ni alloy	3	< 0.1
	(E) Co-Cr alloy	3	< 0.1
	(F) Co-Cr alloy	5	< 0.1
(c) cobalt-chromium alloy	(D) Co-Cr alloy	0.4	0.8
	(E) Co-Cr alloy	0.5	0.8
	(F) Co-Cr alloy	0.1	0.5
(d) cobalt-chromium alloy	(F) Co-Cr alloy	0.4	1.5
(e) phenolic resins + fillers + solid lubricant	(A) Cr-Ni alloy	4	< 0.1
	(B) Cu-Ni alloy	2	< 0.1
(f) phenolic resins + fillers + solid lubricant	(B) Cu-Ni alloy	3	< 0.1
(g) phenolic resins + fillers + solid lubricant	(B) Cu-Ni alloy	2	< 0.1
(h) phenolic resins + fillers + solid lubricant	(B) Cu-Ni alloy	2	< 0.1

Table 8.6 Summary of results of pin and ring testing. (Note that each figure given is the highest figure found in triplicate testing)

8.5.4 LUBRICANT SELECTION FOR INSTRUMENT PIVOT BEARING

(a) PROBLEM

In a cost-benefit analysis of an instrument, the use of two types of rather expensive special lubricant for the lubrication of a steel-sapphire pivot bearing was criticized. Although the bearing system worked very satisfactorily, for the sake of cost reduction the special lubricants had to be replaced by more conventional lubricants. The objective was to select a suitable lubricant which meets the requirement of cost reduction but maintains proper lubrication without deterioration of the bearing's performance. In particular, the wear rate of the pivot should not increase as a consequence of the replacement of the special lubricant.

(b) APPROACH

The problem demands a comparison of the functional behaviour of the pivot bearing lubricated with either the practically proven special lubricants or the new lubricants to be selected on both economic and technological grounds.

A couple of lubricants (mineral oils) were preselected on the basis of criteria like price, chemical constitution, ageing behaviour and viscosity (viscosity range at room temperature: 100-300 mPas). In Table 8.7 some characteristics of the preselected lubricants are listed.

Lubricants (A) and (B) are the practically proven special lubricants, and lubricants (C) - (F) are potential candidates for the replacement. The figures C_A, C_N, C_P characterize the percentage of carbon atoms in aromatic groups, naphthenic groups or paraffinic groups, respectively. (For a description of these terms see Section 4.5.5.)

The tribological performance of the practically proven and the preselected lubricants was tested by simulating the behaviour of the pivot bearing in laboratory tests using the identical steel-sapphire bearing materials and adjusting the operational test variables to that of the practical bearing system. The variables relevant for simulative tribo-testing have been checked with the help of the tribological systems data sheet, see Figure 8.5. In Table 8.8 the pertinent data of the practical system and the test system are listed. It can be seen that the laboratory tests were performed under slightly intensified test conditions concerning contact pressure and speed.

Lubricant	Density at 20°C	Mean molecular weight	$\%C_A$	$\%C_N$	$\%C_P$
(A)	0.878	556	9	20	71
(B)	0.893	779	8	21	71
(C)	0.894	585	4	34	62
(D)	0.894	574	2	38	60
(E)	0.896	656	6	30	64
(F)	0.895	595	6	31	63

Table 8.7 Properties of preselected lubricants.

Operating variables	Pivot bearing	Test rig
Type of motion	spin (continuous)	spin (continuous)
Load F_N (N)	2.6	13.4
Pressure p_{oH} (N/m^2)	$2.1 \cdot 10^9$	$2.5 \cdot 10^9$
Speed (rpm)	22	54
Temperature (°C)	25	25
Duration (h)		24
Geometry	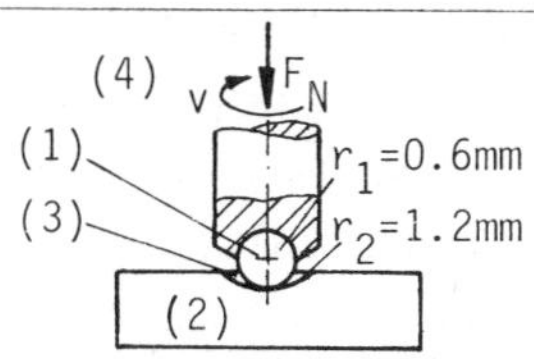	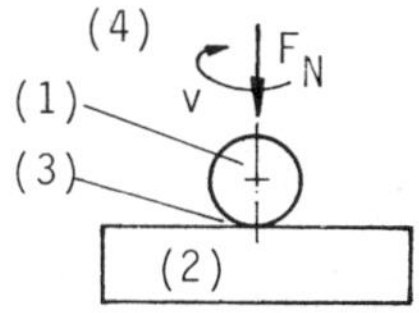
Components of the systems	(1) AISI 52100 steel ball $E = 2 \cdot 10^7$ N/cm^2 H_v = 8000 N/mm^2 R_a = 0.02 µm	
	(2) sapphire $E = 5 \cdot 10^7$ N/cm^2 R_a = 0.02 µm	
	(3) lubricant (see Table 8.7)	
	(4) air	

Table 8.8 Conditions in practice and in simulative tribo-testing.

(c) RESULTS

The functional performance of the pivot bearing is determined mainly by boundary lubrication at the steel-sapphire interface and the limits of proper functional behaviour are reached if the amount of wear of the pivot reaches a certain critical limit. Thus, the wear coefficient of the pivot served as the ultimate lubricant selection criterion. In Table 8.9, the measured wear coefficients for the various systems studied are compiled.

Lubricant used		Wear coefficient of pivot K' (mm^3/Nm)
practically proven special lubricants	(A)	$2.9 \cdot 10^{-8}$
	(B)	$1.5 \cdot 10^{-8}$
lubricants to be used	(C)	$8.9 \cdot 10^{-9}$
	(D)	$8.2 \cdot 10^{-9}$
	(E)	$6.6 \cdot 10^{-9}$
	(F)	$7.8 \cdot 10^{-9}$

Table 8.9 Results of wear testing.

The comparison of these systems shows that no deterioration of the bearing's performance is to be expected if the practically proven special lubricants (A), (B) are replaced by one of the lubricants (C) - (F). Moreover, from the data of Table 8.8, it may be concluded that even an improvement in wear behaviour of the pivot bearing may be obtained by using one of the lubricants (C) - (F).

(d) FOLLOW-UP

On the basis of the above results it was suggested that the special (and expensive) lubricants (A) and (B) should be replaced by the lubricant (E). This has been done in practice and over a period of three years (during which the proper functional behaviour of the pivot bearings has been checked) a satisfactory functional performance of the pivot bearing was observed.

8.5.5 Reduction of severe wear of cam-tappet design

(a) PROBLEM

Cam and tappet combinations (parts of textile machinery) which function at irregular intervals to transmit motion, suffer from severe adhesive wear, frequently within 1 month after putting into use. In particular, the following questions were formulated:

(i) Is failure of the phosphated surfaces force-dependent or pressure-dependent?

(ii) What is the maximum allowable force or pressure?

(iii) Does the application of resin-bonded MoS_2 improve the situation?

In practice, if the system is operated, normal force F_N is built up from 0 to 500 N in 4s, after which F_N remains at the 500 N level for 1s and F_N decreases again to 0 in another 4s. Thus, a full contact cycle requires 9s. The problem is considered to be solved if the cam can perform 15000 movement cycles, irregularly divided over a period of 5 years, without severe adhesive wear occurring. The total contact time is then equal to 37.5 hrs.

(b) APPROACH

Since, from a systems point of view, wear of a given tribo-engineering system is a function of both the operating variables, X, and the system's structure, S, i.e.,

$$\text{wear} = f(X, S),$$

one considers solving the problem by one or more of the following curative actions:

(I) Changing operating variables, $\{X\}$:
- decrease of F_{Nmax}

(II) Changing system's structure, $S = \{A, P, R\}$:
- application of a resin-bonded MoS_2 film
- increase in radius of curvature of the cam (thus lowering the Hertzian contact pressure p_H)

In order to find the optimum solution, simulative laboratory tests were performed with a pin-and-ring test rig. Rings of radius of curvature r_2 = 38mm (representing the tappets) and pins with hemispherical tips with radii r_1 of 1.5, 5 and 11mm (representing cams) were made from 15 CrNi6 steel. Prior to use, the specimens were treated in a manganese-phosphating bath similar to the one used in practice. During test, the specimen assembly was completely immersed in 20W30 oil of 50^{0}C. The speed of rotation of the rings was adjusted at 200 rev./min, which yielded a linear speed in the friction interface of 0.8 m/s.

Operating variables	
Type of motion	sliding
Load $F_N(t)$ (N)	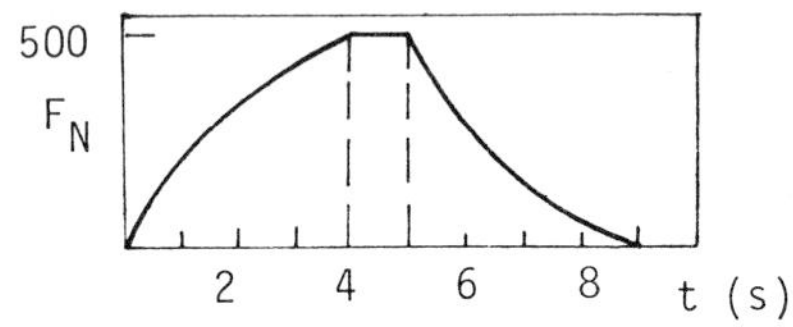
Velocity v (m/s)	0.8
Temperature T (^{0}C)	50
Operating duration	15,000 cycles (37.5 h)

System's structure				
Elements / Properties	Tribo-element (1) (stationary pin)	Tribo-element (2) (rotating ring)	Lubricant (3)	Atmosphere (4)
Material	steel 15 CrNi6	steel 15 CrNi6	oil SAE 20W30	air
Dimensions (mm)	radii 1,5; 5; 11	diameter 38	—	—
Hardness (Rockwell C)	62	62	—	—
Surface treatment	manganese-iron phosphate	manganese-iron phosphate	—	—
Surface roughness R_a (μm)	0.2	0.2	—	—
Ratio ε (%)	100	< 1	—	—

Table 8.10 Conditions of wear test.

During a test the normal force F_N-time characteristic was similar to that in practice, i.e. build up of F_N from 0 to maximum value in 4s, constant level $F_N = F_{max}$ during 1s, followed by decrease to 0 N in another 4s. Contrary to practice, however, the load cycles followed in immediate succession, 15000 cycles thus corresponding to a testing time of 37.5 hrs. In Table 8.10 the simulative test conditions are listed.

(c) RESULTS

r	F_N	p_{Hertz}	endurance			
			with MoS_2		without MoS_2	
(mm)	(N)	(10^7 N/m^2)	min	hrs	min	hrs
1.5	10	217	> 2400	> 40	> 2400	> 40
	50	370	90	1.5	60	1
	100	467	0	-	0	-
	250	633	0	-	0	-
5	100	215	> 2400	> 40	> 2400	> 40
	250	292	1500	25	1200	20
	500	368	110	2	70	1
	750	422	0	-	0	-
11	500	228	> 2400	> 40	> 2400	> 40
	1000	289	1010	17	980	16
	3000	415	0	-	0	-
	4000	457	0	-	0	-

Table 8.11 Results of tribo-testing.

Table 8.11 shows results, expressed in terms of endurance, i.e. time to failure by adhesive wear effects. These results show that the surfaces failed immediately at $p_{Hertz} > 400 \cdot 10^7$ N/m^2. At lower values of p_{Hertz}, increasingly long endurances were found. Microscopic observation directly before and after failure indicated that the occurrence of adhesive-wear effects was associated with failure of the phosphate film.

Table 8.11 shows that the presence of a resin-bonded MoS_2 film, although discernible, has only little influence on the endurance of the phosphate film. In additional experiments it has been shown that this is due to the 20W30 oil rapidly dissolving the resin, thus destroying the cohesive integrity of the solid lubricant film. Under conditions of dry rubbing, a resin-bonded MoS_2 film protects phosphated surfaces quite effectively, but in the present case this is of no practical value.

Clearly, failure of the phosphate film is pressure-dependent rather than force-dependent. Further, the results show that the endurance is more than 40 hrs at $p_{Hertz} \leqslant 230 \cdot 10^7 N/m^2$. At r = 5, to reach this p_{Hertz} value would require a drastic decrease in normal force, i.e. from 500 N to some 150 N. Thus it was recommended to increase the radius of curvature of the cam to 11 mm instead.

(d) FOLLOW-UP

Since performing the experiments, the radius of the cam has indeed been increased, i.e. to 10.5 mm. Since then, severe adhesive wear has not been reported for a period which now amounts to approximately 1 year.

8.5.6 REDUCTION OF FRICTION-INDUCED NOISE OF WHEEL/RAIL SYSTEM

(a) PROBLEM

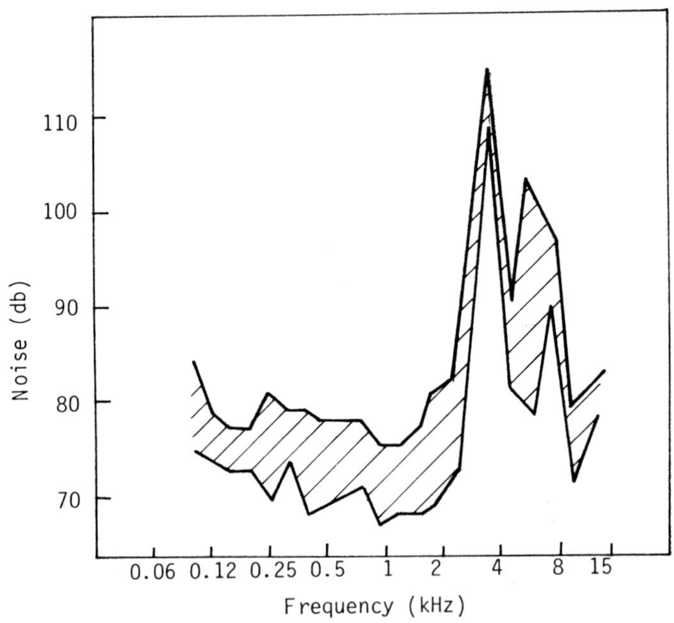

Figure 8.16 Diagram of noise spectrum emitted by wheel-rail system.

City trains running through narrow curves often produce a penetrating squealing noise which reach noise levels up to 120 db at frequencies ranges from 4000 to 8000 Hz. As a typical example in Figure 8.16, a noise diagram is shown measured in six runs of a carriage running through a curve of 193m radius. The causes of friction-induced noise generation are to be studied and methods for noise reduction are to be suggested.

(b) APPROACH

A theoretical consideration of the forces and movements acting between the rail and a rigid wheel-frame running through a curve shows that a transverse slip motion between wheel and rail occurs if the rail-track curve radius is less than 500 m for a railway track width of 1435 mm (Ref. 8.13). For a train speed of 15 m/sec and rail curves 200-500 m, the transverse sliding velocity is in the range of v_t = 0.1 to 0.3 m/sec. The transverse sliding motion and the interfacial friction lead to stick-slip-induced vibrations of the wheel/rail system which is manifest in the generation of noise.

Having discussed in brief the reason for the generation of noise, the question of how to mitigate this "pollutant" must be answered. Interpreting the noise emission as tribo-induced "loss-output" of the tribological wheel-rail system, the general system description can be used as a convenient starting point for a systematic consideration of the parameters that may be changed in order to relieve the noise emission.

According to Section 8.2.4, the tribo-induced noise emission can be described formally as:

$$\text{noise} = f\ (\text{Operating variables, System's structure})$$

$$\text{noise} = f\ (X;\ S)$$

Thus, the main influencing factors can be classified under the headings:

(I) Operating variables, $\{X\}$:

- Load (weight of carriage plus passengers)
- Speed (train speed, transverse slip speed)
- Temperature (depending on season)

(II) System's structure, $S = \{A, P, R\}$

{A} Elements:

(1) Type of wheel
(2) Type of rail
(3) Environmental atmosphere

{P} Properties of elements:

- Material of wheel and rail
- Geometric design of wheel and rail
- Geometric design of rail track
- Surface properties of wheel and rail

{R} Interrelations between elements:
Tribological interrelations between (1), (2), (3)

Clearly, the parameters (I) and also most of the parameters (II) are fixed for a given commercial wheel-rail-train system. Therefore, since changes in design cannot be made, an attempt must be made to influence directly the interfacial tribological processes {R} which are the origin of the noise generation.

Empirically, it was found that if certain phosphate solutions are sprayed on the rails, the noise vanished. Theoretically, from the analogue computer studies of stick-slip effects (see Section 6.2.2), it follows that the stick-slip effects depend on the friction-velocity gradient. In order to clarify the possibilities of relief of stick-slip-induced noise and to optimize the conditions of an appropriate surface treatment of the rails, stick-slip studies in the laboratory were made (Ref. 8.14). For these studies, a pin-on-disc and a ball-on-disc tribometer were used. The tests were performed with the original wheel-rail materials under operating conditions similar to that in practice ($p_H = 50 \cdot 10^7 N/m^2$; $v = 2 - 20$ cm/s).

(c) RESULTS

The experimental studies show that the stick-slip effects are markedly influenced by different phosphate treatments of the surfaces. It was found that through the surface treatment, the static friction coefficient f_s is slightly decreased and the dynamic friction coefficient f_d is considerable increased, so that the difference $\Delta f = f_s - f_d$ is approximately zero. Therefore, the requirement for vanishing stick-slip effects is given (see Section 6.2.2). Indeed, it has been found that under these conditions the stick-slip amplitude is drastically reduced. A typical result of the influence of the surface treatment on the stick-slip behaviour is shown in Figure 8.17.

(d) FOLLOW-UP

On the basis of the systematic laboratory stick-slip investigations, a phosphate solution, which appeared to be optimally suitable for practical purposes could be specified. By treating the rails at regular intervals with this solution, a satisfactory reduction of the stick-slip-induced noise of the practical wheel rail system was obtained.

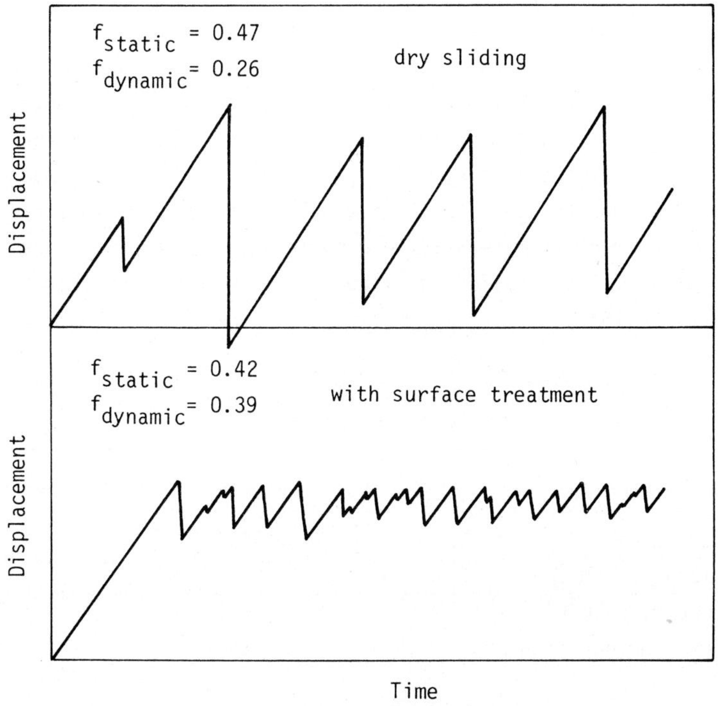

Figure 8.17 Influence of surface treatment on stick-slip behaviour.

8.5.7 FAILURE INVESTIGATION OF ELECTRICAL CONTACTS

(a) PROBLEM

Electrical contacts in telephone exchange units show sudden, inexplicable failures due to severe adhesive wear effects. There is some indication that at low relative humidity (units located in desert country) the failure rate is higher than elsewhere. The causes of failure are to be studied and methods for avoiding failure are to be suggested.

(b) APPROACH

From a systems viewpoint, failure of a given tribo-engineering system occurs through the action of the operating variables on the structure of the system indicated by a severe disturbance of the functional input-output relations (see Section 6.4). Checking the potential influence of the various system's parameters with the help of the tribological systems data sheet (see Figure 8.5), it appears that the failure of the electrical contacts is due to a breakdown of "protective" surface layers influenced differently by contact load F_N and atmospheric humidity R.H. for various contact-material combinations. In order to study these findings further, laboratory failure tests with various contact materials were performed with a pin-on-disc tribometer. In Table 8.12 the failure test conditions are listed. (For further details see Ref. 8.15).

Operating variables			
Type of motion		sliding	
Load F_N (N)		0.01 - 1	
Velocity v (m/s)		0.2	
Temperature (^oC)		18 $\pm$ 1	
Operating duration (h)		up to 72	
System's structure			
Elements / Properties	Contact element (1) (stationary pin)	Contact element (2) (rotating pin)	Atmosphere (3)
Material	70 Ag, 20 Au, 10 Cu	99.9 Cu, 85 Cu, 15 Zn, 90 Cu, 9.5 Sn, 0.5P, 55 Cu, 18 Ni, 27 Zn	air rel. humidity: 0-100%
Dimensions (mm) (diameter)	2	75	—
Surface roughness R_a (μm)	0.1 - 0.2	0.1 - 0.2	—
Ratio ε (%)	100	< 1	—

Table 8.12 Conditions of failure tests.

(c) RESULTS

Results obtained with the material combination Ag-Au-Cu against Cu-Ni-Zn are given in Table 8.13. Obviously, at zero humidity, a transition occurs at a load $F_N = (8 - 10) \cdot 10^{-2}$N. The onset of wear is marked clearly by a deflection in the height versus time recording. The fourth column of Table 8.13 gives the time period which expires before the pin begins to wear. This period of time is called "incubation time". Visual inspection and microscopic observation of the disc surfaces reveal that at $8 \cdot 10^{-2}$N and below, the disc surface remains virtually undamaged.(As usual for contacts sliding in air, the "wear track" becomes marked by a slight discoloration, due to formation of a thin oxide film.) At loads of 0.1 N and higher, no visible oxide films are formed on the disc surface, presumably because fresh metal surfaces form continuously, due to the constant removal of particles in the form of wear debris. The coefficient of friction has a constant value of 0.8 at loads below transition. Above transition, friction rapidly drops to values corresponding to f = 0.6, as soon as the system starts to wear. At 60% relative humidity, the transition has shifted to a much higher load value, viz. between 0.4 and 0.5 N, as a result of introducing water vapour containing air. Friction is lower than at 0% R.H., below as well as above transition.

The results obtained at 0% and 60% humidity suggest that the transition load continuously increases with increasing humidity. This was tested by performing additional experiments at intermediate and higher humidity values. The results of all experiments are summarized in Table 8.14 and show that there are two discrete transition values, one occurring at loads between 0.08 and 0.1 N at humidities of 50% and lower, and the other occurring at loads between 0.4 and 0.5 N, at humidities of 60% and higher. Apparently, for this material combination, the shift in transition load occurs at approximately 55% R.H. To determine whether or not the critical relative humidity of 55% is specific for the material combination at hand, critical humidity determinations were performed with the other disc materials, viz. Cu-Zn, Cu-Sn and Cu. The results are summarized in Table 8.15. It can be seen, that the endurance of the sliding couple increases abruptly at a critical value of relative humidity. This critical value is a function of the alloy composition, probably being determined by the water adsorption characteristics of the different surface oxides.

The analyses clearly show that the proper functional behaviour of the electrical contact system is critically influenced by the relative humidity

Relative humidity (%)	Load (10^{-2} N)	Wear rate of Ag-Au-Cu pin (μm/h)	Incubation time (h)	coefficient of friction f
0	7	< 0.01	> 72	
	8	< 0.01	> 72	
	10	0.7	3	0.8 → 0.6
	15	1.8	1	0.8 → 0.6
	40	2.4	< 0.02	0.6
	60	4.1	< 0.02	0.6
60	20	< 0.01	> 72	0.6
	30	< 0.01	> 72	0.6
	40	< 0.01	> 72	0.6
	50	42	9	0.6 → 0.5
	60	50	0.5	0.5
	70	62	0.5	0.5

Table 8.13 Results obtained with Ag-Au-Cu sliding against Cu-Ni-Zn.

R.H. (%)	Critical load (10^{-2} N)
0	8 - 10
20	8 - 10
40	8 - 10
50	8 - 10
60	40 - 50
80	40 - 50

Table 8.14 Transition load as a function of relative humidity for Ag-Au-Cu sliding against Cu-Ni-Zn.

Disc material (2)	Critical R.H. (%)
Cu-Ni-Zn	55
Cu-Zn	35
Cu-Sn	45
Cu	30

Table 8.15 Critical relative humidity for various material combinations (F_N = 0.4 N, pin material Ag-Au-Cu).

of the surrounding atmosphere, i.e., by a property of the element (3) of the tribo-engineering system under study.

(d) FOLLOW-UP

Some telephone exchange units, operating in extremely dry environments, were provided with air conditioning. This (rather expensive) solution proved to be entirely satisfactorily. In other cases, gold plating of the electrical contacts (also expensive) effectively eliminated the problem. (With gold plated contacts, adhesive material transfer always occurs; however formation of loose wear particles is a relatively rare event.) Finally, in replacing exchange units with mechanical contacts by fully electronic units, precedence was given to replace units, operating in extremely dry environments.

8.5.8 COMPILATION OF CHARACTERISTICS OF METAL-WORKING PROCESSES

(a) PROBLEM

An important area of tribology which has not been treated explicitly in this volume so far concerns the tribology of manufacturing processes. Also in this field, progress is possible only through a better understanding of the great number of controlling variables. While it is not possible to enter into details of this field within the scope of this volume, at least a brief overall view of its main "systems parameters" should be attempted here.

(b) APPROACH

In the Appendix (A), a compilation of the various manufacturing processes is given. Clearly, in all these processes the tribological mechanisms, described in Chapter 4, i.e., the contact, friction, wear and lubrication processes, act all at the workpiece-tool interface. In manufacturing, the tribological processes play a "dual" role:

(i) On one hand, tribological processes influence directly the surface properties of the manufactured workpiece. Thus, tribological processes determine to a great extent the quality of the manufactured workpiece, i.e., the quality of the "use-output" of the manufacturing system.

(ii) On the other hand, tribological processes at the workpiece-tool interface are responsible for tool-wear and tool-workpiece friction. Thus, as in any tribo-engineering system, tribological processes are also responsible, of course, for the energy and material "loss-outputs" of the manufacturing system.

Manufacturing processes can be classified broadly in two groups: "cutting" or "machining" processes, and "forming" processes. In machining, material is removed from a crude piece of stock to leave behind a precision-shaped useful part. Manufacturing by this method is, however, wasteful of material; sometimes as much as 90 percent of the original material is removed and discarded in the process. In future, such waste may become intolerable as a consequence of the increasing efforts of conservation of materials. Thus, the ultimate method of manufacture in the future will be "consolidative processing" - the manufacture of goods by forming or assembling material in a controlled precise manner (Ref. 8.16). One of the major areas of consolidative manufacturing processes is the field of metal-working processes. In the following, the main systems parameters relevant to the tribology of metal-working processes will be studied in brief.

(c) RESULTS

A comprehensive review of the tribology of metal-working processes was published by Schey in 1970 (Ref. 8.17). Further reviews of the overall field or special important aspects can be found in Refs 8.18 to 8.22. If on the basis of this work the main systems parameters of the tribology of metal-working processes are compiled with the help of the tribological systems data sheet (see Figure 8.5), the following compilation results:

I. Technical function of metal-working systems

Forming of materials through:

- drawing
- pressing
- rolling
- forging
- extrusion

II. Operating variables of metal-working process

- type of motion at the workpiece-die interface
- load and/or interfacial pressure, strain rate
- surface and/or deformation velocity, velocity distribution

- working temperature (ambient temperature and friction-induced temperature)
- operating duration
- material flow and/or deformation rate

III. Structure of metal-working system

(a) Components of metal-working system
   (1) Workpiece
   (2) Die
   (3) Lubricant/coolant
   (4) Surrounding atmosphere

(b) Properties of metal-working components
   (1) Workpiece: material, geometry and dimensions, composition, ductility, hardness, surface roughness, pretreatment
   (2) Die: material, geometry and dimensions, composition, surface coating (if any), hardness, surface roughness
   (3) Lubricant: type, chemical composition, viscosity, η-T- and η-p-behaviour, η-strain or velocity dependence
   (4) Atmosphere: composition, humidity

(c) Tribological processes at the workpiece-tool interface
   - contact deformation processes: type, state of stress and strain, state of deformation
   - adhesion and tribochemical processes
   - friction processes
   - wear processes, with respect to (1) workpiece and (2) die
   - lubrication modes

IV. Tribological "loss-output" characteristics
   - friction data (also amount of friction-induced power loss or heat generation)
   - wear and pickup data and surface condition of die
   - surface condition of workpiece

In addition to this general compilation of effects and parameters of the tribology of metal-working processes, typical ranges of the basic operating variables of pressure, product speed and temperature are outlined in Table 8.16 for the main metal-working processes.

Metalworking process	Basic operating variables		
	pressure ($10^7 N/m^2$)	speed (m/s)	temperature (^oC)
Drawing 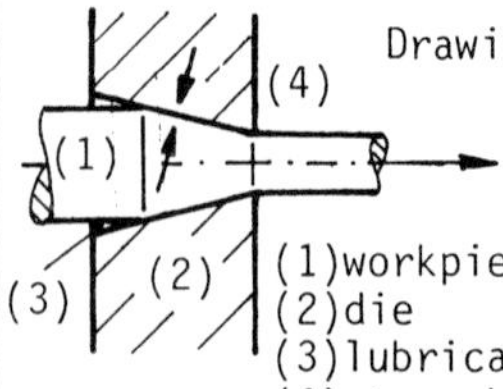 (1)workpiece (2)die (3)lubricant (4)atmosphere	...20	0.0005 (bar and tube) ...1 (heavier wire) ...15-30 (fine wire)	cold: room temperature + ΔT=200-300 warm: 400-600 hot : 800 for Mo,W
Pressing 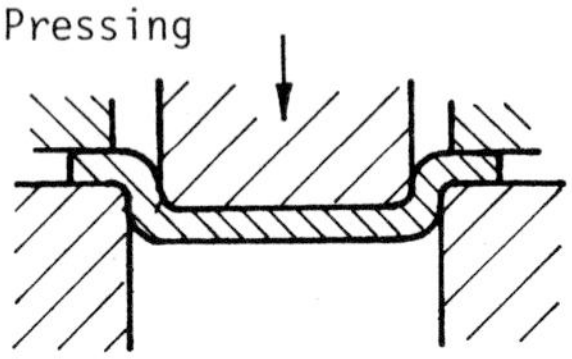	...80	0.02-0.5	room or elevated temperature (e.g. 500-800 for Ti)
Rolling 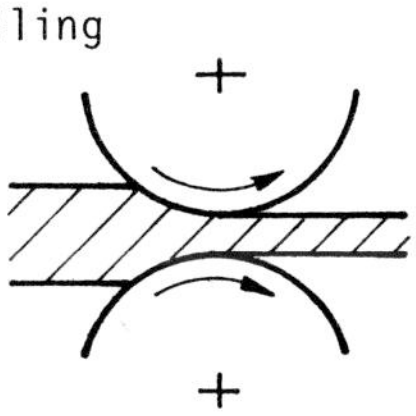	20-150	5-25 (cold rolling) 0.1-0.5 (hot rolling)	room temperature or elevated temperature (400-1250, hot rolling)
Forging 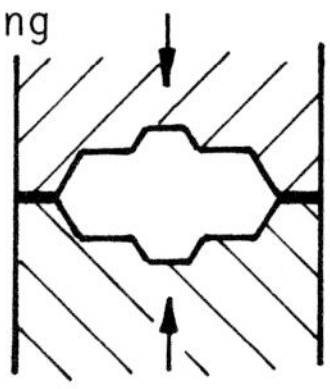	30-130	0.02-0.2 (hydr. presses) ...10 (hammers)	100-200 (cold) 500-600 (warm) ...1200 (hot forging)
Extrusion 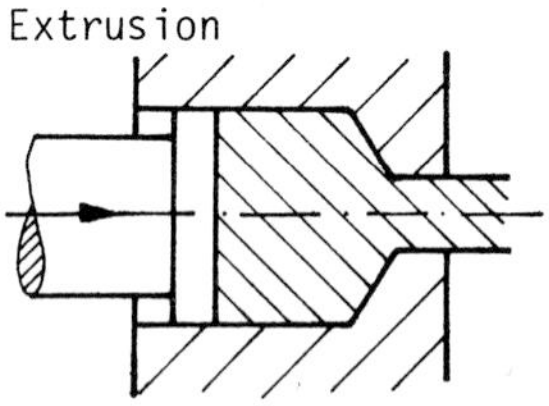	100-200	0.05 ("hot shot") 0.5-5	room or elevated temperature (1000-1200 hot extrusion) 600 non-ferrous

Table 8.16 Some characteristics of metal-working processes.

(d) FOLLOW-UP

Clearly, the given compilation of characteristics of metal-working processes must be specified and extended for a given actual situation. However, the systematic compilation outlined in brief may be used for purposes of material and lubricant selection as well as a framework for reporting experimental variables or performing simulative tribo-testing of metal-working processes.

Appendix (A) A compilation of basic tribo-engineering systems

The treatise on the tribology of mechanical systems, presented in this volume in a generalized manner, applies to a great variety of actual tribo-engineering systems. As a supplement to the general treatise on the tribology of mechanical systems, a compilation of basic engineering systems in which tribological processes, i.e., contact, friction, lubrication and wear processes, occur is given in this section of the appendix.

In general terms, a tribo-engineering system is defined as an entity whose technical function is connected with "the interaction of material surfaces in relative motion". It was described in Chapter 3 how the technical functions realized through interacting material surfaces can be broadly classified in the following groups:

(a) Transmission of MOTION
(b) Transmission of WORK (or POWER)
(c) Generation or reproduction of INFORMATION
(d) Transportation of MATERIALS
(e) Forming of MATERIALS

Using this classification, a compilation of basic tribo-mechanical systems of technology and biology is given in the following. Most of the tribo-systems named can be additionally classified. For example, the group of "sliding bearings" can be classified further into journal bearings, thrust bearings, dry bearings, fluid bearings, etc., or the group of "human joints" can be classified further into hip joints, knee joints, etc. While not claiming to be complete, this listing shows the extremely wide relevance of the tribology of mechanical systems. It should be emphasized that the systems listed below are the basic, i.e., the "simplest", tribo-mechanical systems which may be combined to systems of higher ranking order - i.e., complete machines, instruments, vehicles, etc. - as explained in Chapter 2, Figure 2.1.

Primary technical function	Tribo-engineering system (or tribo-engineering processes)
(a1) Guidance or transmission of MOTION	Sliding bearings Ball and roller bearings Needle bearings Pivot bearings Gyroscopic bearings Guides Slideways Ball-and-screw mechanisms Spindles Hinges Engineering joints Human and animal joints Shoe (or foot) on ground
(a2) Coupling of MOTION	Flanges Gear couplings Splines Cone clutches Disc clutches Band clutches
(a3) Annihilation of MOTION	Block brakes Band brakes Disc brakes Fastener sets Bolt sets Friction dampers
(b) Transmission of WORK or POWER	Rack-and-pinions Spur gears Bevel gears Hypoid gears Planetary gears Worm gears Power screws Chain drives Belt drives Rope drives Crank drives Cam drives Actuators Friction drives Hydraulic drives

Primary technical function	Tribo-engineering system (or tribo-engineering processes)
(c1) Generation of INFORMATION	Clock-works Cam-and-followers Electrical contacts Switches Relays Printing units Typewriter units Writing units
(c2) Reproduction of INFORMATION	Tape-recorder heads Slip-ring assemblies Audio pick-ups Video pick-ups
(d1) Transportation of MATERIALS	Wheel and rail Tyre and road Pipelines Conveyors Earth movement units Human and animal veins
(d2) Control of flow of MATERIALS	Seals Valves Packings Washers Piston-cylinder assemblies
(e1) Forming of MATERIALS	Drawing Deep drawing Pressing Extrusion Forging Rolling Casting Injection moulding Spinning Weaving
(e2) Tearing of MATERIALS	Mining Well drilling Quarrying Comminution Dredging Human and animal teeth

Primary technical function	Tribo-engineering system (or tribo-engineering processes)
(e3) Machining of MATERIALS	Cutting Punching Sawing Filing Turning Shaping and Planing Boring Drilling Grinding Abrasive cutting Chipless cutting Lapping Polishing Blasting Brushing

Supplementing remark:

The compilation of basic tribo-engineering systems shows the great variety of mechanical systems in which friction, lubrication, and wear processes occur. While it is not possible to deal with specific details of the individual tribo-engineering systems within the scope of this volume, the tribological fundamentals relevant to the various mechanical systems are presented in the preceding chapters under the headings:

→ Contact processes, Section 4.2, page 47
→ Friction processes, Section 4.3, page 69
→ Wear processes, Section 4.4, page 97
→ Lubrication modes, Section 4.5, page 130

The influences of friction and wear processes on mechanical engineering systems are discussed in general terms under the headings:

Structural aspects:

→ Surface topography changes, Section 5.2.1, page 179
→ Surface composition changes, Section 5.2.2, page 186
→ Surface strength properties changes, Section 5.2.3, page 191
→ Solutions for no-wear conditions, Section 5.4, page 199

Functional aspects:

→ Dynamic behaviour, Section 6.2.1, page 215
→ Stick-slip effects, Section 6.2.2, page 218
→ Efficiency, Section 6.3, page 221
→ Failure and safety, Section 6.4, page 227
→ Reliability, Section 6.5, page 234
→ Machinery condition monitoring, Section 7.5, page 277
→ Requirements for proper functional behaviour, Section 6.6, page 240

Appendix (B) Elements of tribo-engineering systems

The tribo-engineering systems compiled in Appendix (A) consist of (material) elements directly involved in the contact, friction, lubrication and wear processes. The components forming the "interacting surfaces in relative motion" of a given tribo-engineering system are termed tribo-element (1) and tribo-element (2). It was emphasized in Chapters 4 and 5 that for an understanding of friction, lubrication and wear, the interaction of all the material elements involved in the interfacial tribological processes must be taken into consideration. Thus, the interfacial medium (3) (if any) between components (1) and (2) and the surrounding medium (4) are integral constituent of any tribo-engineering system. In the following, the elementary elements (1), (2), (3), (4) of tribo-engineering systems are listed as examples from every group of the basic tribo-engineering systems compiled in Appendix (A).

Tribo-engineering system (or process)	Elements of the system			
	Tribo-element (1) (moving or stationary)	Tribo-element (2) (moving or stationary)	Interfacial medium (3)	Surrounding medium (4)
(a1) Sliding bearing Human hip joint	 Shaft Femur	 Bushing Acetabulum	 Lubricant Synovia	 Air —
(a2) Band clutch Spline	 Shaft Inner spline	 Band Outer spline	 — Grease	 Air Air
(a3) Disc brake Fastener set	 Disc Bolt	 Pad Nut	 Contaminant —	 Air Air
(b) Worm gear set Belt drive	 Worm Shaft	 Gear Belt	 Gear oil —	 Air Air
(c1) Cam and follower Printing unit	 Cam Print-head	 Follower Paper	 Lubricant Dye	 Air Air
(c2) Electrical contact Audio pick-up	 Ring Record	 Brush Sapphire tip	 Spray —	 Cover gas Air
(d1) Locomotion Pipeline	 Wheel Fluid	 Rail Pipe	 Contaminant —	 Air —
(d2) Valve Piston assembly	 Plug Piston ring	 Valve-body Cylinder	 Fluid Lubricant	 Fluid Fluid
(e1) Wiredrawing Hot extrusion	 Wire Billet	 Die Die	 Borax Glass	 Air Air
(e2) Well drilling Dredging	 Borer Dredge	 Soil Earth	 — —	 — Air
(e3) Turning Grinding	 Workpiece Workpiece	 Cutting tool Grinding wheel	 Cutting fluid Cutting fluid	 Air Air

Appendix (C)
Bibliographic work:
DOCUMENTATION TRIBOLOGY

The literature in the field of tribology is extremely voluminous owing to the vast number of topics included. For example, as mentioned already in Section 1.3, in the period 1966 to 1977 some 55,000 papers were published in this field. An international and comprehensive bibliography is provided by the DOCUMENTATION TRIBOLOGY:

- an annual systematic bibliography of titles covering the entire field of wear, friction and lubrication, including literature from all over the world (English translations of the titles are given in brackets wherever the titles are Non-English),
- a comprehensive source material, also referring to parallel publications, translations and abstract material, edited and published by the

  Bundesanstalt für Materialprüfung (BAM)
in Berlin (West), Germany

- providing direct and easy access to required references from various points of interest by a detailed and logical classification of the field, consisting of ten main sections and more than 100 subsections,
- relieving the individual scientist or engineer of the pains of currently scanning the literature and registering the relevant titles referring to his/her field of interest,
- total number of references at present 6000-8000 titles per year, among these are:

materials and pairings	650
wear-resistant coatings	450
lubricants, lubricant testing	600
bearings	1250
transmissions	250
manufacturing processes, tools	1200

- may be ordered through a bookstore or directly from the publisher:

  BAM, Unter den Eichen 87,
  D-1000 Berlin 45

In the following, a summary of the classification of main sections and sub-sections of the DOCUMENTATION TRIBOLOGY is given:

1. General, monographs, reviews
2. Experimental methods and equipment
3. Tribological fundamentals, friction and wear: laws, mechanisms, effects
   - 3.1 Mechanical and acoustical phenomena
   - 3.2 Chemical, physico-chemical and thermal phenomena
   - 3.3 Electrical and optical phenomena
4. Appearance of damage
5. Wear and friction under specific stressing conditions
   - 5.0 Resting and preliminary displacement in static friction
   - 5.1 Sliding, including stick-slip
   - 5.2 Rolling, rolling with sliding
   - 5.3 Spin
   - 5.4 Impact
   - 5.5 Grain sliding
   - 5.6 Fretting, fretting corrosion
   - 5.7 Wear caused by flowing media
   - 5.8 Thermal wear, ablation
6. Materials and combinations of materials (with and without intermediate materials)
   - 6.1 Metals
   - 6.2 Minerals, ceramics
   - 6.3 Polymers (plastics and rubber)
   - 6.4 Composites (glass-fibre reinforced materials, cermets, laminated materials etc.)

6.8 Textile materials, leather

6.9 Other materials and combinations of materials (including wood, paper, shoe soles, floorings, granular materials, biological materials, dental materials)

7. Influential factors and control parameters

7.1 Material-dependent factors

7.1.1 Strength

7.1.2 Composition, texture, structure

7.1.9 Other material characteristics (including compatibility)

7.2 Surface-dependent factors

7.2.1 Topography

7.2.2 Surface layers (adsorbed films, oxide layers)

7.2.3 Surface treatment (mechanical and heat treatment)

7.2.4 Surface coatings

7.3 Operating conditions

7.3.1 Motion and velocity

7.3.2 Shear force, load

7.3.3 Temperature

7.3.4 Environment

7.4 Other influential factors

7.4.1 Electric potential, current passage

7.4.2 Solid intermediate particles, dust, contaminants

7.4.3 Irradiation

7.4.4 Geometric factors, effects of design

8. Lubrication

8.1 Thick-film lubrication, especially hydrodynamic lubrication

8.2 Thin-film lubrication (boundary lubrication, elasto-hydrodynamic lubrication)

8.3 Lubrication in special environments

8.4 Lubrication at extreme temperatures

8.5 Lubrication under radiation and/or in reactor technique

8.6 Extreme pressure lubrication

8.7 Lubricants, lubricant testing

8.8 Lubrication systems and methods

8.9 Hygiene and microbiological problems

9. Fields of application. Technical designs and processes
   - 9.0 Life, reliability and maintenance of technical systems
   - 9.1 Machinery and components: selection and performance
     - 9.1.1 Bearings
     - 9.1.2 Transmissions
     - 9.1.3 Couplings, clutches and brakes
     - 9.1.4 Seals
     - 9.1.5 Machines, sets. Vehicles and vessels
     - 9.1.6 Wheel and rail
     - 9.1.7 Tyre and road
     - 9.1.8 Specific industries
     - 9.1.9 Other machine elements and products
   - 9.2 Manufacturing processes. Tools
     - 9.2.1 Forming
     - 9.2.2 Chipless cutting
     - 9.2.3 Machining
     - 9.2.4 Surface finishing
     - 9.2.5 Joining
     - 9.2.6 Coating
   - 9.3 Materials testing
   - 9.4 Rock mechanics, geomechanics

10. Basic information on solid surfaces and contacts
   - 10.1 Solid surfaces
     - 10.1.1 Morphology
     - 10.1.2 Strength, surface tension and energy, chemical reactions, environmental effects
   - 10.2 Contacts
     - 10.2.1 Area of contact, deformation of contact, heat transfer
     - 10.2.2 Joint character, adhesion

References

Chapter 1

1.1 R.H. Cannon, Jr., Dynamics of Physical Systems, Mc Graw-Hill, New York, 1967.

1.2 Tribology - A Glossary of Terms and Definitions, Organisation for Economic Co-operation and Development, OECD, Paris, 1969.

1.3 Lubrication (Tribology) Education and Research - A report on the present position and industry's needs, Her Majesty's Stationery Office, London, 1966.

1.4 F.P. Bowden and D. Tabor, The Friction and Lubrication of Solids, Clarendon Press, Oxford, 1964.

1.5 E. Rabinowicz, Friction and Wear of Materials, Wiley, New York, 1965.

1.6 I.V. Kragelskii, Friction and Wear, Butterworths, London, 1965.

1.7 A. Cameron, The Principles of Lubrication, Longmans, London, 1966.

1.8 J. Halling (Editor), Principles of Tribology, MacMillan, London, 1975.

1.9 G. Amontons, De la Resistance Causée dans les Machines, Histoire Acad. Roy. Sci., Paris, 12(1699) 206.

1.10 E. Coulomb, Théorie des machines simples, Mem. Math. Phys., Paris, 10(1785) 161.

1.11 J.T. Desagulier, A Course of Experimental Philosophy, London, 1734.

1.12 L. Reti (Editor), Leonardo-Künstler, Forscher, Magier, S. Fischer Verlag, Frankfurt, 1974.

1.13 W.B. Hardy and J.K. Hardy, Note on static friction and on the lubricating properties of certain chemical substances, Phil. Mag., 6(1919) 32.

1.14 G.A. Tomlinson, A molecular theory of friction, Phil. Mag., 7(1929) 905.

1.15 R. Holm, Die Reibungskraft in der wahren Kontaktfläche, Wiss. Veröff. Siemens Werke, 17(1938) 38.

1.16 H. Ernst and M.E. Merchant, Surface friction between metals - A basic factor in the metal cutting process, Proc. Special Summer Conf. Friction and Surface Finish, MIT Press, Cambridge, Mass., 1940, p. 76.

1.17 F.P. Bowden and D. Tabor, The theory of metallic friction and the role of shearing and ploughing, Council Sci. and Ind. Research, Comm. of Australia, Bulletin 145, 1942.

1.18 F.P. Bowden and D. Tabor, The Friction and Lubrication of Solids, Clarendon Press, Oxford, 1954.

1.19 M. Füchsel, Über Verschleißbarkeit der Werkstoffe bei trockener Reibung, Organ Fortschr. Eisenbahnwes., 84(1929) 413.

1.20 M. Fink und U. Hofmann, Zur Theorie der Reiboxydation, Archiv Eisenhüttenwes., 6(1932) 161.

1.21 R. Mailänder und K. Dies, Beitrag zur Erforschung der Vorgänge beim Verschleiß, Tech. Mitt. Krupp Forsch.-Ber., 5(1942) 209.

1.22 E. Siebel, Über die praktische Bewährung der mit Verschleißversuchen gewonnenen Ergebnisse, Tagungsband VDI-Verschleißtagung, Stuttgart, 1938, S.4.

1.23 I-Meng Feng, Metal transfer and wear, J. Appl. Phys., 23(1952) 1011.
J.T. Burwell and C.D. Strang, On the empirical law of adhesive wear, J. Appl. Phys., 23(1952) 18.

1.24 M. Kerridge, Metal transfer and the wear process, Proc. Phys. Soc. B, London, 68(1956) 400.

1.25 J.F. Archard, Contact and rubbing of flat surfaces, J. Appl. Phys. 24(1953) 24.
J.F. Archard and W. Hirst, The wear of metals under unlubricated conditions, Proc. Roy. Soc., London, A 236(1956) 397.

1.26 J.T. Burwell, Survey of possible wear mechanisms, Wear, 1(1957) 119.

1.27 cited in:
R.P. Feynman, R.B. Leighton, M. Sands, The Feynman Lectures on Physics, Vol. II, Addison-Wesley, Reading, 1964, p. 40-3.

1.28 O. Reynolds, On the theory of lubrication and its application to Mr. Beauchamp Tower's experiments, including an experimental determination of the viscosity of olive oil, Phil. Trans., 177(1886) 157.

1.29 R. Stribeck, Die wesentlichen Eigenschaften der Gleit- und Rollenlager, VDI-Zeitschrift, 46(1902) 1341, 1432, 1463.

1.30 A. Sommerfeld, Zur hydrodynamischen Theorie der Schmiermittelreibung, Z. Math. Phys., 50(1904) 97.

1.31 C. Biel, Die Reibung in Gleitlagern bei Zusatz von Voltoöl zu Mineralöl und bei Veränderung der Umlaufzahl und der Temperatur, VDI-Zeitschrift, 64(1920) 449, 483.

1.32 G. Vogelpohl, Betriebssichere Gleitlager, Springer Verlag, Berlin, 1958.

1.33 W.B. Hardy and J. Doubleday, Boundary lubrication, The paraffin series, Proc. Roy. Soc., London, A 100(1922) 550.

1.34 K. Kirschke, H. Czichos, K.-H. Habig, P. Studt, Tribologie - Forschungsbericht zum Schwerpunktprogramm Verschleiß, Reibung, Schmierung, Franz Steiner Verlag, Wiesbaden, 1971.

1.35 The introduction of a new technology, Report 1966-1972 of the Committee on Tribology, Her Majesty's Stationery Office, London, 1973.

1.36 G.H. Göttner, Tribologie - Begriff, Wesen und Bedeutung, Schmiertechnik und Tribologie, 17(1970) 285.

1.37 P. Jost, Some economic factors of tribology, Proc. JSLE-ASLE Internat. Lubr. Conf. (Editor: T. Sakurai), Elsevier, Amsterdam, 1976, p. 2.

1.38 D. Summers-Smith, 10 years after Jost: the effect on industry, Proc. Tribology Convention, Inst. Mech. Engrs, 1976, p. 21.

1.39 DOCUMENTATION TRIBOLOGY - Wear, Friction and Lubrication, 11 Volumes, 1967-1974, Edited by Bundesanstalt für Materialprüfung (BAM), Fachgruppe Rheologie und Tribologie, D-1 Berlin 45.

Chapter 2

2.1 L. von Bertalanffy, General System Theory, Penguin, London, 1971.

2.2 K.E. Boulding, General systems theory - the skeleton of science, Management Science, 2(1956) 197.

2.3 G. Ropohl, Systemtechnik - Grundlagen und Anwendung, Hanser Verlag, München, 1975.

2.4 M.D. Mesarovic, Views on General Systems Theory, Wiley, New York, 1964.

2.5 G.J. Klir, An Approach to General Systems Theory, Van Nostrand Reinhold, New York, 1969.

2.6 J.J. van Dixhoorn and F.J. Evans, (Editors), Physical Structure in Systems Theory, Academic Press, London, 1974.

2.7 P. Faurre and M. Depeyrot, Elements of System Theory, North Holland, Amsterdam, 1977.

2.8 S. Seely, An Introduction to Engineering Systems, Pergamon, New York, 1972, p. 55.

2.9 J.U. Thoma, Introduction tc Bond Graphs and Their Application, Pergamon, Oxford, 1975, p. 167.

2.10 H.M. Paynter, Analysis and Design of Engineering Systems (Class notes for M.I.T. course 2.751), The M.I.T. Press, Cambridge, Mass., 1960.

2.11 D. Karnopp and R. Rosenberg, Systems Dynamics: A Unified Approach, Wiley, New York, 1975.

2.12 N. Wiener, Cybernetics or Control and Communication in the Animal and the Machine, The M.I.T. Press, Cambridge, Mass., 1948.

2.13 A.D. Hall, A Methodology for Systems Engineering, Van Nostrand, New York, 1962.

CHAPTER 3

3.1 G. Fleischer, Systembetrachtungen zur Tribologie, Wiss. Z. TH Magdeburg, 14(1970) 415.

3.2 G. Salomon, Application of systems thinking to tribology, ASLE Trans, 17(1974) 295.

3.3 H. Czichos, The principles of systems analysis and their application to tribology, ASLE Trans, 17(1974) 300.

3.4 H. Czichos, Systemanalyse und Physik tribologischer Vorgänge, Teil 1: Grundlagen, Teil 2: Anwendungen, Schmiertechnik und Tribologie, 22(1975) 126 und 23(1976) 6.

3.5 H. Czichos, and J. Mølgaard, Towards a general theory of tribological systems, Wear, 44(1977) 247.

3.6 J. Mølgaard, The dry wear of metals as a process in an open system, Wear, 32(1975) 353.

3.7 W.M.J. Schlösser, A contribution to the study of analogies of power transmission in machines, Proc. J. Mech. E. London, 188(1974) 1.

3.8 J.U. Thoma, Bondgraph design for hydrostatic bearings, Tribology, 6(1973) 97.

3.9 J.N. Brønsted, On the concept of heat, Det Kgl. Danske Videnskabernes Selskab, Mat.-fysk. Meddelelser, 19(1941) No. 8, 79p.

CHAPTER 4

4.1 A.D. de Pater and J.J. Kalker (Editors), The Mechanics of the Contact between Deformable Bodies, Delft University Press, Delft, 1975.

4.2 F.F. Ling, Surface Mechanics, Wiley, New York, 1973.

4.3 H. Hertz, Über die Berührung fester elastischer Körper, J. für die reine u. angew. Mathem., 92(1881) 156.

4.4 M.H. Jones, R.I.L. Howells, and S.D. Probert, Solids in static contact - a review, Wear, 12(1968) 225.

4.5 J.F. Archard, Surface topography and tribology, Tribology, 7(1974) 213.

4.6 D.J. Whitehouse, The measurement and analysis of surfaces, Tribology, 7(1974) 249.

4.7 T.R. Thomas, Recent advances in the measurement and analysis of surface microgeometry, Wear, 33(1975) 205.

4.8 H. v. Weingraber, Möglichkeiten für eine statistische Auswertung technischer Oberflächen, Feingerätetechnik, 19(1970) 416.

4.9 J. Peklenik, Grundlagen zur Korrelationstheorie technischer Oberflächen, Ind. Anz., 87(1965) 456.

4.10 A.M. Jones, P.W. O'Callaghan and S.D. Probert, Prediction of contact parameters from the topographies of contacting surfaces, Wear, 31(1975) 89.

4.11 P.T. Suratkar, S.M. Pandit and S.M. Wu, A stochastic approach to the mode of deformation and contact between rough surfaces, Wear, 39(1976) 239.

4.12 J.F. Archard, Contact and rubbing of flat surfaces, J. Appl. Phys., 24(1953) 981.

4.13 J.F. Archard, Elastic deformation and the laws of friction, Proc. Roy. Soc., London, A 243(1957) 190.

4.14 J.A. Greenwood and J.H. Tripp, The elastic contact of rough spheres, Trans ASME: J. Appl. Mech., 89(1967) 153.

4.15 A. Seireg and E.J. Weiter, Behaviour of frictional Hertzian contacts under impulsive loading, Wear, 8(1965) 208.

4.16 K.C. Ludema and D. Tabor, The friction and visco-elastic properties of polymeric solids, Wear, 9(1966) 329.

4.17 R. Stribeck, Kugellager für beliebige Belastungen, VDI-Zeitschrift, 45(1901) 73.

4.18 J.A. Greenwood and J.B.P. Williamson, The contact of nominally flat surfaces, Proc. Roy. Soc., London, A 295(1966) 300.

4.19 D.J. Whitehouse and J.F. Archard, The properties of random surface of significance in their contact, Proc. Roy. Soc., London, A 316(1970) 97.

4.20 D. Tabor, A simplified account of surface topography and the contact between solids, Wear, 32(1975) 269.

4.21 J.F. Nicholas, The dissipation of energy during plastic deformation, Acta Met., 7(1959) 544.

4.22 H.C. Gatos, Structure of surfaces and their interactions, in: Interdisciplinary Approach to Friction and Wear, (Editor: P.M. Ku), NASA SP-181, Washington, 1968, p.7.

4.23 G. Schmaltz, Technische Oberflächenkunde, Springer, Berlin, 1936.

4.24 H.B.G. Casimir, Proc. Koninkl. Ned. Akad. Wetenshap, B 51(1948) 795.

4.25 B.J. Briscoe and D. Tabor, Surface forces in friction and adhesion, Faraday Spec. Disc. Chem. Soc., No. 2(1972) 7.

4.26 J.M. Ziman, Electrons in Metals - A Short Guide to the Fermi Surface, Taylor and Francis, London, 1963.

4.27 J. Ferrante and J.R. Smith, A theory of adhesion at a bimetallic interface: overlap effects, Surface Science, 38(1973) 77.

4.28 J. Ferrante and J.R. Smith, Metal interfaces: adhesive energies and electronic barriers, Solid state comm., 20(1976) 393.

4.29 H. Czichos, The mechanism of the metallic adhesion bond, J. Phys. D: Appl. Phys., 5(1972) 1890.

4.30 D.H. Buckley, Adhesion of various metals to a clean iron (011) surface studied with LEED and Auger emission spectroscopy, NASA TN D-7018, Washington, 1971.

4.31 F.P. Bowden and G.W. Rowe, The adhesion of clean metals, Proc. Roy. Soc. London, A 233(1956) 429.

4.32 M.E. Sikorski, Correlation of the coefficient of adhesion with various physical and mechanical properties of metals, Trans. ASME, D 85(1963) 279.

4.33 D.H. Buckley, The influence of crystal structure, orientation and solubility on the adhesion and sliding of various metal single crystals in vacuum (10^{-11} Torr), ASTM STP No. 431, 1967, p.248.

4.34 D.H. Buckley and R.L. Johnson, The influence of crystal structure and some properties of hexagonal metals on friction and adhesion, Wear, 11(1968) 405.

4.35 K.-H. Habig, Der Einfluß der gegenseitigen Löslichkeit von Metallen auf einige ihrer tribologischen Eigenschaften, Metalloberfläche, 24(1970) 375.

4.36 D.H. Buckley, Friction, Wear and Lubrication in Vacuum, NASA SP-277, Washington, 1971, p.66.

4.37 D.H. Buckley, The influence of the atomic nature of crystalline materials on friction, Trans. ASLE, 11(1968) 89.

4.38 K.-H. Habig, Zur Struktur- und Orientierungsabhängigkeit der Adhäsion und der trockenen Gleitreibung von Metallen, Materialprüfung, 10(1968) 417.

4.39 D.H. Buckley, Effect of various material properties on the adhesive stage of fretting, AGARD Conf. Proc. No. 161, 1975, p. 13-1.

4.40 N. Gane, P.F. Pfaelzer and D. Tabor, Adhesion between clean surfaces at light loads, Proc. Roy. Soc., London, A 340(1974) 495.

4.41 K.L. Johnson, K. Kendall and A.D. Roberts, Surface energy and the contact of elastic solids, Proc. Roy. Soc., London, A 324(1971) 301.

4.42 K.N.G. Fuller and D. Tabor, The effect of surface roughness on the adhesion of elastic solids, Proc. Roy. Soc., London, A 345(1975) 327.

4.43 R.D. Mindlin, Compliance of elastic bodies in contact, Trans. ASME, J. Appl. Mech., 71(1949) 259.

4.44 J.S. Mc Farlane and D. Tabor, Relation between friction and adhesion, Proc. Roy. Soc., London, A 202(1950) 244.

4.45 J.S. Courtney-Pratt and E. Eisner, The effect of a tangential force on the contact of metallic bodies, Proc. Roy. Soc., London, A 238(1957) 529.

4.46 H. Czichos, Festkörperreibung - Teilgebiet der Tribologie, Umschau (1971) 116.

4.47 M. Godet, Surface and shape effects in the measurement of friction forces, Wear, 28(1974) 115.

4.48 R. Courtel, Modes d'interpretation de la troisième dimensions dans les processus de frottement intermétallique. Role du bourrelet frontal, Bull. Inst. Sci. Techn. du Centre Energie At., 90(1965) 3.

4.49 L.A. Mitchell and C. Osgood, A theory of friction and wear based on a new characterization of asperity interactions, Wear, 40(1976) 203.

4.50 D.H. Buckley, The metal-to-metal interface and its effect on adhesion and friction, J. Coll. Interface Sci., 58(1977) 36.

4.51 D. Tabor, Interaction between Surfaces: Adhesion and Friction, in: Surface Physics of Materials, Vol. II (Editor: J.M. Blakely) Academic Press, New York, 1975, p. 475.

4.52 H. Czichos, Die Energieverlustmechanismen der Rollreibung, Schmiertechnik und Tribologie, 16(1969) 62.

4.53 O. Reynolds, On rolling friction, Phil. Trans. Roy. Soc., London, 116(1876) 155.

4.54 H.L. Heathcote, The ball bearing, Proc. Instn. Automotive Engrs, 15(1921) 1569.

4.55 F.W. Carter, On the action of a locomotive driving wheel, Proc. Roy. Soc., London, A 112(1926) 151.

4.56 H. Poritsky, Stresses and deflections of cylindrical bodies in contact with application to contact of gears and of locomotive wheel, Trans. ASME, J. Appl. Mech., 72(1950) 191.

4.57 L. Föppl, Die strenge Lösung für die rollende Reibung, München, 1947.

4.58 D. Tabor, The mechanism of rolling friction, II, the elastic range, Proc. Roy. Soc., London, A 229(1955) 198.

4.59 B.G. Brothers and G.R. Bremble, The effect of geometric conformity between a ball and its track on the free rolling resistance, Wear, 20(1972) 175.

4.60 D. Tabor, The mechanism of rolling friction, Phil. Mag., 43(1952) 1055.

4.61 D.G. Flom and A.M. Bueche, Theory of rolling friction for spheres, J. Appl. Phys., 30(1959) 1725.

4.62 J.R. Halaunbrenner, Elasto-optical investigations of rolling of a rigid cylinder on a visco-elastic base, Wear, 8(1965) 30.

4.63 K.R. Eldredge and D. Tabor, The mechanism of rolling friction, I, the plastic range, Proc. Roy. Soc., London, A 229(1955) 181.

4.64 L.D. Dyer, Rolling friction on single crystals of copper in the plastic range, Acta Met., 9(1961) 928.

4.65 K.L. Johnson, A review of the theory of rolling contact stresses, Wear, 9(1966) 4.

4.66 A.W. Crook, Simulated gear-tooth contacts: some experiments upon their lubrication and subsurface deformations, Proc. Conf. Lubr. Wear, I, Mech. Eng., 1957, 701.

4.67 G. M. Hamilton, Plastic flow in rollers loaded above the yield point, Proc. Inst. Mech. Engrs, 177(1963) 667.

4.68 J.W. Merwin and K.L. Johnson, Analysis of plastic deformation rolling contact, Proc. Inst. Mech. Engrs, 177(1963) 676.

4.69 H. Czichos, Über den Zusammenhang zwischen Adhäsion und Elektronenstruktur von Metallen bei der Rollreibung im elastischen Bereich, Z. angew. Physik, 27(1969) 40.

4.70 K. Kendall, Rolling friction and adhesion between smooth solids, Wear, 33(1975) 351.

4.71 B.V. Derjaguin and V.P. Smilga, Electrostatic component of the rolling friction force moment, Wear, 7(1964) 270.

4.72 J.H. Dismant, Reviving the classical theory of friction by a modern dislocation theory of deformation revision, J. Appl. Phys., 31(1960) 221.

4.73 B.J. Kostetskii and P.V. Nazarenko, Influence of changes of dislocation structure on the relation between friction and normal pressure (The Amontons – Coulomb law), Soviet Physics - Doklady, 9(1965) 1011.

4.74 R. Feder and P. Chaudhari, Transmission electron microscopy of wear tracks, Wear, 19(1972) 109.

4.75 G.E.R. Schultze, Metallphysik, Akademie-Verlag, Berlin, 1967, p. 217.

4.76 N. Gane and J. Skinner, The generation of dislocations in metals under a sliding contact and the dissipation of frictional energy, Wear, 25(1973) 381.

4.77 G. Andarelli, D. Maugis and R. Courtel, Observation of dislocations created by friction on aluminium thin foils, Wear, 23(1973) 21.

4.78 J.G. Bollinger, Noise - an industrial pollutant of international concern, CIRP Annals, 22(1973) 197.

4.79 D.M. Tolstoi, Significance of the normal degree of freedom and natural normal vibrations in contact friction, Wear, 10(1967) 199.

4.80 Triboluminescence, dtv-Lexikon der Physik, Deutscher Taschenbuch-Verlag, München, 1971, Bd. 9, p. 168.

4.81 J. Ferrante, Exoelectron emission from a clean, annealed magnesium single crystal during oxygen adsorption, ASLE Trans., 20(1977).

4.82 J. Wortmann, Zur Exoelektronenemission (EEE), in "Verhalten und Eigenschaften von metallischen Oberflächen", TU Berlin, Fachbereich 17 (Fachseminar WS 75/76) 1976, p. 280.

4.83 H.S. Carslaw and J.C. Jaeger, Conduction of Heat in Solids, Clarendon Press, Oxford, 1947, p. 56.

4.84 R. Holm, Calculation of the temperature development in a contact heated in the contact surface, and application to the problem of the temperature rise in a sliding contact, J. Appl. Phys., 19(1948) 361.

4.85 H. Blok, Theoretical study of temperature rise at surfaces of actual contact under oiliness lubricating conditions, Proc. Gen. Disc. Lubr., Inst. Mech. Engrs, London, 1937, vol. 2, p. 14.

4.86 J.F. Archard, The temperature of rubbing surfaces, Wear, 2(1958/59) 438.

4.87 P.S. Kounas, A.D. Dimarogonas and G.N. Sandor, The distribution of friction heat between a stationary pin and a rotating cylinder, Wear, 19(1972) 415.

4.88 F. Charron, Partage de la chaleur entre deux corps frottants, Editions Blondel la Rougery, Paris, 1943.

4.89 J.F. Archard, Wear, in: Interdisciplinary Approach to Friction and Wear (Editor: P.M. Ku), NASA SP-181, Washington, 1968, p. 267.

4.90 H. Czichos und K.-H. Habig, Grundvorgänge des Verschleisses metallischer Werkstoffe - Neuere Ergebnisse der Forschung, VDI-Ber. Nr. 194, 1973, S. 23.

4.91 P.A. Engel, Impact Wear of Materials, Elsevier, Amsterdam, 1976.

4.92 O.R. Lang, Surface fatigue of plain bearings, Wear, 43(1977) 25.

4.93 K.H. Kloos und E. Broszeit, Zur Frage der Dauerwälzfestigkeit, Z. f. Werkstofftechnik, 5(1974) 181.

4.94 M.J.W. Schouten, Der Einfluß elastohydrodynamischer Schmierung auf Reibung, Verschleiß und Lebensdauer von Getrieben, Schmiertechnik und Tribologie, 20(1973) 147.

4.95 W.E. Littmann, The mechanism of contact fatigue, in: Interdisciplinary approach to the lubrication of concentrated contacts, NASA SP-237, Washington, 1970, p. 309.

4.96 Y.P. Chiu, T.E. Tallian and J.I. Mc Cool, An engineering model of spalling fatigue failure in rolling contact, I, The subsurface model, Wear, 17(1971) 433.

T.E. Tallian and J.I. Mc Cool, II, The surface model, Wear, 17(1971) 447.

T.E. Tallian, III, Engineering discussion and illustrative examples, Wear, 17(1971) 463.

4.97 J.C. Tyler, R.A. Burton and P.M. Ku, Contact fatigue under oscillatory normal load, ASLE Trans., 6(1963) 255.

4.98 P. Haasen, Physikalische Metallkunde, Springer-Verlag, Berlin, 1974, p. 298.

4.99 D. Kuhlmann-Wilsdorf and C. Laird, Dislocation behaviour in fatigue, Mat. Sci. Eng., 27(1977) 137.

4.100 E. Hornbogen, The role of fracture toughness in the wear of metals, Wear, 33(1975) 251.

4.101 N.P. Suh, The delamination theory of wear, Wear, 25(1973) 111.

4.102 J.P. Hirth and D.A. Rigney, Crystal plasticity and the delamination theory of wear, Wear, 39(1976) 133.

4.103 T.S. Eyre, Wear characteristics of metals, Tribology International, (1976) 203.

4.104 K. Wellinger und H. Uetz, Gleitverschleiß, Spülverschleiß, Strahlverschleiß unter Wirkung von körnigen Stoffen, VDI-Forschungsheft 449, Beilage zu Forsch. Ing. Wes. 21(1955) 1.

4.105 K. Wellinger, H. Uetz und M. Gürleyik, Gleitverschleißuntersuchungen an Metallen und nichtmetallischen Hartstoffen unter Wirkung körniger Stoffe, Wear, 11(1968) 173.

4.106 M.M. Khrushov, Resistance of metals to wear by abrasion; related to hardness, Proc. Conf. Lubr. and Wear, Instn. Mech. Engrs, London, 1957, p. 655.

4.107 M.M. Khrushov, Principles of abrasive wear, Wear, 28(1974) 69.

4.108 R.C.D. Richardson, The wear of metals by relatively soft abrasives, Wear, 11(1968) 245.

4.109 P.L. Hurricks, Some metallurgical factors controlling the adhesive and abrasive wear resistance of steel, a review, Wear, 26(1973) 285.

4.110 H. Uetz und J. Föhl, Verschleiß und Reibung als komplexe mechanisch-physikalisch-chemische Prozesse, Erzmetall, 23(1970) 205.

4.111 D.H. Buckley, Wear and interfacial transport of material, NASA TMX-71781, 1975.

4.112 M. Cocks, Shearing of junctions between metal surfaces, Wear, 9(1966) 320.

4.113 M. Antler, Stages of wear of a prow forming metal, ASLE Trans., 13(1970) 79.

4.114 D. Landheer and J.H. Zaat, The mechanism of metal transfer in sliding friction, Wear, 27(1974) 129.

4.115 D.H. Buckley, Effect of various material properties on the adhesive stage of fretting, NASA TMX-71582, 1974.

4.116 P.A. Thiessen, K. Meyer und G. Heinicke, Grundlagen der Tribochemie, Akademie-Verlag, Berlin, 1967.

4.117 T.F.J. Quinn, The effect of "hot spot" temperatures on the unlubricated wear of steel, ASLE Trans., 10(1967) 158.

4.118 P.L. Hurricks, The mechanism of fretting - a review, Wear, 15(1970) 389.

4.119 R.B. Waterhouse, Fretting Corrosion, Pergamon, Oxford, 1972.

4.120 R.B. Waterhouse, Physics and metallurgy of fretting, AGARD-CP-161, 1975, p. 8-1.

4.121 H. Czichos, Discussion on the adhesive stage of fretting, AGARD-CP-161, 1975, p. 13-16.

4.122 J. Mølgaard, Die Entwicklung von Verschleißprozeßthesen, Schmiertechnik und Tribologie, 23(1976) 126.

4.123 see Ref. 1.29.

4.124 see Ref. 1.28.

4.125 see Ref. 1.27, Chapter 40.

4.126 M.D. Hersey, Theory and Research in Lubrication, Wiley, New York, 1966, p. 34.

4.127 H.C. Rippel, O. Decker and Z. Zudans, NASA Contributions to Fluid-Film Lubrication - A Survey, NASA SP-5058, Washington, 1969.

4.128 J. Holland, Die Ermittlung der Kenngrößen für zylindrische Gleitlager, Konstruktion, 13(1961) 100.

4.129 G. Vogelpohl, Betriebssichere Gleitlager, Springer-Verlag, Berlin, 1958, p. 73.

4.130 D.F. Moore, Principles and Applications of Tribology, Pergamon, Oxford, 1975, p. 346.

4.131 J.J. O'Connor, J. Boyd, E.A. Avallone, (Editors), Standard Handbook of Lubrication Engineering, McGraw-Hill, New York, 1968.

4.132 L.D. Wedeven, What is EHD?, Lubr. Engng, 31(1975) 291.

4.133 F.H. Theyse, Elastohydrodynamik, Schmiertechnik, 15(1967) 15.

4.134 J.W. Kannel, J.C. Bell and C.M. Allen, Methods for determining pressure distributions in lubricated rolling contact, ASLE Trans., 8(1965) 250.

4.135 see Ref. 1.7, p. 203.

4.136 D. Dowson and G.R. Higginson, New roller-bearing lubrication formula, Engineering, London, 192(1961) 158.

4.137 H.S. Cheng and F.K. Orcutt, A correlation between the theoretical and experimental results on the elastohydrodynamic lubrication of rolling and sliding contact, Proc. I. Mech. E., London, Vol. 180, Part 3B, (1965/66) 158.

4.138 A.J. Petrousevitch, D.S. Kodnir, R.G. Salukvadze, D.L. Bakashvili and V.Sh. Schwarzman, The investigation of oil film thickness in lubricated ball-race rolling contact, Wear, 19(1972) 369.

4.139 F.J. Westlake and A. Cameron, Optical elastohydrodynamic fluid testing, ASLE Trans., 15(1972) 81.

4.140 A. Jackson, Film thickness in EHL point contacts: revision of Westlake and Cameron's equations, ASLE Trans., 19(1976) 48.

4.141 J. Tevaarwerk and K.L. Johnson, A simple non-linear constitutive equation for elastohydrodynamic oil films, Wear, 35(1975) 345.

4.142 D. Dowson, Transition to boundary lubrication from elastohydrodynamic lubrication, in: Boundary Lubrication, (Editors: F.F. Ling, E.E. Klaus and R.S. Fein), ASME, New York, 1969, p. 229.

4.143 A. Dyson, discussion in: Interdisciplinary approach to the lubrication of concentrated contacts, NASA SP-237, Washington, 1970, p. 58.

4.144 J.F. Archard, Elastohydrodynamic lubrication of real surfaces, Tribology, 6(1973) 8.

4.145 L.D. Wedeven, Traction and film thickness measurements under starved elastohydrodynamic conditions, Trans. ASME, JOLT, 97(1975) 321.

4.146 H. Münnich, Einfluß der Schmierung auf Lebensdauer, Reibung und Verschleiß von Wälzlagern, Schmiertechnik und Tribologie, 16(1968) 87.

4.147 V.K. Ausherman, H.S. Nagaraj, D.M. Sanborn and W.O. Winer, Infrared temperature mapping in elastohydrodynamic lubrication, Trans. ASME, F, 98(1976) 236.

4.148 L. Gümbel, Der Einfluß der Schmierung auf die Konstruktion, Jb. schiffbautechn. Ges., 18(1917) 299.

4.149 G. Vogelpohl, Die Stribeck-Kurve als Kennzeichen des allgemeinen Reibungsverhaltens geschmierter Gleitflächen, VDI-Z., 96(1954) 261.

4.150 H. Christensen, A theory of mixed lubrication, Proc. Instn. Mech. Engrs, London, 186(1972) 421.

4.151 D. Berthe and M. Godet, A more general form of Reynolds equation - application to rough surfaces, Wear, 27(1974) 345.

4.152 K.L. Johnson, J.A. Greenwood and S.Y. Poon, A simple theory of asperity contact in elastohydrodynamic lubrication, Wear, 19(1972) 91.

4.153 T.E. Tallian, The theory of partial elastohydrodynamic contacts, Wear, 21(1972) 49.

4.154 E.S. Forbes, Antiwear and extreme pressure additives for lubricants, Tribology, 3(1970) 145.

4.155 P. Studt, Schmieröladditive für hohe Flächenpressung und ihre Wirkung, VDI-Berichte, 156(1970).

4.156 D. Godfrey, Boundary lubrication, in Ref. 4.131, Chapter 2.

4.157 D. Godfrey, Boundary lubrication, in: Interdisciplinary Approach to Friction and Wear, (Editor: P.M. Ku), NASA SP-181, Washington, 1968, p. 335.

4.158 J.S. Courtney-Pratt, An optical method of measuring the thickness of adsorbed monolayers, Proc. Roy. Soc., London, 212(1952) 505.

4.159 R. Holm, Electric Contacts, H. Gebers Förlag, Stockholm, 1946, p. 198.

4.160 F.P. Bowden and D. Tabor, Friction - An Introduction to Tribology, Heinemann, London, 1973, p. 128.

4.161 W.E. Campbell, Boundary lubrication, in: Boundary Lubrication, (Editors: F.F. Ling, E.E. Klaus and R.S. Fein), ASME, New York, 1969, p. 87.

4.162 F.G. Rounds, Influence of steel composition on additive performance, ASLE Trans., 15(1972) 54.

4.163 R.S. Fein, Chemistry in concentrated - conjunction lubrication, in: Interdisciplinary Approach to the Lubrication of Concentrated Contacts, NASA SP-237, Washington, 1970, p. 489.

4.164 C.N. Rowe, Some aspects of the heat of adsorption in the function of a boundary lubricant, ASLE Trans., 9(1966) 100.

4.165 C.N. Rowe, A relation between adhesive wear and heat of adsorption for the vapor lubrication of graphite, ASLE Trans., 10(1967) 10.

4.166 A. Dyson, Scuffing - a review, Tribology, 8(1975) 77 and 117.

4.167 T.v. Kârmân, Über den Mechanismus des Widerstandes, den ein bewegter Körper in einer Flüssigkeit erfährt, Nachrichten der K. Ges. d. Wissenschaften, Math-phys. Klasse, Göttingen, 1911.

4.168 G.I. Taylor, Stability of a viscous liquid contained between two rotating cylinders, Trans. Roy. Soc., London, A 223(1923) 289.

4.169 R.C. di Prima, A note on the stability of flow in loaded journal bearing, ASLE Trans., 6(1963) 249.

4.170 V.N. Constantinescu, Analysis of bearings operating in turbulent regime, Trans. ASME, D, 84(1962) 139.

4.171 E. Saibel, Turbulence in lubrication, in Ref. 1.7, p. 441.

4.172 T.E. Tallian, Y.P. Chiu, D.F. Huttenlocher, J.A. Kamenshine, L.B. Sibley and N.E. Sindlinger, Lubricant films in rolling contact of rough surfaces, ASLE Trans., 7(1964) 109.

4.173 H. Czichos, Untersuchungen über die Verteilung metallischer und nichtmetallischer Kontaktanteile bei Mischreibung, Wear, 17(1971) 209.

4.174 H. Czichos, Influence of asperity contact conditions on the failure of sliding elastohydrodynamic contacts, Wear, 41(1977) 1.

4.175 R.S. Fein, C.N. Rowe and K.L. Kreuz, Transition temperatures in sliding systems, ASLE Trans., 2(1959) 50.

4.176 R.S. Fein, Operating procedure effects on critical temperatures, ASLE Trans., 10(1967) 373.

4.177 H. Czichos and K. Kirschke, Investigations into film failure (transition point) of lubricated concentrated contacts, Wear, 22(1872) 321.

4.178 H. Czichos, Failure modes of sliding lubricated concentrated contacts, Wear, 28(1974) 95.

4.179 A. Begelinger and A.W.J. de Gee, Thin film lubrication of sliding point contacts of AISI 52100 steel, Wear, 28(1974) 103.

4.180 H. Czichos, Failure criteria in thin film lubrication: the concept of a failure surface, Tribology, 7(1974) 14.

4.181 G. Salomon, Failure criteria in thin film lubrication - the IRG program, Wear, 36(1976) 1.

4.182 H. Czichos, Tribometrische Untersuchung der Versagensgrenzen von Gleitreibungssystemen mit Hertzschem Kontakt im Mischreibungsgebiet, Materialprüfung, 16(1974) 32.

4.183 A. Sethuramiah, H. Okabe, T. Sakurai, Critical temperatures in EP lubrication, Wear, 26(1973) 187.

4.184 M. Tomaru, S. Hironaka and T. Sakurai, Effects of oxygen on the load-carrying action of some additives, Wear, 41(1977) 117.

4.185 M. Tomaru, S. Hironaka and T. Sakurai, Effects of some chemical factors on film failure under EP conditions, Wear, 41(1977) 141.

4.186 A. Begelinger and A.W.J. de Gee, Lubrication of sliding point contacts of AISI 52100 steel - the influence of curvature, Wear, 36(1976) 7.

4.187 H. Czichos, Film failure of sliding Hertzian contacts: The influence of contact geometry, Proc. JSLE-ASLE Internat. Lubr. Conf. Tokyo 1975, (Editor: T. Sakurai), Elsevier, Amsterdam, 1976, p. 368.

Chapter 5

5.1 J.B.P. Williamson, J. Pullen and R.T. Hunt, The shape of solid surfaces, in: Surface Mechanics (Editor F.F. Ling), The American Soc. of Mech. Engrs, New York, 1969, p. 32.

5.2 M.J. Neale (Editor), Tribology Handbook, Butterworths, London, 1973, p. A31.

5.3 K.J. Stout, T.G. King and D.J. Whitehouse, Analytical techniques in surface topography and their application to a running-in experiment, Wear, 43(1977) 99.

5.4 J.E. Willn, Characterization of cylinder bore surface finish - a review of profile analysis, Wear, 19(1972) 143.

5.5 J.C. Campbell, Cylinder bore surface roughness in internal combustion engines: its appreciation and control, Wear, 19(1972) 163.

5.6 J.R. Dowd and F.T. Barwell, Tribological interaction between piston and cylinder of a model high pressure hydraulic pump, ASLE Trans., 18(1975) 21.

5.7 R. Østvik and H. Christensen, Changes in surface topography with running-in, Proc. Instn. Mech. Engrs, London, Vol. 183 Pt 3P, (1968/69) 57.

5.8 G.W. Rowe, Surface topographic changes at breakdown of thin-film lubrication; a short account of a collaborative C.I.R.P. investigation, Wear, 28(1974) 125.

5.9 E. Saibel, A statistical approach to run-in and the dependence of the coefficient of friction on velocity, Wear, 35(1975) 383.

5.10 J. Volz, Erstellung optimierter Einlaufprogramme von Dieselmotoren (Ein System unter Anwendung von Radionukliden) Kernforschungszentrum Karlsruhe, Laboratorium für Isotopentechnik, KFK 2432, März 1977.

5.11 D.H. Buckley, Wear and interfacial transport of material, J. Vac. Sci. Technol., 13(1976) 88.

5.12 M. Fink and U. Hofmann, Zur Theorie der Reiboxydation, Archiv f.d. Eisenhüttenwesen, 6(1932) 161.

5.13 K. Dies, Die Vorgänge beim Verschleiß bei rein gleitender trockener Reibung, VDI-Z., 83(1939) 307.

5.14 T.S. Eyre and D. Maynard, Surface aspects of unlubricated metal-to-metal wear, Wear, 18(1971) 301.

5.15 S. Hogmark, O. Vingsbo and S. Friedström, Mechanisms of dry wear of some martensitic steels, Wear, 31(1975) 39.

5.16 T.F.J. Quinn and J.L. Wooley, The unlubricated wear of 3% Cr-1/2% Mo steel, Lubr. Eng., 26(1970) 226.

5.17 B.A. Baldwin, Chemical characterization of wear surfaces using X-ray photoelectron spectroscopy, Lubr. Engng, 32(1976) 125.

5.18 R.J. Bird and G.D. Galvin, The application of photoelectron spectroscopy to the study of E.P. films on lubricated surfaces, Wear, 37(1976) 143.

5.19 R.C. Coy and T.F.J. Quinn, The use of physical methods of analysis to identify surface layers formed by organosulphur compounds in wear tests, ASLE Trans, 18(1975) 163.

5.20 G. Levy, R.G. Linford and L.A. Mitchell, Wear behaviour and mechanical properties: the similarity of seemingly unrelated approaches, Wear, 21(1972) 167.

5.21 see Ref. 1.5, p. 18.

5.22 J.T. Burwell and C.D. Strang, On the empirical law of adhesive wear, J. Appl. Phys., 23(1952) 18.

5.23 R.D. Arnell, A.P. Herod and D.G. Teer, The effect of combined stresses on the transition from mild to severe wear, Wear, 31(1975) 237.

5.24 K.-H. Habig, K. Kirschke, W.-W. Maennig, H. Tischer, Festkörpergleitreibung und Verschleiß von Eisen, Kobalt, Kupfer, Silber, Magnesium und Aluminium in einem Sauerstoff-Stickstoff-Gemisch zwischen 760 und $2 \cdot 10^{-7}$ Torr, BAM-Berichte Nr. 13, Berlin, Juni 1972.

5.25 K.-H. Habig, K. Kirschke, W.-W. Maennig, H. Tischer, Festkörpergleitreibung und Verschleiß von Eisen, Kobalt, Kupfer, Silber, Magnesium und Aluminium in einem Sauerstoff-Stickstoff-Gemisch zwischen 760 und $2 \cdot 10^{-7}$ Torr, Wear, 22(1972) 373.

5.26 N.C. Welsh, The dry wear of steel. I. The general pattern of behaviour; II. Interpretation and special features, Phil. Trans. Roy. Soc., London, A 257(1965) 31.

5.27 see Ref. 4.89.

5.28 H. Uetz and J. Föhl, Prüftechnik bei einem Verschleißsystem auf Grund der Verschleißanalyse, insbesondere der thermischen Analyse, VDI-Berichte, Nr. 194, 1973, p. 57.

5.29 see Ref. 1.5, p. 151 ff.

5.30 A.A. Reda, R. Bowen and V.C. Westcott, Characteristics of particles generated at the interface between sliding steel surfaces, Wear, 34(1975) 261.

5.31 T.E. Tallian, G.H. Baile, H. Dalal, O.G. Gustafsson, Rolling Bearing Damage - A Morphological Atlas, SKF Industries, Technology Center, King of Prussia, Pa, 1974.
E.R. Bowen, V.C. Westcott, Wear Particle Atlas, Report No N00156-74-c-1682, prepared for Naval Air Engineering Center, Lakehurst, N.J., July 1976.

5.32 see Ref. 4.129, p. 252 ff.

5.33 H.C. Rippel, Designing fluid-film bearings by computer, Mech. Engng, (1970) 30.

5.34 P.B. Davies and R.B. Howarth, Hydrostatic lubrication, see Ref. 1.8, p. 308 ff.

5.35 C.M. Taylor, Unusual bearings: air and magnetic suspension, Ind. Lubrication and Tribology, 24(1972) 234.

5.36 M.I. Anwar and C.M. Rodkiewicz, Slider bearing with segmented electric field, Wear, 29(1974) 173.

5.37 R. Katterloher, Magnetlager auch im Maschinenbau: Aufbau, Eigenschaften, Anwendungen, Maschinenmarkt, 81(1975) 315.

5.38 J.L. Potter, The elastomeric bearing: what it can do for you, Mech. Engng, (1973) Dec, 22.

5.39 see Ref. 4.131, p. 5-110.

5.40 S. Hildebrand, Feinmechanische Bauelemente, VEB Verlag Technik, Berlin, 1967, p. 456 f.

5.41 see Ref. 5.2, section C2.

5.42 M. Donovan, J.L. Sanders, R. Hazzard, G.R. Bell, R.E. Mansford, Surface coatings (Part I to IV), Tribology, 5(1972) 205.

5.43 S. Jahanmir, N.P. Suh and E.P. Abrahamson, II, The delamination theory of wear and the wear of composite surface, Wear, 32(1975) 33.

5.44 see Ref. 4.90.

5.45 see Ref. 1.6, p. 301 ff.

5.46 G. Cros and J.-L. Polti, Einige theoretische und praktische Gesichtspunkte der Reibung, Schweizer Archiv, 36(1970) 147.

5.47 E. Roemer, Werkstoffe und Schichtaufbau bei Gleitlagern, Z. f. Werkstofftechnik, 4(1973) 2.

CHAPTER 6

6.1 W.M.J. Schlösser, A contribution to the study of analogies of power transmissions in machines, Power drives, (1974) 48 and 49.

6.2 C.T. Walters, The dynamics of ball bearings, Trans ASME, JOLT, (1971) 1.

6.3 D.G. Stewart, J.B. Hunt, Relaxation oscillations on a machine tool slideway, Proc. Inst. Mech. Engrs, London, Vol. 184 Pt 3L, (1969/70) 33.

6.4 J.A. Bailey, Friction in metal machining - mechanical aspects, Wear, 31(1975) 243.

6.5 L. Eaton, Friction instability, in Ref. 1.8, p.147.

6.6 H.-U. Mittmann and H. Czichos, Analogue computer studies of stick-slip effects, BAM, Berlin-Dahlem, (1976) unpublished.

6.7 G. Niemann, Maschinenelemente, Bd 1 und Bd 2, Springer-Verlag, Berlin, 1961.

6.8 F. Sass, Ch. Bouché, A. Leitner, DUBBEL Taschenbuch für den Maschinenbau, Bd 1, Springer-Verlag, Berlin, 1970, p. 232.

6.9 R.H. Creamer, Machine Design, Adison-Wesley Publ. Co. Reading, 1976, p. 281 f.

6.10 E. Radzimovsky and A. Mirarefi, Dynamic behaviour of gear systems and variation of coefficient of friction and efficiency during the engagement cycle, Trans ASME, B, 97(1975) 1274.

6.11 G.B. Stanton, Jr., The future of safety and health: a challenge to engineering managers, Trans ASME, J. of Engng for Industry, (1976) 761.

6.12 H.W. Heinrich, Industrial accident prevention: a scientific approach, Mc Graw-Hill, New York, 1950, p. 11 ff.

6.13 W. Haddon, Jr, The prevention of accidents, in Preventive Medicine (Editors: D.M. Clark and B. Mac Mahon) Little, Brown, Boston, 1967, chapter 33.

6.14 J.V. Grimaldi, Applying systems analysis techniques in safety situations, The Center of Safety, New York University, New York, Apr. 1968, p. 31 ff.

6.15 R.A. Collacott, Mechanical fault diagnosis and condition monitoring, Chapman and Hall, London, 1977, chapter 2.

6.16 Allianz, Handbuch der Schadenverhütung, Allianz Versicherungs AG, München und Berlin, 1. Auflage, 1972, p. 361 and p. 376; 2. erweiterte und überarbeitete Auflage, 1976, p. 645 and p. 695.

6.17 P.M. Ku, Gear gailure modes - importance of lubrication and mechanics, ASLE Trans, 19(1976) 239.

6.18 S.O. Rafique, Failures of plain bearings and their causes, Proc. Inst. Mech. Engrs, London, Vol. 178, Pt 3N, (1963/64) 180.

6.19 F.F. Simpson, Failure of rolling contact bearings, Proc. Inst. Mech. Engrs, London, Vol. 178, Pt 3N, (1963/64) 215.

6.20 D. Scott, A. Smith, J. Tait, G.R. Tremain, Materials and metallurgical aspects of piston ring scuffing - a literature survey, Wear, 33(1975) 293.

6.21 E. Dombrowski, Einführung in die Zuverlässigkeit elektronischer Geräte und Systeme, AEG-TELEFUNKEN, Berlin, 1970.

6.22 J.C. Cluley, Electronic equipment reliability, Mac Millan Press, London, 1974.

6.23 R.A. Bones, Designing for reliability, Engineering, 216(1976) 798.

6.24 W. Gerisch, Zuverlässigkeit aus philosophischer, mathematischer und ingenieurwissenschaftlicher Sicht, Material und Technik, (1973) 51.

6.25 Messerschmitt-Bölkow-Blohm (Herausgeber), Technische Zuverlässigkeit, Springer-Verlag, Berlin, 1971.

6.26 G. Mathhäi, Über den Gleitverschleiß ungeschmierter Oberflächen, Forsch. Ing. Wes., 30(1964) 1.

6.27 H. Thum, Zuverlässigkeit und Verschleiß von mechanischen Baugruppen, Schmierungstechnik, 3(1972) 139.

6.28 H. Yoshikawa, Fundamentals of mechanical reliability and its application to computer aided machine design, CIRP Annals, 24(1975) 297.

6.29 G. Fleischer, Probleme der Zuverlässigkeit von Maschinen, Wiss. Z. TH Magdeburg, 16(1972) 289.

6.30 G. Bergling, Betriebszuverlässigkeit von Wälzlagern, Kugellager-Zeitschrift, 51(1976) 1.

6.31 P. Eschmann, Betriebssicherheit und Gebrauchsdauer von Wälzlagern, Wälzlagertechnik, 13(1974) 3.

6.32 H. Pittroff, Funktionssichere Wälzlagerungen für Elektromaschinen, Kugellager-Zeitschrift, 51(1977) 10.

6.33 W.J. Bartz, Tribotechnik als Grundlage zur Vermeidung von Schäden an Maschinenelementen, Schmiertechnik und Tribologie, 20(1973) 50.

6.34 G. Fleischer, Der Einfluß der Schmierungstechnik auf die Minderung der Verschleißintensität und die Sicherung einer hohen Zuverlässigkeit, Standardisierung und Qualität, 21(1975) 83.

6.35 M.L. Shooman, Probabilistic reliability - an engineering approach, Mc Graw-Hill, New York, 1968.

6.36 N.N., "Darunter kommt alles ins Rutschen": Moderne Elektronik schafft Arbeitslosigkeit - sowohl bei Herstellerfirmen als auch bei den Abnehmern, Der Spiegel, 30(1976) Nr. 51, 87.

6.37 B. Cramer, Der elektronische Fernschreiber als Beispiel für den Strukturwandel in der Feinwerktechnik, Feinwerktechnik und Meßtechnik, 85(1977) 1.

6.38 D.F. Wilcock and E.R. Booser, Bearing design and application, Mc Graw-Hill, New York, 1957.

6.39 E.E. Bisson and W.J. Anderson, Advanced bearing technology, NASA, SP-38, Washington, 1964.

6.40 F.E. Kennedy and H.S. Cheng (Editors), Computer-aided Design of Bearings and Seals, The American Soc. of Mech. Engrs, New York, 1976.

6.41 H. Peeken, Die tribologisch richtige Konstruktion, VDI-Z., 118(1976) 201.

6.42 A.E. Green and A.J. Bourne, Reliability, Technology, Wiley - Interscience, London, 1972, p. 19.

6.43 R.G. Bayer and T.C. Ku, Handbook of analytical design for wear, Plenum Press, New York, 1964.

6.44 J.H. Dumbleton and S.R. Rhee, The application of a zero wear model to metal/polyethylene sliding pairs, Wear, 35(1975) 233.

6.45 see Ref. 4.129.

6.46 J.v. Neumann, The general and logical theory of automata, in: Modern systems research for the behavioral scientist (Editor: W. Buckley), Aldine Publ. Co, Chicago, 1968, p. 97.

Chapter 7

7.1 J. Blouet, Tribométrie, Le Frottement et l'Usure, Journées d'Etude, 5/6 déc 1966, 1.

7.2 G. Salomon, The structure of tribological systems, Lubr. Engng, 32(1976) 458.

7.3 H. Czichos, Experimentelle Methoden zur Untersuchung tribologischer Effekte im Mischreibungsgebiet, VDI-Berichte, 156, 1970, 5.

7.4 H. Wiegand u. E. Broszeit, Zur Frage der Beanspruchungszeit bei Verschleißuntersuchungen unter Mischreibungsbedingungen, Motortechn. Zeitschrift, 34(1973) 33.

7.5 H. Czichos, Tribologische Materialprüfung - Physikalische Grundlagen und technisch-wirtschaftliche Bedeutung, Amts- u. Mitteilungsblatt der BAM 7(1977) 5.

7.6 R. Benzing, I. Goldblatt, V. Hopkins, W. Jamison, K. Mecklenburg, M. Peterson, Friction and Wear Devices, 2nd edition, ASLE, Park Ridge, 1976.

7.7 K.R. Mecklenburg, The effect of wear on the compressive stress in the sphere-on-plane configuration, ASLE Trans, 17(1974) 149.

7.8 G.H. Göttner, Definition und Erfassung einiger Begriffe zur Beschreibung des Geschehens in Reibstellen bei Mischreibung, Schmiertechnik und Tribologie, 20(1973) 22.

7.9 E.D. Braun, Modellierung - eine objektive Methode zur Beurteilung der Reibungseigenschaften von Werkstoffen, Schmierungstechnik, 2(1971) 12.

7.10 J.A. Elder, Jr. and N.S. Eiss, Jr., A study of the effect of normal stiffness on kinetic friction forces between two bodies in sliding contact, ASLE Trans, 12(1969) 234.

7.11 G. Salomon, The simulation of marginal control, Lubr. Engng, 32(1976) 570.

7.12 H.-U. Mittmann, N. Czaika and H. Czichos, A new device for simultaneous measurement of friction force, normal force and friction coefficient, Wear, 31(1975) 179.

7.13 E. Broszeit, K.H. Kloos, F.J. Hess und E. Wagner, Einfluß von Schwingungen auf die Verschleißeigenschaften eines Reibsystems, Wear, 28(1974) 395.

7.14 H. Czichos und K. Kaffanke, Zur Bestimmung von Grenzflächentemperaturen bei tribologischen Vorgängen, VDI-Z., 112(1970) 1491 und 1643.

7.15 see Ref. 4.147.

7.16 K.-H. Habig, Möglichkeiten der Modell-Verschleißprüfung, Materialprüfung, 17(1975) 358.

7.17 R.G. Bayer (Editor), Selection and Use of Wear Tests for Metals, ASTM Technical Publication 615, Philadelphia, 1976.

7.18 K.-H. Habig, On the determination of wear rates, Wear, 28(1974) 135.

7.19 R. Holm, Electrical contacts, Springer-Verlag, Berlin, 1967.

7.20 H. Czichos, W. Grimmer and H.-U. Mittmann, Rapid measuring techniques for electrical contact resistance applied to lubricant additives studies, Wear, 40(1976) 265.

7.21 B.B. Seth and T. Willis, Techniques for film thickness measurements in elastohydrodynamic lubrication, Trans ASME, Paper 76-Det-79.

7.22 A.K. Pogosian, Forcecasting the viability of friction pairs in accelerated tests, Wear, 26(1973) 175.

7.23 S.V. Pinegin, A.V. Cicinadze und E.D. Braun, Entwicklungsrichtungen der Modellierung zur Beurteilung von Reibungs- und Verschleißvorgängen, Schmierungstechnik, 6(1975) 5.

7.24 D. Play and M. Godet, Design of high performance dry bearings, Wear, 41(1977) 25.

7.25 A.W.J. de Gee, Selection criteria for lubricated journal bearings, Wear, 36(1976) 33.

7.26 G. Heinke, Verschleiß - eine Systemeigenschaft; Auswirkungen auf die Verschleißprüfung, Z. Werkstofftechnik, 6(1975) 164.

7.27 R.D. Brown, Test methods, in: Boundary lubrication, edited by F.F. Ling, E.E. Klaus and R.S. Fein, ASME, New York, 1969, p. 241.

7.28 K. Kutzbach, Reibung und Abnutzung von Zahnrädern, VDI-Z., 70(1926) 999.

7.29 see Ref. 4.131, chapter 26.

7.30 R.B. Howarth, Lubricant properties and testing, in Ref. 1.8, p. 202.

7.31 C. Zerbe, Mineralöle und verwandte Produkte, Springer-Verlag, Berlin, 1969.

7.32 H. Jünemann, Mechanische Prüfverfahren für Schmierstoffe, Erdöl und Kohle, 29(1976) 259.

7.33 see Ref. 4.131, p. 20-17.

7.34 S.C. Dodson and L.A. Godsave, Lubricating oils, in Ref. 1.7, p. 501.

7.35 see Ref. 4.131, p. 27-8.

7.36 D. Scott, Failure diagnosis and investigation, Tribology, (1970) 22.

7.37 see Ref. 6.15.

7.38 F.K. Orcutt, Das Auffinden von Störungen an mechanischen Komponenten, verursacht durch Reibung und Verschleiß, Technica, (1970) 883.

7.39 G.H. Göttner, Zur Funktionsdiagnose von Maschinenbauteilen und -anlagen geeignete Meßgrößen, Schmiertechnik und Tribologie, 24(1977) 3.

7.40 K. Ziegler, Funktionsüberwachung von Maschinen und Bauteilen mit Hilfe der Schallmeßtechnik, Schmiertechnik und Tribologie, 24(1977) 5.

7.41 K.A. Bowen and T.S. Graham, Noise and analysis: a maintenance indicator, Mech. Engng, (Oct. 1967) 31.

7.42 ALLIANZ, Handbuch der Schadenverhütung, 2. erweiterte und überarbeitete Auflage, München, 1976, p. 672.

7.43 see Ref. 7.36.

7.44 A. Beerbower, Spectrometry and other analysis tools for failure prognosis, Lubr. Engng, 32(1976) 285.

7.45 see Ref. 6.15, p. 258.

7.46 A. Gervé, Einsatzmöglichkeiten von Radionukliden zur Untersuchung konstruktiver und schmierstoffabhängiger Einflüsse auf den Verschleiß von Maschinenteilen, VDI-Berichte, Nr. 196(1973) 43.

7.47 W.W. Seifert and V.C. Westcott, A method for the study of wear particles in lubricating oil, Wear, 21(1972) 27.
A.W. Ruff, Characterization of debris particles recovered from wearing systems, Wear, 42(1977) 49.

7.48 D. Scott, W.W. Seifert and V.C. Westcott, The particles of wear, Scientific American, 230(1974) 88.

7.49 H. Czichos, Meßtechnische Methoden zur Untersuchung tribologisch beanspruchter Werkstoffoberflächen, Schmiertechnik und Tribologie, 21(1974) 25.

7.50 G. Nomarski et A.R. Weill, Application à la métallographie des méthodes interférentielles à deux oudes polarisées, Revue de Metallurgie, 52(1955) 121.

7.51 H.-U. Mittmann und H. Czichos, Reibungsmessungen und Oberflächenuntersuchungen an Kunststoff Metall-Gleitpaarungen. Materialprüfung, 17(1975) 366.

7.52 S. Tolansky, Multiple-beam Interferometry of Surfaces and Films, Clarendon Press, Oxford, 1948.

7.53 H. Trumpold, Die Anwendungsgrenzen der Lichtinterferenzen bei Rauheitsmessungen, Feingerätetechnik, 18(1969) 247.

7.54 K.H. Habig, Der Einsatz des Raster-Elektronenmikroskops zur Aufklärung von Verschleißmechanismen, Beitr. Elektronenmikroskop Direktabb. Oberfl. 3(1970) 235.

7.55 L. Reimer und G. Pfefferkorn, Rasterelektronenmikroskopie, Springer-Verlag, Berlin, 1973.

7.56 R. Stickler, C.W. Hughes, G.R. Booker, Application of the selected area ECP method to deformation studies, Proc. 4 Annual SEM Symp., JJTRJ, Chicago, April 1971, p. 473.

7.57 H. Czichos und D. Klaffke, Deformationsuntersuchungen an tribologisch beanspruchten Metalloberflächen mit der Methode der electron chaneling pattern, BAM, Berlin-Dahlem, (1977) unveröffentlicht.

7.58 J.B.P. Williamson, Microtopography of surfaces, in: Properties and Metrology of Surfaces, Proc. Inst. Mech. Engrs, London, Vol. 182, 3K(1968) 21.

7.59 T.F.J. Quinn, The application of modern physical techniques to tribology, Newnes - Butterworth, London, 1971, p. 221.

7.60 A. Benninghoven, New developments in the surface analysis of solids, Appl. Phys. 1(1973) 3.

7.61 J. Ferrante, D.H. Buckley, S.V. Pepper, W.A. Brainard, Use of LEED, Auger emission spectroscopy and field ion microscopy in microstructural studies, in: Microstructural Analysis-Tools and Techniques, (Edited by J.L. Mc Call and W.M. Mueller), Plenum Publ. Co, New York, 1973, 241 ff.

7.62 R. Holm und S. Storp, Monolagenanalyse an "schmutzigen" Oberflächen?, Phys. Blätter, 32(1976) 342.

7.63 D.H. Buckley, The use of analytical surface tools in the fundamental study of wear, Proc. Int. Conf., Wear of Materials, St. Louis, Apr. 1977, ASME, New York, 1977, p. 12.

7.64 D.R. Wheeler, X-ray photoelectron spectroscopic study of surface chemistry of dibenzyl-disulphide on steel under mild and severe wear conditions, submitted to Wear.

7.65 A. Benninghoven und S. Storp, Untersuchung der Grenzflächen und des Volumens dünner Schichten mit Hilfe der Sekundärionen - Massenspektroskopie, Z. angew. Physik, 31(1971) 31.

7.66 P.L. Stöhr, Nachweis und Analyse von Reaktionsschichten bei metallischen Reibpartnern im Reibverschleiß-Modellgerät und im Verbrennungsmotor, Dissertation, Eberhard-Karls-Universität, Fachbereich Physik, Tübingen, 1976.

7.67 E.W. Müller, Field ion microscopy, American Elsevier, New York, 1969.

Chapter 8

8.1 A. Pieuchot, J. Blouet, R. Gras, M. Alfred and R. Courtel, Méthodologie et classement des essais de frottement et de leurs résultats, Méc. Electr., Oct. 1969, p. 8.

8.2 H. Czichos, A systems analysis data sheet for friction and wear tests and an outline for simulative testing, Wear, 41(1977) 45.

8.3 M.N. Gardos, Labyrinth sealing of aerospace mechanisms - theory and practice, ASLE Trans., 17(1974) 237.

8.4 G. Katzenmeier, Das Verschleißverhalten und die Tragfähigkeit von Gleitlagern im Übergangsbereich von der Vollschmierung zu partiellem Tragen (Untersuchungen mit Hilfe von Radioisotopen), Kernforschungszentrum Karlsruhe, KFK, 1569, Febr. 1972.

8.5 D.H. Buckley and R.L. Johnson, Marked influence of crystal structure on the friction and wear characteristics of cobalt and cobalt-base alloys in vacuum to 10^{-9} millimeter mercury, NASA TND-2523, Washington, 1964.

8.6 J.K. Appeldorn, J.B. Goldmann and F.F. Tao, Corrosive wear by atmospheric oxygen and moisture, ASLE Trans, 12(1969) 140.

8.7 J. Mølgaard, A simulative wear study critically reviewed, Wear, 41(1977) 57.

8.8 J.K. Lancaster, Dry bearings: a survey of materials and factors affecting their performance, Tribology international, 6(1973) 219.

8.9 L.H. Lee (Editor), Advances in Polymer Friction and Wear, Polymer science and technology, Vol. 5, Parts A and B, Plenum Press, New York, 1974.

8.10 see Ref. 7.51.

8.11 B.J. Briscoe, C.M. Pooley and D. Tabor, Friction and transfer of some polymers in unlubricated sliding, see Ref. 8.9, Part 5A, p. 191.

8.12 K.H. Habig, R. Chaterjee-Fischer und F. Hoffmann, Zum tribologischen Verhalten von Oberflächenschichten, die durch Eindiffusion von Stickstoff, Bor oder Vanadin in Eisenwerkstoffe gebildet werden, Härtereitechn. Mitteilungen, in print.

8.13 H. Hübner, Unterdrückung stick-slip induzierter Kurvengeräusche bei Eisenbahnschienen durch physikalisch-chemische Oberflächenbehandlung, Dissertation, TU-Berlin, 1973, D83.

8.14 H. Czichos und H. Hübner, Modelluntersuchungen über stick-slip Vorgänge, BAM, Berlin-Dahlem, (1973), unveröffentlicht.

8.15 G.H.G. Vaessen and A.W.J. de Gee, Influence of water vapour on the wear of lightly loaded contacts, Wear, 18(1971) 325.

8.16 M.E. Merchant, Friction and Adhesion, in: Interdisciplinary approach to friction and wear (Editor: P.M. Ku), NASA SP-181, Washington, 1968, p. 198.

8.17 J.A. Schey, Metal Deformation Processes. Friction and Lubrication, Marcel Dekker, New York, 1970.

8.18 G.W. Rowe, Principles of Industrial Metalworking Processes, Arnold, London, 1977.

8.19 A. Gileji, Bildsame Formgebung der Metalle, Akademie-Verlag, Berlin, 1967.

8.20 O. Pawelski, G. Graue und D. Löhr, Reibungsbeiwert und Temperaturverteilung beim Warmformen von Stahl mit verschiedenen Schmiermitteln, Schmiertechnik und Tribologie, 17(1970) 120 und 170.

8.21 H. Krause und A.H. Demirci, Factors influencing the real trend of the coefficient of friction of two elastic bodies rolling over each other in the presence of dry friction, in Ref. 4.1.

8.22 M. Gierzynska, Problem of friction in rolling and forging processes, in: Tribology in iron and steel works, ISI Publication, The Iron and Steel Institute, London, 1970, p. 156.

List of symbols

I Characteristics of a system (Chapter 2)

A	=	$\{a_i\}$	set of the elements of a system
P	=	$\{P(a_i)\}$	set of element's properties
R	=	$\{R(a_i, a_j)\}$	set of interrelations between elements
S	=	$\{A, P, R\}$	structure of a system
X	=	$\{x_i\}$	inputs of a system
Y	=	$\{y_i\}$	use-outputs of a system
Z	=	$\{z_i\}$	loss-outputs of a system
T	=	$T(X,Y)$	input-output relation (transfer function)

II General symbols

a	coefficient of adhesion
a_H	radius of Hertzian contact
A	(see I)
A_c	Arrhenius constant
A_o	geometric area of contact
A_m	mantle surface of a body

A_r	real area of contact
A_W	wear-track area
b	Burgers vector
B	magnetic field
c	specific heat
c_o	velocity of light
c_r	radial clearance
C	general constant
C_d	damper constant
C_s	spring constant
d	diameter
E	modulus of elasticity
ΔE	energy
ΔE^W	mechanical energy or work
ΔE_e	energy of edge dislocation
ΔE_s	energy of screw dislocation
$\dot{E}$	power
$\dot{E}^W$	mechanical power
f	coefficient of friction
F	force
F_A	force of adhesion
F_F	friction force
F_N	normal force
F_S	separation force
F_T	tangential force
F(t)	failure distribution function
g	acceleration due to gravity
G	shear modulus
h	film thickness

$\hbar$	Planck's constant / 2π
H	hardness
ΔH	enthalpy
i	electric current
i, j	counter
J	mechanical equivalent of heat
k	thermal conductivity
k_1, k_2	Hamaker constants
k_p	parabolic oxidation rate constant
K, K'	wear coefficients
K_c	crack parameter
l	length or distance
m	mass
M	torque
M_F	friction torque
n	asperity density per unit area
n_1, n_2	electron number density
N	gear ratio
o	ratio: $\frac{\text{heat emitting area}}{\text{heat absorbing volume}}$
p	pressure
$\overline{p}$	load/bearing projected area
p_H	maximum Hertz pressure
$\overline{p}_H$	mean Hertz pressure
p_p	pump pressure
p_y	yield pressure
P	(see I)
q_e	electric charge
q	flow rate
q^m	flow rate of mass

q_l	flow rate of lubricant
Q	amount of heat
r	radius
R	(see I)
$\overline{R}$	mean roughness
R_a	center-line-average roughness
R_t	peak-to-valley-height roughness
R_c	gas constant
R(t)	reliability function
s	standard deviation
S	(see I)
ΔS	entropy
$\dot{S}$	enthalpy flow generation rate
So	Sommerfeld number
t	time
T	(see I)
T	temperature
u	voltage
ΔU	potential energy
v	velocity
V	volume
V_w	wear volume
w	wear rate
W	work
x_i	coordinates (i = 1, 2, 3)
X	(see I)
y	surface profile height
Y	(see I)
z	distance of approach or separation

Z	(see I)
$Z(\Delta)$	auto-correlation function of profile
α	pressure-viscosity coefficient
β	radius of asperity curvature
γ	surface free energy
γ_R	constant (zero-wear model)
$\Gamma(x)$	gamma function
δ	friction angle
Δ	difference of quantities
ε	ratio: $\frac{\text{contact area}}{\text{wear track area}}$
ε_w	wear resistance
ζ	no-contact time fraction
η	viscosity
η	efficiency
ϑ	convection heat-transfer coefficient
Θ	adhesion parameter
$\varkappa$	thermal diffusivity
λ	film thickness to roughness ratio
$\lambda(t)$	failure rate
μ	mean of normal distribution
ν	Poisson number
ξ	critical oxide film thickness
ρ	density
σ	stress or strength (general)
σ_a	adhesion bond strength
σ_y	yield strength
σ^*	mean deviation of asperity height
τ	shear stress

τ_s	shear strength
φ	angle (general)
Φ	potential
ψ	plasticity index
ψ_r	ratio: $\frac{\text{radial clearance}}{\text{shaft radius}}$
ω	angular velocity

$\vec{\nabla}=(\frac{\partial}{\partial x},\frac{\partial}{\partial y},\frac{\partial}{\partial z})$ Nabla operator

$\nabla^2=\frac{\partial^2}{\partial x^2}+\frac{\partial^2}{\partial y^2}+\frac{\partial^2}{\partial z^2}$ Laplace operator

Author index

The numbers are reference numbers, composed of the number of the chapter in which an author's work is referred to and the running number. The complete references are listed on pages 360 to 381.

Subject index